BEYOND CONVERGENCE

For sale by the Superintendent of Documents, U.S. Government Publishing Office
Internet: bookstore.gpo.gov Phone: toll free (866) 512-1800; DC area (202) 512-1800
Fax: (202) 512-2104 Mail: Stop IDCC, Washington, DC 20402-0001

ISBN 978-0-16-093613-5

BEYOND CONVERGENCE
World Without Order

Edited by
Hilary Matfess and Michael Miklaucic

Center for Complex Operations
Institute for National Strategic Studies
National Defense University
Washington, D.C.

Cover: Yayoi Kusama, *Infinity Mirror Room-Filled with the Brilliance of Life* (2011)

Layout Design: Viviana Edwards

First printing, October 2016

For current publications of the Center for Complex Operations, consult the CCO website at http://www.cco.ndu.edu.

Contents

III. Pandora

IV. A Toolbox for the 21st Century

Acknowledgments

In 2013, National Defense University Press published *Convergence: Illicit Networks and National Security in the Age of Globalization*. It was an effort to map the many issues arising from the accelerating interactions among international terrorist, transnational criminal, and networked insurgent organizations. To meet demand, over 10,000 copies were printed. It generated much discussion—including much criticism—and led the editors to conclude that the debate over how to understand the emerging threat environment was only beginning. Therefore, we would like to acknowledge and thank the many readers and critics of *Convergence*, for taking up the challenge and for laying out a gauntlet.

Ongoing research on the scope and dynamics of illicit network convergence indicates that this phenomenon is penetrating new domains and markets, taking on new characteristics, and continuing to morph at a velocity very hard to match. It is evident that continuing empirical analysis and thought-provoking discussion of these new dimensions are critical if the United States is to meet this challenge to both national and international security. The Center for Complex Operations (CCO) at NDU and the editors wish to acknowledge and thank the Office of the Secretary of Defense—in particular, the leadership of the Office of Counternarcotics and Global Threats, for recognizing the national security implications of this phenomenon and supporting our effort to take up the gauntlet laid out. Without their sponsorship, this project would not have materialized. Deputy Assistant Secretary of Defense Caryn Hollis is a visionary in the struggle to counter transnational criminal organizations and emerging global threats. We also wish to recognize David Sobyra for his encouragement and early and consistent support for this NDU/CCO effort.

CCO benefits enormously from our outstanding team of volunteer interns in all of its work. The editors are especially indebted to Samantha Fletcher, Oliver Vaughn, Shawn McFall, and Chris Johnson for their assistance in proofreading and formatting these chapters. We also thank other members of the CCO staff, who provided unwavering support. In particular, the untiring efforts of Becky Harper, Dale Erickson, and Dr. Joseph Collins have been critical in enabling us to succeed. Additionally, Tamara Tanso has again proven herself a peerless professional and a perfectionist in the final preparation of *Beyond Convergence*. We wish also to express our deep gratitude to Vivian Edwards, Michael Mann, and the William J. Perry Center for Hemispheric Defense Studies at NDU for heroic support in the production of the first printing.

Many have advised us along the way. If we have left any unacknowledged, it is either by their request, or our oversight, which we regret. There are some who would like to have contributed but were prevented by circumstances. Nevertheless, we thank them all and look forward to engaging with them again in the future. However, our greatest appreciation goes to the authors for their enthusiasm, patience, and commitment to this project.

Introduction: World Order or Disorder?

Hilary Matfess and Michael Miklaucic

The world order built upon the Peace of Westphalia is faltering. State fragility or failure are endemic, with no fewer than one-third of the states in the United Nations earning a "high warning"—or worse—in the Fragile States Index, and an equal number suffering a decline in sustainability over the past decade.[1] State weakness invites a range of illicit actors, including international terrorists, globally networked insurgents, and transnational criminal organizations (TCOs). The presence and operations of these entities keep states weak and incapable of effective governance, and limit the possibility of fruitful partnerships with the United States and its allies. Illicit organizations and their networks fuel corruption, eroding state legitimacy among the governed, and sowing doubt that the state is a genuine guardian of the public interest. These networks can penetrate the state, leading to state capture, and even criminal sovereignty.[2] A growing number of weak and corrupt states is creating gaping holes in the global rule-based system of states that we depend on for our security and prosperity. Indeed, the chapters of this book suggest the emergence of a highly adaptive and parasitic alternative ecosystem, based on criminal commerce and extreme violence, with little regard for what we commonly conceive of as the public interest or the public good.

The last 10 years have seen unprecedented growth in interactivity between and among a wide range of illicit networks, as well as the emergence of hybrid organizations that use methods characteristic of both terrorist and criminal groups. In a convergence of interests, terrorist organizations collaborate with cartels, and trafficking organizations collude with insurgents. International terrorist organizations, such as al-Qaeda and Hezbollah, engage energetically in transnational crime to raise funds for their operations. Prominent criminal organizations like Los Zetas in Mexico and D-Company in Pakistan have adopted the symbolic violence of terrorists—the propaganda of the deed—to secure their "turf." And networked insurgents, such as the Islamic State of Iraq and the Levant (ISIL), the Revolutionary Armed Forces of Colombia (FARC), and the Liberation Tigers of Tamil Eelam (LTTE), have adopted the techniques of both crime and terror.[3]

An Emerging Criminal Ecosystem

The unimpeded trajectory of these trends—convergence, hybridization, and state capture—poses substantial risks to the national security interests of the United States, and threatens international security. Illicit networked organizations are challenging the fundamental principles of sovereignty that undergird the international system. Fragile and failing states are both prey to such organizations, which feed on them like parasites, and Petri dishes for them, incapable of supporting effective security partnerships. The Westphalian, rule-based system of sovereign polities itself is at risk of fraying, as fewer and fewer capable states survive to meet these challenges, and populations around the world lose faith in the Westphalian paradigm. The emergence of an alternative ecosystem of crime and violence threatens us all and much of the progress we have seen in recent centuries. This dark underworld weakens national sovereignty and erodes international partnerships.

We should not take for granted the long-term durability of the Westphalian system. It was preceded by millennia of much less benign forms of governance, and alternative futures are imaginable. This book describes "convergence" (the interactivity and hybridization of diverse illicit networks), the emergence of new networks and new domains or "battlespaces," and the threat illicit networks pose to national and international security. It examines dystopian visions of a world in which these trajectories go indefinitely unimpeded, and concludes by discussing possible countermeasures to be explored.

While some recognize the growing threat to the global system of governance that these new phenomena impose, others are skeptical. According to the conventional wisdom, TCOs and international terrorist organizations are unlikely candidates for partnership. Such analysis suggests that criminals are motivated by the pursuit of wealth in defiance of law, morality, or ideology. They typically prefer to remain undetected, and have little interest in the violence committed by, or risks taken by, international terrorists. Already pursued by law enforcement, criminals are not keen to receive the attention of the Central Intelligence Agency or SEAL Team Six. International terrorists and insurgents, on the other hand, are politically motivated; driven by ideological, religious, or nationalistic motives; and repelled by the vulgar materialism and greed of criminals. They have no desire to get on the radar of the Drug Enforcement Administration (DEA), or other national or international law enforcement agencies. This logic is understandable, and may have prevailed in previous times, but the evidence of extensive interconnectivity—if not explicit partnership—between TCOs, international terrorists, and globally networked insurgents is compelling. Recent research undertaken by the Combating Terrorism Center at West Point reveals that, "criminals and terrorists are largely subsumed (98 percent) in a single network as opposed to operating in numerous smaller networks."[4] In its Performance Budget Congressional Submission for FY 2014, the DEA stated that by "the end of the first quarter of FY 2013, 25 of the 67 organizations on the Attorney General's Consolidated Priority Organization Target (CPOT) List are associated with terrorist organizations."[5] According to a more recent DEA statement, roughly half of the Department of State's 59 officially designated foreign terrorist organizations have been linked to the global drug trade.[6] The six degrees of separation that may have once divided people is a relic of the past—today, international terrorists, insurgents, and criminals are merely a click away from each other.

It might be argued that terrorism, insurgency, and organized crime have existed since time immemorial, and that their modern iterations represent nothing new. Such an argument naively discounts modern enablers such as information and communication technology, transportation advances, and the unprecedented volumes of money generated in illicit markets. These are game changers. They permit illicit actors to avail themselves of lethal technology, military-grade weaponry, real-time information, and professional services of the highest quality, including legal, accounting, technological, security, and paramilitary services. Cartels and gangs, as well as terrorists and some insurgents, can now outman, outspend, and outgun the governments of the countries where they reside. They can communicate across the globe in real time, using widely available and inexpensive technology. The November 2008 Mumbai terrorist attackers used satellite phones, internet communications, and global positioning systems, under the direction of Pakistan-based handlers to carry out an atrocious binge of murder and terror.[7] The string of ISIL attacks across Europe in 2015 and 2016 further illustrates the global consequences of this technological acceleration. International travel has never been easier or cheaper than it is today, and would-be terrorists, traffickers, launderers, and even assassins can fly nearly undetected from continent to continent, in the sea of traveling humanity.

Though it is clear that this connectivity is widespread and threatens global security, the details of the agreements or arrangements between terrorist, insurgent, and transnational criminal organizations remain murky. A partial exception to this is in instances where both organizations wish for new relationships to be known, such as the 1998 merger of Ayman al-Zawahiri's Egyptian Islamic Jihad organization with Osama bin Laden's al-Qaeda.[8] Other relationships, such as between the FARC and al-Qaeda in the Islamic Maghreb (AQIM), are opaque as neither organization has an interest in revealing the relationship. It is unclear in the majority of cases what kinds of partnerships these are, and we are often unable to discern whether such instances of cooperation are one-time affairs or longer-term arrangements. This lack of information handicaps our response and threatens global security.

The purposeful opacity of illicit organizations presents a vexing challenge to mapping and understanding these actors. Operating by intention outside the vision of regulators or researchers, their activities and revenues are hidden. So how do we determine the magnitude of their operations, or the harm they inflict? How do we know the value of their transactions? We extrapolate from extremely inexact evidence, such as seizures, arrests, convictions, and the associated testimony of witnesses, often themselves members of such organizations and motivated to dissemble. Analysts still rely on the nearly 20-year-old "International Monetary Fund (IMF) consensus range," of "$1 to $3 trillion" or "two to five percent" of global product. In 1998, Michel Camdessus, then managing director of the IMF, provided that estimate of the amount of money laundered annually across the globe. Given what we know about global trafficking in drugs, persons, weapons, counterfeits, and other contraband it seems unlikely that the value of illicit trade has decreased over the past 20 years. Even at a "mere" two to five percent of global product, Camdessus described the magnitude of the problem as "almost beyond imagination...."[9]

Less difficult, but still challenging and far more visceral to calculate, is the cost of global terrorism in human lives. At publication, the most recent estimates suggest that 2014 saw an increase of 35 percent in the number of terrorist attacks globally, with total fatalities rising to nearly 33,000 by some counts; 2015 is likely to mark another increase, as ISIL continues its brutal global campaign, and Boko Haram terrorizes the Lake Chad

Basin.[10] This does not take into account nonfatal injuries, the destruction of families and communities, and the economic costs. These cannot be monetized, but few would deny that the opportunity cost of the "global war on terror" (GWOT) has been huge. A 2008 estimate by Nobel Laureate Joseph Stiglitz and Linda J. Bilmes put the long-term costs of the Iraq War at $3 trillion.[11] The Cost of War Project puts the total economic cost of America's post-9/11 campaigns at $4.4 trillion through FY 2014.[12]

These two sets of costs—the global illicit market plus the costs associated with the GWOT—comprise a staggering portion of global product, and give a plausible indication of the magnitude of the emerging alternative ecosystem. Consider the drag on global productivity and development if so much of human activity is dedicated to transnational crime and terrorism. Adding to this, the cost of networked insurgencies in countries such as Afghanistan, Colombia, Sri Lanka, and South Sudan, suggests that an unconscionable proportion of global resources is being expended by efforts to undermine the well-being of citizens worldwide. Imagine what might be accomplished for all mankind if those resources were available for more constructive investment.

Net Systemic Costs

Not only do these networks divert economic resources globally, but they also reduce the capacity of states to govern, rendering them incapable of effectively governing their territory or borders, let alone exercising a monopoly of the legitimate use of force, or providing other vital public services. The net systemic harm is imposed at four levels:

- the inability of states to govern their populations and territories, which creates seedbeds for international terrorism, networked insurgency, and transnational crime, causing immense human suffering;
- the regional spillover effect from state fragility and instability, that sometimes penetrates key U.S. allies and partners;
- the growing feral regions that serve as launch pads for attacks against U.S. national security interests worldwide, as well as potentially direct attacks on the homeland, as occurred on September 11, 2001; and
- the cost associated with the decline of the global, rule-based system and the shrinking Westphalian domain.

A cursory examination of a few key states shows the toll illicit networks take on our national security interests. Though Mexico's death rate has subsided somewhat over the past two years, the wars between the narcotics cartels and state authorities, and between the cartels themselves, are thought to have caused as many as 130,000 deaths between 2007 and 2013, or over 20,000 per year.[13] Mexican cartels today work hand-in-hand with the criminal gangs of Central America's Northern Triangle—comprised of El Salvador, Honduras, and Guatemala—resulting in some of the highest homicide rates in the world. El Salvador's official forensic unit estimated the homicide rate in 2014 at nearly 70 per

100,000.[14] Despite their collaborative intentions, these countries are under such duress that their security partnership contributions cannot yet inspire confidence. Indeed, in 2014, nearly 70,000 unaccompanied children from Central America and Mexico made their way through Mexico to the United States to escape the tormented lands of their births.[15]

Another key security partner, Nigeria is the most populous African state with the largest economy, and a major oil producer. Nigeria could and should play a stabilizing role throughout the continent. In fact, Nigerian forces were critical in staunching the civil wars that hemorrhaged West Africa in the 1990s through the 2000s. Yet today, Nigeria is hobbled by the burgeoning Boko Haram insurgency in the north, and resurgent gang insurgency in the Niger Delta. Moreover, the Boko Haram scourge has bled into the neighboring countries of the Lake Chad Basin.

The once-hopeful suppositions that Iraq and Afghanistan could act as U.S. security partners now seem to be wishful thinking. Despite the investment of hundreds of billions of dollars to bolster the capacity of these two potential partners, effective collaboration seems extremely unlikely for the foreseeable future. Afghanistan today struggles to survive the attacks of al-Qaeda, the Taliban and Haqqani networks, and more recently ISIL. Though the Government of Afghanistan welcomes U.S. engagement, its effectiveness as a security partner remains questionable. Similarly, Iraq struggles to survive as an autonomous state, depending on Kurdish and Shia nonstate militias in its fight with ISIL. Afghanistan and Iraq may continue to act as incubators for terrorist groups planning attacks against the United States well into the future.

Though the nature or extent of the connections between these terrorist and criminal organizations is not transparent, what is clear is that when they desire to interact, they are able to do so. Joint training, learning, and sharing of experience are certainly likely, if not yet joint operations. While states unwillingly and unwittingly act as safe havens for destabilizing global actors, even more troubling are instances in which there is clear collusion between such groups and elements of sovereign states. For example, Iran's Quds Force, a special forces unit of the Islamic Revolutionary Guards Corps, has been both directly engaged in terrorist acts around the world, and is supportive of other terrorist organizations. Ominously, in 2011, an attempt by the Quds Force to collaborate with the Los Zetas cartel to assassinate the Saudi Arabian ambassador to the United States was intercepted.[16] That this effort was interdicted by the vigilant DEA is extremely fortunate—at that particular moment in time, with the combustible tension between Iran and Saudi Arabia, and between Sunni and Shia throughout the Islamic world, the consequences of the intended assassination are difficult to imagine. One need only consider the consequences of the assassination of Archduke Ferdinand in Sarajevo just a century ago to put this into perspective. This effort by the Quds Force to conspire with Los Zetas, now fully documented in U.S. case law, demonstrates beyond a reasonable doubt the potential collusion of sovereign states and terrorist organizations with criminal organizations.

This type of collusion is not limited to the Middle East. As Douglas Farah has written, Venezuela has utilized the state's diplomatic tools to support criminal and terrorist activity.[17] North Korea has long been known as a hub of illicit activity, allegedly including smuggling, counterfeit trade, production of controlled substances, illegal weapons trafficking, and money laundering. Pyongyang's infamous Bureau 39 is thought to generate between $500 million and $1 billion per year from such illicit activities.[18]

The Stakes Are High

To succeed in meeting the international security challenges of the 21st century, the United States and its allies need capable and legitimate partners. Today, ISIL's assault on Syria and Iraq is being vigorously resisted by a coalition that includes many American partner countries, including Saudi Arabia, the United Arab Emirates, Jordan, Qatar, and Bahrain, among others. Imagine a world in which the United States had no partners. No partners in the Middle East or Africa would leave only U.S. boots on the ground to combat ISIL, al-Qaeda, al-Shabaab, and Boko Haram. But capable and legitimate partners are hard to find, and getting harder.

The global community of democracies from which America prefers to choose its partners has shrunk, as the domain of freedom is much diminished in recent years.[19] Many potential partners have shown deep fault lines leading to instability. Consider for example Egypt, Mali, and Thailand; each so consumed with internal fissures that effective partnership is beyond their current capability. Though some argue the world has actually become gradually safer over time, many countries once thought to be stable and safe have recently experienced the trauma of indiscriminate terrorism.[20] The Global Peace Index reports that, "The world has become less peaceful every year since 2008."[21] Attacks in countries as diverse as Kenya, India, France, Belgium, and the United States show that there is no nationality, religion, or terrain immune from this onslaught.

The Westphalian system of global governance has always been an aspirational model—and a geographically limited one at that. Despite its limitations, however, it is unclear if a better model of governance exists. Under the Westphalian system, economic growth has surged and the quality of life has flourished. In the 368 years since the Peace of Westphalia established this rule-based system based on sovereign equality, the world has experienced an unprecedented surge across a range of quality of life indicators: life expectancy has surged from below 40 to over 70 years, per capita gross domestic product increased from around $600 to over $10,000 per year, and literacy has increased from less than 10 percent to over 80 percent of the global population. Rather than abandoning the Westphalian system in favor of an untested, and likely less capable, system, we must cultivate global partnerships to reform and strengthen the system.

A chain is only as strong as its weakest link, and a growing number of weak and corrupt states leaves alarming gaps in the global rule-based system of states. This book aspires to act as a roadmap for those seeking to understand the forces—both the external pressures and the internal failings—that have led us to the current global crisis of governance. The text is organized in four sections.

The first section, "Slouching Toward Dystopia," offers a vision of a world unmoored from the organizational principles of the Westphalian order. This part imagines the worst-case scenarios if current assaults on the international system go unchecked. It includes Phil

Williams' discussion of the crisis of the international order, arguing that global governance has failed because of the inability of states to govern themselves. Nils Gilman describes the state under pressure from "twin insurgencies," plutocrats and criminal networks, both detached from any loyalty to the state, and both limiting the capacity of the state. Scott Atran reveals the profound and widespread alienation from the global status quo that leads to violent extremism as a redeeming virtue. Francis Fukuyama and Hilary Matfess examine emerging alternative forms of governance, emulating antidemocratic norms, that are cropping up globally, complicating the American search for willing partners abroad. Jay Chittooran and Scott Helfstein explain how criminality has affected equity market returns, illustrating the tangible economic effects that new criminal actors have had on economic stability and development.

Section II, "One Network," examines the expansion of existing criminal networks and explores their operational characteristics and policy implications. Christopher Dishman describes the extent and interconnectivity of criminal networks, terrorist groups, and other violent actors that enable the external corrosion of the state. Matthew Levitt explores the global reach of Hezbollah, detailing how the group's global networks have allowed it to exploit "Useful Idiots, Henchmen, and Organized Criminal Facilitators." Douglas Farah discusses the spread of criminality and anti-system norms in Latin America, suggesting that an anti-American coalition is on the rise south of the border. Jessica Stern describes the rise of ISIL as a global threat—highlighting the group's mixture of ideological and material interests that has propelled it to the forefront of national and international security discussions.

Section III, "Pandora," describes recent innovations that complicate the global threat landscape. Tuesday Reitano and Andrew Trabulsi discuss the role of social media in bolstering the appeal of antistate actors, allowing them to establish "cult-like" followings and to facilitate "intimate connections to an individual which can be used to raise funds, identify and cultivate associates and victims."[22] Mark Shaw describes the rise of "protection economies," particularly in West Africa, where jihadist networks are increasingly a part of the drug smuggling business in the region. Describing a massive and nefarious parallel economy, Karl Lallerstedt adds to this discussion through an exploration of global counterfeit and smuggling networks, and the growing gray space between licit and illicit commerce. Weak state capacity hampers efforts to counter this trend, threatening to allow the region to descend into alternatively governed spaces. Raj Samani shows how the technological innovations that have made our lives and work so much easier have produced disconcerting vulnerabilities in the cyber domain that are increasingly being exploited by criminal groups, terrorists, and hostile states alike.

Section IV, "A Toolbox for the 21st Century," offers responses to these challenges; the authors offer tangible policy options to mitigate the threats. Clare Lockhart and Michael Miklaucic discuss the critical role of state-building as a remedy to the rise of illicit actors and tempting but toxic ideologies. Celina Realuyo explains how public-private partnerships (P³) can be leveraged to form effective alliances against antistate forces. Sebastian Gorka details the remarkable appeal of ISIL's destructive ideology, the ways in which this affects the nature of the fight to counter violent extremism worldwide, and how armed forces can adapt. Christopher Fussell and D.W. Lee build upon General (Ret.) Stanley McChrystal's "Team of Teams" approach to offer an organizational solution to the rigidity of governmental bureaucracy, rendering it nimbler and more effective in the face of a metastasizing threat.

Throughout the book, a number of common threads emerge, which should be considered by leaders seeking to preserve and strengthen the liberal world order. The first is that American confidence that the end of the Cold War also denoted the end of the global ideological struggle was premature. As this collection shows, across the globe, the contemporary paradigm of governance consisting of democracy and liberalization is being challenged. This challenge emanates not only from China, despite the media attention focused on this purported rivalry, but also from gangs and cartels in Latin America and nonstate actors in Africa, the Middle East, and elsewhere. The "new brand" of global jihadist terrorism traffics not just in weapons, oil, and people, but also in a profound sense of communal marginalization. And it has global reach—the arc of connectivity spans from the cartels in Mexico to the insurgents in Mindanao, and encompasses the gangs of Central America, the cartels of Colombia, al-Qaeda affiliates in the Maghreb, Boko Haram in Nigeria, al-Shabaab in Somalia, al-Qaeda in the Arabian Peninsula, the Taliban in Afghanistan, Lashkar-e-Taiba in Pakistan, the LTTE in Sri Lanka, and the Jemaah Islamiyah in Indonesia. Hovering over many of these is the specter of a new caliphate, ISIL.

The second is that new technology not only reduces the "capacity gap" between conventional and unconventional forces, but also introduces new vulnerabilities to America's security and that of its allies. Communications technology, which has been a force for democratic change, has also proven to be a powerful enabler for recruitment to groups like ISIL, and facilitated its ability to coordinate attacks in Paris, Brussels, Istanbul, and across the Levant. The rise of social media has allowed remote groups to have a global presence; consider, in an age before propaganda videos could "go viral," would Abubakar Shekau, the leader of Boko Haram, have a presence outside of the Lake Chad Basin? Further, the innovations that have made life easier for affluent Westerners, including personal computers and web- or cloud-based technologies, are increasingly being exploited by criminal groups to gather funds and collect valuable personal information. Even more troubling than the rise of internet scams, however, is the looming possibility of major hacks and cyber warfare. The "Sony Hack" in the fall of 2014 was quickly relegated to a late-night punch line, overlooking the significance of North Korean operatives having the capacity to hack into a multibillion dollar company. The following summer, news broke that the U.S. Office of Personnel Management had been hacked. The records of an estimated 21.5 million people who had worked for, or had applied for positions within the U.S. federal government were compromised in the breach, which was traced to China.

Third, many of the states within the international community are at a severe handicap in their efforts to mitigate the unprecedented threats to their sovereignty. Their weakness is exacerbated by networked adversaries, of either the terrorist, insurgent, or criminal types, which eat away at state institutions—and more importantly, erode the social contract between governments and the governed. The proliferation of weak, fragile, and failed states leaves big holes in the rule-based system of sovereign states, thus weakening the system, and rendering vulnerable all the gains that flow from that system. An alternative model of global disorder is emerging in which the public good or public interest is a constant casualty. In this alternate global disorder, pure self-interest, violence, and deceit are the major currencies, and the vulnerable of the earth are the constant victims.

Finally, strong states, led by trusted, capable governments that are accountable to their populations, are the most effective line of defense against these threats. While "state-building" has become anathema in some circles, it is clear that improving state governance is a necessary corrective measure in the fight against endemic insecurity. Learning from our previous endeavors and identifying effective means of building partner capacity is necessary if the United States is to remain a global leader. Exporting democracy, defined merely by elections, without corresponding rule of law and economic development, will likely exacerbate the disruptive dynamics already at play.

Though these themes are addressed by all of the authors, the correct "solution" to the problems described is elusive, and remains a source of disagreement even among the most reasonable people. What role should America play in global ideological conflicts, how legal and regulatory systems should adapt to technology, and how best to promote state-building globally are all thorny questions with no obvious answers. Undeniable, though, is that the discussions surrounding how America and its allies must respond to these challenges should be well-informed, nuanced, and timely.

Ultimately, our purpose is not to offer a comprehensive review of every threat facing the United States and its allies or to prescribe solutions. Rather, it is to provide insight for understanding the contemporary threat environment, and offer strategies to mitigate the accelerating trends and forces that threaten us. If this book contributes to a reorientation away from our siloed, traditional approach to analyzing national security, to a more holistic understanding of the threat landscape we face in the 21st century, we will count ourselves successful. Most importantly, we hope that this book will generate discussions recognizing the gravity of these threats among those with the power to affect change in our current policies for addressing the challenges on the horizon, in our backyards, in our bureaucracies, and in our future.

Notes

[1] The Fund for Peace, a nongovernmental organization, ranks the stability of 178 countries each year based on 12 key political, social, and economic indicators (which in turn include over 100 sub-indicators). The ranking categories are "sustainable," "stable," "less stable," "low warning," "warning," "high warning," "alert," "high alert," and "very high alert." The Fund for Peace, "Fragile States Index 2015," available at <http://library.fundforpeace.org/library/fragilestatesindex-2015.pdf>.

[2] Michael Miklaucic and Moises Naim, "The Criminal State," in *Convergence: Illicit Networks and National Security in the Age of Globalization*, ed. Michael Miklaucic and Jacqueline Brewer (Washington, DC: National Defense University Press, 2013).

[3] Stanford University, "Liberation Tigers of Tamil Eelam," *Mapping Militant Organizations Project*, available at <https://web.stanford.edu/group/mappingmilitants/cgi-bin/groups/view/225>.

[4] Scott Helfstein with John Solomon, "Risky Business: The Global Threat Network and the Politics of

Contraband," *The Combating Terrorism Center at West Point*, May 2014, available at <https://www.ctc.usma. edu/v2/wp-content/uploads/2014/05/RiskyBusiness_final.pdf>.

[5] U.S. Department of Justice Drug Enforcement Administration, "FY 2014 Performance Budget Congressional Submission," available at <https://www.justice.gov/sites/default/files/jmd/legacy/2014/05/16/dea-justification.pdf>.

[6] Remi L. Roy, "Dissecting the Complicated Relationship Between Drug Operations and Terrorism," *The Fix*, October 8, 2014, available at <http://www.thefix.com/content/dissecting-confounding-nexus-drugs-and-terror>.

[7] Jeremy Kahn, "Mumbai Terrorists Relied on New Technology for Attacks," *New York Times*, December 8, 2008, available at <http://www.nytimes.com/2008/12/09/world/asia/09mumbai.html?_r=0>.

[8] "Al-Qaida / Al-Qaeda (The Base)," *GlobalSecurity.org*, last modified on December 17, 2015, available at <http://www.globalsecurity.org/military/world/para/al-qaida.htm>.

[9] Michel Camdessus, "Money Laundering: The Importance of International Countermeasures" (address given at the Plenary Meeting of the Financial Action Task Force on Money Laundering, Paris, February 10, 1998).

[10] U.S. Department of State, "Country Reports on Terrorism 2014," June 2015, available at <http://www. state.gov/documents/organization/239631.pdf>.

[11] Joseph Stiglitz and Linda J. Bilmes, *The Three Trillion Dollar War: The True Cost of the Iraq Conflict* (New York, NY: W.W. Norton & Company, 2008).

[12] "Economic Costs," in *Lessons Encountered: Learning from the Long War*, ed. Richard D. Hooker, Jr. and Joseph J. Collins (Washington, D.C.: National Defense University Press, 2015).

[13] Molly Molloy, "The Mexican Undead: Toward a New History of the 'Drug War' Killing Fields," *Small Wars Journal* 9, no. 8 (August 2013).

[14] David Gagne, "InSight Crime 2014 Homicide Round-up," *InSight Crime*, January 12, 2015, available at <http://www.insightcrime.org/news-analysis/insight-crime-2014-homicide-round-up>.

[15] American Immigration Council, "A Guide to Children Arriving at the Border: Laws, Policies and Responses," June 6, 2015, available at <http://immigrationpolicy.org/special-reports/guide-children-arriving-border-laws-policies-and-responses>.

[16] Daniel Valencia, "The Evolving Dynamics of Terrorism: The Terrorist-Criminal Nexus of Hezbollah and The Los Zetas Drug Cartel" (capstone project, Institute for National Strategic Studies, 2014), available at <http://academics.utep.edu/Portals/4302/Student%20research/Capstone%20projects/Valencia_Evolving%20 Dynamics%20of%20Terrorism.pdf>.

[17] Douglas Farah, "Terrorist-Criminal Pipelines and Criminalized States: Emerging Alliances," *PRISM* 2, no. 3 (2011): 15-32.

[18] Kelly Olsen, "North Korea's Secret: Room 39," *The Salt Lake Tribune*, June 11, 2009, available at <http://www.sltrib.com/ci_12566697>.

[19] Freedom House, "Freedom in the World 2016," January 27, 2016, available at <https://freedomhouse. org/sites/default/files/FH_FITW_Report_2016.pdf>.

[20] For an argument that the world has become less violent and more secure over time, see Steven Pinker, *The Better Angels of Our Nature: Why Violence has Declined* (New York: Viking Press, 2011).

[21] The Institute for Economics and Peace, "Global Peace Index Report 2015," available at <http://static. visionofhumanity.org/sites/default/files/Global%20Peace%20Index%20Report%202015_0.pdf>.

[22] See Tuesday Reitano and Andrew Trabulsi's chapter entitled, "Virtually Illicit: The Use of Social Media in a Hyper-Connected World," in this volume.

I. Slouching Toward Dystopia

1
The Global Crisis of Governance

Phil Williams

The world has entered a period of kaleidoscopic, irregular conflicts in which the reassertion of traditional geopolitical rivalries is inextricably linked with the activities of a bewildering assortment of violent nonstate actors (VNSAs). States in the Middle East, for example, increasingly define national interests in terms of sectarianism; however, the civil war in Islam is being played out not only in the direct, competitive dynamic of Saudi Arabia and Iran, but also through the proxies these two states use, including sectarian factions, tribes, warlords, insurgents, and transnational criminal organizations (TCOs). These VNSAs pursue their own agendas, yet interact and ally with states when it is convenient and advantageous to do so. They might, on occasion, act as state proxies; but, they are not pawns. On the contrary, they generate their own conflict dynamics and follow strategic imperatives that sometimes complement the actions of their state allies, but, on other occasions, can equally well confound them.

In South Asia, D-Company, the criminal organization led by Dawood Ibrahim, is closely allied with Pakistan's intelligence agency, Inter-Service Intelligence (ISI). Indeed, D-Company is used by ISI to provide money and logistic support for terrorist actions against India, and for assistance in introducing counterfeit currency into India. The organization also provides plausible deniability for ISI and for Pakistan, in return for which Pakistan provides sanctuary and protection.[1] The relationship is symbiotic and D-Company enjoys a high degree of impunity, while continuing to profit from its extensive portfolio of transnational criminal activities. Moreover, at times, its close relationship with Pakistani intelligence and military services has hindered efforts to improve relations between India and Pakistan.

In other words, traditional geopolitics is alive and well, but is sharing the stage with a variety of new players that are useful to states but are not necessarily or not fully under state control. This is a complex picture in which states, at the very least, remain the major players, and often set the frameworks within which VNSAs operate. At the same time, VNSAs add elements of fluidity and unpredictability, complicating state calculations and rendering desired outcomes uncertain. Both Iraq and Afghanistan have revealed that favorable power asymmetries do not guarantee victory, that military power often matters less than political resilience, and that even political success can prove impossible to sustain. Such complexities and uncertainties are increasingly reflected in current U.S. military planning with its focus on gray zones, irregular operations, and hybrid enemies, as well as its reliance on technological superiority to provide what has been characterized as the "third offset."[2]

Underlying this panorama of actors, and somewhat obscured by the current crises and tensions, is a fundamental global trend in which the nature of governance provided by many states is inadequate and unable to meet the needs, demands, and expectations of their citizens. In other words, the Westphalian order is undergoing a long-term secular decline that is bringing with it a series of convulsions along with what Nathan Freier terms, "prolific insecurity."[3] This does not mean that the state is going away anytime soon; the state remains critical in defining political order, and will continue to play much of that role for the foreseeable future.

Yet there is an important, albeit often unrecognized, distinction between failed states and failed governance. Governance can fail dismally even while the formal state remains intact. In many cases around the world, state governance is failing to meet the needs of citizens even as states continue to meet all the formalities of statehood and are recognized by their peers as part of the international community. Sovereignty as a formal legal status in which the state recognizes no higher authority than itself and mandates nonintervention in its domestic affairs is alive and thriving; sovereignty as exclusive and full territorial control and protection of citizens within the area of the state's jurisdiction, however, is increasingly illusionary. As a result, other actors are stepping in, both to challenge the state directly and to provide governance where the state has limited presence or is simply absent. In many countries, especially in the developing world, the traditional equation between the state and governance has broken down. This has four major consequences: high levels of violence in many societies, the rise of alternative loyalties that supersede loyalty to the state, the emergence of alternative governance mechanisms, and large numbers of refugees and migrant flows from countries where governance—but not the state—has effectively failed. These consequences are discussed more fully below.

Although there is considerable hand-wringing over what is often seen as the failure of global governance, it should be emphasized that this is not what is being discussed here. This chapter is not about global governance; it is concerned with governance at the state level, which is becoming a problem of truly global proportions and one that can only exacerbate the shortcomings of global institutions and thereby underline the paucity of effective global governance. If the constituent units that make up the global community of states cannot effectively govern the territory and populations they nominally control, then developing common solutions to global challenges such as climate change will likely prove impossible.

In other words, the continued absence or, charitably, the weakness of global governance is likely to be a major consequence of the global crisis of governance at the state level. Although we still typically refer to the state as "Westphalian," more often than not contemporary states can best be described by terms such as "weak," "fragile," "anorexic," "truncated," "predatory," "corrupt," "criminal," "mafia," "parasitic," "vampire," "exploitative," or "kleptocratic," to name just a few appropriate terms.[4] These descriptors refer either to the absence of certain positive, desirable and, perhaps necessary, state characteristics or the presence of undesirable or negative characteristics. In effect, they reflect the realities of many states that differ significantly from the traditional notion of the Westphalian state that emerged primarily in Europe and North America and reached its zenith with the total wars of the 20th century. These realities can be summed up as the qualified state, with the word "qualified" suggesting two things: (1) that the state does not exhibit the attributes traditionally—if sometimes erroneously—associated with the modern nation-state in the 20th century; and (2) that many contemporary states cannot be understood without the use as qualifiers for some of the descriptors identified above.

In other words, the key issue is not the weakness, frailty, or inadequacy of institutions or norms designed to promote collective security and provide governance at the global level. The crisis of governance that is most disturbing and corrosive is occurring at the state level and reflects a world of more and more perennially weak, corrupt, or captured states that are unable or unwilling to meet the needs of their citizens, to provide an inclusive fold of protection and provision, to evoke the continued loyalty of their citizenry, to maintain the rule of law, to impose and maintain order in their major cities, and to control their borders.

Although this crisis manifests itself differently in different parts of the world, it has become a crisis of near global proportions. This is not to suggest that all states are undergoing a crisis of governance. There is a significant minority of states, most of them associated with the U.S. alliance systems of the Cold War, and located largely but not exclusively in the Northern Atlantic, that retain high levels of authority, legitimacy, and effectiveness. To talk about a global crisis of governance, therefore, is to be somewhat guilty of hyperbole. Yet it is worth noting that, in recent years, even the United States has suffered from intermittent bouts of institutional paralysis and has pursued economic and fiscal policies that have accentuated rather than reduced the inequalities in American society. Moreover, even European states have been fraying at the edges, and the economic problems of Greece, Spain, and—to a lesser degree—Italy suggest that many of the characteristics of states in the Global South are creeping northwards.

Yet, partly because of the dominance of the state system and the attractive fictions associated with territorial sovereignty, the scope and nature of the crisis of the state are largely unrecognized. Neither denial nor avoidance of the issues, however, is an adequate strategy. By clinging to fictitious notions such as the sovereign equality of states, denial and avoidance somehow become not only more palatable, but a powerful inhibitor to the development of coherent responses. Unfortunately, the results of denial are likely to be even more far-reaching and corrosive. After all, states are the building blocks of regional or global governance; consequently, crises of governance at the state level inevitably undermine the already modest and flawed efforts to provide more effective governance at the global level.

This argument not only runs against the orthodoxies of much international relations theory, but also against the valuable and enduring artifacts of the interstate system— embassies, national armies, national intelligence agencies—as well as the explicit rules of international law and the implicit or tacit codes of conduct that serve to constrain great power behavior. The central proposition here, however, is that this elaborately constructed edifice is built on fragile foundations. Put somewhat differently, the somewhat idealized concept of the modern sovereign state system obscures a reality that is much more complex, partial, incomplete, and uneven. Unfortunately, the deviations from the idealized norm are often treated as anomalies, that are typically (and often somewhat glibly) explained by the fact that states in much of the developing world are at a different (and less advanced) level of modernization than the advanced postindustrial states, or have been hijacked by groups and individuals, intent on exploiting the state for their own purposes. If looked at empirically, however, the relatively few legitimate and effective states are the real anomalies and these less legitimate and less effective, or "qualified" states are much more prevalent. Unfortunately, this perspective is also obscured by a fixation with failed states—that are in reality few and far between. The real problem is not state failure, but a state that, in many cases, has to be qualified with terms that are pejorative.

Instead of explaining the qualified states, therefore, it might be useful to focus briefly on why the relatively successful nation-states developed the way they did. Drawing on the seminal work of Charles Tilly, it is clear that this evolution was a result of war making and state making going hand in hand.[5] Moreover, the most salient feature of this juxtaposition was that it gave the state and the society a degree of congruence that emerged organically out of the challenges of fighting wars in an era, and in regions, characterized by high levels of nationalism and industrialization. The imperatives of fighting total wars ensured that the state became good at resource extraction. Perhaps equally significant, however, is that the state became good at resource provision, reflecting an implicit social contract based on a common experience: in effect, the state protected its citizens, who in turn identified with the state and supported it with their lives if necessary. The concomitant was a strong sense that the collective sacrifice of the citizens needed to be rewarded through social provision. This started in Germany in the 1880s when Chancellor Otto von Bismarck introduced social insurance, which was designed to reduce migration to the United States and to increase support among the populace for the new German state. A few other European states began to emulate this in the first decade of the 20th century, but it was World Wars I and II that provided the impetus for both the deepening and the widening of this system. Indeed, it was no coincidence that the welfare state developed most fully in Europe in the aftermath of World War II. Providing economic opportunities and, where possible, minimizing economic inequalities helped to ensure that the relationship between the state and the society remained copasetic. Moreover, states forged by warfare became very inclusive. The collective effort required for survival encouraged rather than discouraged inclusion in the society. It also provided a basis for social cohesion that was reflected in the family and other components of civil society. As Herbst notes:

> in Europe there was an almost symbiotic relationship between the state's extractive capacity and nationalism: war increased both as the population was convinced by external

threat that they should pay more to the state, and as, at the same time, the population united around common symbols and memories that were important components of nationalism. Fighting wars may be the only way whereby it is possible to have people pay more taxes and at the same time feel more closely associated with the state.[6]

The concomitant of this is that where major interstate war was absent, the state failed to develop a comparable capacity either for extraction or for social provision. This is true of many states in Africa, which suffered from what Herbst terms, "the incompleteness of state consolidation."[7] In Latin America, too, the state was much weaker because of the absence of the large—and sometimes—prolonged interstate wars that helped shape, extend, and consolidate the state in Europe and, by extension, North America.

Another way of thinking about the prototypical Westphalian state is in terms of the sophisticated and successful management of a series of complex balancing acts:

- the balance between resource extraction and the provision of services;
- the balance between the state and the society;
- the balance between the exercise of political power on one side, and the social contract between governors and the governed on the other;
- the balance between top-down rule and bottom-up expressions of needs and preferences;
- the balance between security and welfare;
- the balance between responsibility and deference; and
- the balance between multiple roles and identities, as citizens juggle their allegiance to the state with their allegiance to nonstate entities and organizations.

Expressed in this way, it is clear that well-balanced states are relatively rare. Moreover, even when balance has been attained it is often difficult to maintain. In many instances, it proves highly elusive. Consequently, the relationship between state and society is all too often characterized by disequilibrium rather than equilibrium, by jarring and fractious imbalances that are highly corrosive of good governance. As discussed more fully below, since the 1970s, both globalization and the dominance of the neoliberal ethos that relegates the role of the state to the promotion of free markets, have created and perpetuated these imbalances.

Against this background, this chapter sets out to consider the causes of disequilibrium. It then looks at the manifestations of the crisis of governance at the state level. In the final section, it suggests that priority should be given to reestablishing good governance and that this is not synonymous with strengthening the state. In fact, it requires what might be described as shared governance, in which the state is only one of several actors providing governance.

The Crisis of the State

The idea of the Westphalian state has such powerful connotations as both an abstract concept and as an organizing principle for world politics that all too many international relations scholars have been reluctant to look critically at the underlying realities. One of the first and most important exceptions was Robert H. Jackson who developed the notion of quasi-states, a particularly powerful qualifier, if ever there was one.[8] Jackson recognized that not all states are truly Westphalian in origin, and argued that many of the states that emerged from the decolonization process "are independent largely by international courtesy. They exist by virtue of an external right of self-determination—negative sovereignty—without yet demonstrating much internal capacity for effective and civil government—positive sovereignty."[9] This was, and arguably still is, particularly true of many African states that, in his view, "frequently lack the characteristics of a common or public realm: state offices possess uncertain authority, government organizations are ineffective and plagued by corruption and the political community is highly segmented ethnically into several 'publics' rather than one."[10] Herbst, too, has described African states in a similar manner, noting that the incomplete consolidation process created a fundamental "contradiction of states with only incomplete control over the hinterlands but full claims to sovereignty."[11] In turn, this led to a series of political pathologies, including systems in which "leaders who steal so much from the state that they kill off the productive sources of the economy; a tremendous bias in deference and the delivery of services toward the relatively small urban population; and the absence of government in large parts of some countries."[12]

Yet the problem goes beyond this. As Jackson also notes, the state in Africa is regarded primarily as "an exploitable treasure trove devoid of moral value.... Corruption is integral rather than incidental to African politics. Self-enrichment and personal or factional aggrandizement constitute politics."[13] Other observers have been even more explicit, characterizing the state in much of Africa as a predatory state, where "anyone with an official designation can pillage at will.... Their over-arching [sic] obsession is to amass personal wealth...."[14] Some progress has obviously been made since the 1990s, and at least some of the most egregious and blatant forms of corruption and exploitation have been replaced by more subtle and less overt examples. Moreover, under pressure from donors and groups like Transparency International, mechanisms of participation, accountability, and transparency have been put in place. The difficulty, however, is that these mechanisms are circumvented and undermined by what one scholar has summarized as patrimonial structures and practices, personal rule and "clientelism," and a significant disconnect between the state and society.[15]

In other words, all too often the state in Africa is for the rulers rather than the ruled. As William Reno pointed out almost 20 years ago, private or selfish interests often take precedence over collective interest or the public good.[16] Straightforward in its conception, Reno's observation is profound and far-reaching in its implications. It suggests that the state in Africa is a vehicle for predation and corruption rather than for serving the citizens and is widely seen as such. The state in Africa provides opportunities for private resource acquisition and that, rather than a genuine conception of public service, is the main attraction of political life. Indeed, controlling the state has become the strategic equivalent for political entrepreneurship of military control of the high ground in traditional forms of warfare.

Underlying such a system is the subordination of the public good to selfish interests—whether personal, familial, tribal, or ethnic—that are motivated by greed and expediency. This is not peculiar to Africa. Indeed, if the state in Africa is far from the Westphalian model, the same can also be said about the state in parts of Latin America. The Central American state, for example, has been variously described as "improvisational," "truncated," and as a paper Leviathan.[17] Specific country studies have sometimes yielded similar characterizations, with the state in Guatemala, for example, being described as "anorexic."[18]

Michael Mann, an eminent sociologist who has written extensively about the state has summarized the reasons why "Latin American states developed and will develop according to their own rhythms." He puts it thus:[19]

Two distinctive features delayed the emergence of true nation-states. (1) The military/ fiscal pressures were much weaker…. Wars were fewer and smaller, and so states and their militaries also remained small. Taxation rates were much lower than in Europe…. Since provincial elites were not bothered much by the state, they retained their local controls. States continued to rule their provinces indirectly, through the caciques, the local bosses. The rich paid virtually no taxes, and even the poor paid less than they did in Europe. States remained weak and ruled through rural landed oligarchies, which stifled pressures for land reform and for greater equality. (2) Greater ethnic differences remained for much longer. Most colonies in Latin America did not almost completely exterminate their indigenous peoples…. Racial differences between whites, mestizos, blacks, mulattos and indios generally reinforced class differences. In many areas the upper classes/castes considered themselves to possess an altogether superior "civilization" to the indios. Some still believe this. This means that the continent has long possessed unusually steep and deeply entrenched class/caste hierarchies. These profound differences were also expressed regionally. Regions settled by whites dominated regions populated by indios.[20]

Mann goes on to argue that this combination of factors ensured that infrastructure penetration of their territories by Latin American states remained feeble, while "levels of class, ethnic and regional inequality among the citizen body" remained high.[21] At the same time, the dominance of the elites was self-perpetuating and inhibited the emergence of powerful state institutions. Indeed, it should be emphasized that state structures, processes, and activities—such as low levels of direct taxation on income or wealth—that are sometimes interpreted as evidence of state weakness are in large part a consequence of choices made by elites whose primary concern is with extending and perpetuating their own wealth and privilege. As Sarah Chayes argues, "acute corruption should be understood not as a failure or distortion of government but as a functioning system in which ruling networks use selected levers of power to capture specific revenue streams. This effort often overshadows ordinary activities routinely connected with running a state."[22] From this perspective, the state in Latin America has, until recently, often been very effective in providing protection, cover, and access to resources for the few while ignoring the needs of the many. In the past, the growing concern over corruption and inequality, has generated left-wing insurgencies based on the desire for greater social justice and equality. Moreover, in 2015, popular protest at egregious corruption led to the ouster of the Guatemalan president, and the early months of 2016 saw global outrage over the revelations of the "Panama Papers."[23] Although such developments have created dents in the existing systems, they have done little to remedy the fundamental shortcomings associated with pervasive and systemic corruption. Indeed, in all too many cases, "governments have been repurposed to serve an objective that has little to do with public administration: the personal enrichment of ruling networks. And they achieve this aim quite effectively. Capacity deficits and other weaknesses may be part of the way the system functions, rather than reflecting a breakdown."[24]

In other words, even states that are weak in terms of provision of services might actually be both strong and adept at the extraction of rents and corrupt payments. In some instances, this system can be described as one of structured corruption where control is consolidated at high levels.[25] In others, the corruption system is pervasive but lacks "the same degree of consolidation at the top of the pyramid. Monopolies on the instruments of force may be less complete, so elite networks may engage in open, violent competition to capture revenue streams."[26] It is arguable that the first kind of relationship exists in Central Asia, where participants in the state apparatus acquire and control the rents associated with the trafficking of opium and heroin from Afghanistan to Russia. According to David Lewis, drug trafficking in Central Asia:

> is conducted with the active connivance and support of state institutions, controlled by senior security officers, government officials, and parliamentarians who have effectively nationalized drug transit through the region. They have brokered lucrative deals with Turkish and Russian criminal groups and with Afghan suppliers, many of whom also benefit from close relations with state structures in their countries.[27]

This is in stark contrast to West Africa; consider, for example, Guinea-Bissau, where civilian and military elites have fought over control of the rents from cocaine shipments being transshipped to Europe by Colombian and Venezuelan drug trafficking organizations. Whatever the precise arrangements, it is clear that political elites are exploiting state power and position to enrich themselves, while ignoring their responsibilities to citizens.

In Russia, there is a consolidated corruption pyramid centered on Vladimir Putin. As Karen Dawisha notes, Putin "has built a system based on massive predation on a level not seen in Russia since the tsars."[28] He has done this in ways that ensure he and his cronies benefit both politically and financially, and on a scale that would be the envy of infamous looters like General Abacha of Nigeria and President Mobutu of Zaire. Putin has achieved enormous political power and massive wealth through a system of "state capitalism," in which "the state nationalizes the risk but continues to privatize the rewards to those closest to the president in return for their loyalty."[29] Moreover, successful entrepreneurs who are not closely linked to Putin or supportive of his regime become targets of corporate raiding through the justice system.[30] The other result is that, as with neoliberalism (which is discussed more fully below) the citizen becomes the victim of greed, rather than the beneficiary of state protection and support. As Dawisha notes:

> the biggest threat to the success of ordinary Russians occurs…when Russia's all-powerful overlord, or one of his cronies, demolishes a village to build a palace, steals the money intended for health reforms, stymies innovation by maintaining state ownership of patents, or sends waves of tax, fire, and health inspectors as part of a shakedown. The only way for ordinary Russians to avoid state predation is to keep their heads down and believe in fate, or turn into cheerleaders of the system in order to gain insurance and a few crumbs from the table.[31]

Outside the former Soviet Bloc, this kind of predation and exploitation, although less blatant, is often justified by what elites claim is necessary to meet the requirements of competition in a globalized world. Certainly, globalization has created a new set of challenges for states. Yet at its core, globalization can be understood simply as increased connectivity among societies that has resulted from speed, ease, and low cost of global communications. David Held, one of the major theorists of globalization, has emphasized that one of the most salient features of globalization is the vast flow of people, money, commodities, information, messages, digital signals, and services—around the world.[32] Not only are these flows much denser and more rapid than ever before, but the transaction costs have shrunk enormously. The dominant assessment of globalization emphasizes the benefits of all this: the globalization of trade, finance, information, and communications systems is regarded as a benevolent development, and even though the accrued benefits are not distributed evenly, benefits do accrue to all—a high tide lifts all boats.[33] Relative gains are less important than the absolute gains that are made by all countries. Moreover, it is not simply that technological advancement has reduced the costs of global trade and finance; globalization also involves the triumph of liberal democracy and the free market economy and is the natural concomitant of the collapse of the Soviet Bloc and U.S. victory in the Cold War.

Similarly, it is argued, the benefits of globalization for developing economies are very real. In the long term, connectivity to the global system is essential for development and prosperity; being disconnected is a recipe for continued poverty, despair, and instability.[34] Yet globalization has another—rather darker—side that has been well captured by Thomas Friedman. An avowed champion of globalization, Friedman is also sensitive to its adverse effects. As he notes:

> The globalization system…is not static, but a dynamic ongoing process: globalization involves the inexorable integration of markets, nation-states and technologies to a degree never witnessed before—in a way that is enabling individuals, corporations and nation-states to reach around the world farther, faster, deeper and cheaper than ever before, and in a way that is also producing a powerful backlash from those brutalized or left behind by the new system.[35]

In other words, Friedman acknowledges that globalization has winners and losers—and that the pain for the losers can be enormous. For Friedman though, the pain is essential to progress. For critics of globalization, in contrast, the costs exceed the gains. Benefits from globalization are outweighed by its disruptive impact on employment, traditional cultures, and state capacity to govern. To take liberties with the high tide analogy, the globalization critique suggests that some of the boats will be driven onto rocks as the economic gap between the most developed and the least developed countries and regions increases rather than decreases. One astute observer has even termed this "the globalization gap."[36] In his view, "globalization encourages the well-positioned to use tools of economics and politics to exploit market opportunities, boost technological productivity, and maximize short-term material interests in the extreme. The result is a rapid increase in inequality between the affluent and the poor."[37] In short, globalization creates not only losers, but also real victims. Some segments of the population are both excluded from the benefits of globalization and seriously hurt by market dynamics. In Saskia Sassen's judgment, many of these people are brutally and involuntarily expelled from the global economy.[38] They exist in what Manual Castells terms, "zones of social exclusion."[39] Such zones, which are economic as well as social, can be found in parts of Central Asia, in large parts of Africa, in Latin America and the Caribbean, and in South and Southeast Asia.

The downside of globalization is all too often glossed over by champions of the new globalized free market economy. This is not surprising: globalization has its own philosophical underpinnings, ideological justification, and policy rationale in the shape of neoliberalism. As one analysis notes:

> Neoliberalism is a rather broad and general concept referring to an economic model or 'paradigm' that rose to prominence in the 1980s. Built upon the classical liberal ideal of the self-regulating market, neoliberalism comes in several strands and variations. Perhaps the best way to conceptualize neoliberalism is to think of it as three intertwined manifestations: (1) an ideology; (2) a mode of governance; and (3) a policy package.[40]

The ideology emphasizes the power of unfettered markets and the virtues of economic interdependence. For its part, the "neoliberal mode of governance adopts the self-regulating free market as the model for proper government" and supersedes and relegates the more traditional approach "of pursuing the public good (rather than profits) by enhancing civil society and social justice."[41] The policy component focuses on economic deregulation, liberalization of trade and industry, and privatization of state-owned enterprises. Governments emphasize the reduction of social welfare provision and the rise of "new commercial urban spaces shaped by market imperatives."[42]

If globalization directly and indirectly challenges the power and authority of the state, neoliberalism provides an intellectual rationale for the state to relinquish not only power and authority, but also the responsibility for the welfare and security of its citizens. As one study observes, "while the state is far from dissolved, its functioning has been restricted in scope as it is becoming increasingly difficult to legitimately incorporate other values, interests and goals in the policy-making [sic] process than those fitting within neoliberal parameters."[43] Despite the mantra of good governance, which has been widely enunciated in the first decade and a half of the 21st century, globalization and neoliberalism have undermined the foundations of the state, and have led to what can only be understood as a widespread crisis of governance. As Clunan and Trinkunas note, one of the ironies is that "Western liberalism created the criteria for 'good governance' that states are expected to adhere to today, while at the same time undermining the ideological legitimacy and institutional capacity of state authority."[44] Hayek and Milton Friedman have trumped Hobbes. Perhaps nowhere was this more pernicious than in Central America, as notions of public interest or the collective good that seemed to come to the fore briefly in the aftermath of civil wars and the beginning of democratic transitions were subverted by neoliberalism.

Neoliberalism justifies the retreating state with its argument that "human well-being can best be advanced by liberating individual entrepreneurial freedoms and skills."[45] Neoliberal economic and political reforms implemented in Central America from the 1980s onward included "market liberalization, privatization of industry and state services, reductions in public expenditure, and opening to foreign trade."[46] While it is plausible that the contraction of the state and a fundamental loosening of state control over all facets of social and economic life were essential in totalitarian states, such as the former Soviet Union, the same kinds of processes were not necessarily desirable elsewhere. Indeed, in countries where the state was controlled by privileged elites who were rarely responsive to the needs and demands of their citizens, neoliberalism not only neutralized any faint stirrings of responsibility brought on by democratic transitions but also justified the abdication of state responsibilities in the name of the free market. The results in terms of the security and well-being of many ordinary citizens in Honduras, Guatemala, and El Salvador have been catastrophic. In effect, neoliberalism provided a tacit justification for the perpetuation of elite domination on one side, and a deepening of social and economic exclusion on the other.

Robert Mandel has highlighted how neoliberalism has also undermined one of the critical functions of the state as Leviathan: the provision of its security and safety for its citizens. As he has noted, in many cases, the state has privatized security, effectively transferring the security function of the state to nonstate actors.[47] Mandel, however, does not see this as a problem. On the contrary, in an incisive and provocative analysis of the provision of security, he contends that "not all people expect or want a central state government to fulfill this task, and the privatized free-market mentality embedded in globalization suggests that security might well be treated as a service to be bought or sold on the open market...."[48] The difficulty with this approach, however, is that what has been understood as a public good is transformed into a preferential private good that is not equally available to the collective. Moreover, in many countries, the state is still widely regarded as the primary purveyor or provider of security, and when insecurity is endemic then the state becomes the culprit and typically suffers a serious decline in its legitimacy. While private security can be an invaluable supplement for efforts to provide public security, when it becomes a substitute for those efforts, or a luxury enjoyed only by the elite, the concomitant is usually an erosion of the state's political legitimacy.

Pervasive and Wicked Problems

The other problem with the dominance of neoliberalism is that it coincides with a period in which the challenges facing communities from the local to the global level are unprecedented. Continued global population growth, which will likely reach 9 billion people by 2050 (an increase of over 1.6 billion people from the current number), will place enormous demands on food and water supplies, sanitation systems, and employment opportunities. Rapid, unplanned, and often chaotic urbanization will result in larger cities that are potentially fragile and unmanageable. Global climate change will bring both unpredictable transformations and a very predictable but potentially massive increase in environmental refugees, to add to those fleeing violence, repression, and hunger. Moreover, the recurrence of periodic economic and financial crises will require careful and continued management by states and will be ill-served by a blind faith that markets will self-correct in ways that provide optimum outcomes. In other words, the challenges are enormous. They are also best understood as wicked problems, with the following, very distinct characteristics:

- They are multifaceted and typically cross several distinct—but ultimately interdependent—issue areas and policy domains. Moreover, every wicked problem challenging governance is a symptom of other problems, and responding to it has an impact (sometimes positive and sometimes negative) on other problems and the ability to manage them.[49]
- There are multiple stakeholders, both at the national and international levels, who often find it difficult to achieve a consensus on either the nature of the problem or the most appropriate solutions. Wicked problems encourage bureaucratic conflicts and typically require a "whole of government" approach simply to get to the point of agreement on the holistic nature of the problem. And even then effective implementation can prove enormously difficult.
- Measures of effectiveness, let alone success, are very elusive partly because of the dynamism of the problem and partly because "success" can often have inadvertent and unexpected consequences that make the problem even more intractable.
- Wicked problems are dynamic rather than static and constantly evolve and morph, often in ways that make them more resilient in the face of efforts to respond effectively to them. Often there is no obvious end game. Wicked problems can rarely be solved; consequently, they have to be managed and even this is usually highly problematic.

As the world increases in complexity and velocity, and wicked problems continue to fester and confound policymakers, state responsibility both for citizens and for policy innovation and creativity will become indispensable. Yet neoliberalism encourages the state to abdicate responsibilities beyond ensuring the optimum functioning of the market.

Neoliberalism has already been highly pernicious; continued adherence to its precepts and policies in a period of unprecedented change and massive turmoil is likely to be quietly apocalyptic in its consequences. What is already a huge gap between the Westphalian concept and the realities at ground level will turn into a chasm. Indeed, it is arguable that the point of no return has already been passed. Ironically, the financial crisis, which seemed to critics to demand a reappraisal of neoliberalism, has resulted in rigid reaffirmation of economic orthodoxies and the widespread imposition of austerity policies that further relegate the state.[50]

Manifestations of Crisis

The crisis of the state is not an abstract and future problem; it is already here and its consequences are immediate, severe, and far-reaching. The preceding analysis has already touched on social and economic exclusion and the expulsion of people from the licit economy. Yet there are four other immediate consequences of the crisis of the state that need to be discussed: high levels of violence and impunity; a loss of faith in the state as protector and provider and the consequent emergence of alternative loyalties; the emergence of alternative governance entities and mechanisms to compensate for the absence or shortcomings of the state; and massive refugee and migratory flows.

Violence and Impunity

While homicide rates are more complex than they sometimes appear, they provide a basic and telling metric of violence. All other things being equal, the higher the homicide rate in a country, the less well the country is doing in providing safety and security of its citizens. There are, of course, many variables in this, including history, culture, availability of firearms, the role of the family, the quality of community life, as well as governance. Nevertheless, it cannot be ignored that particularly high levels of homicides are most prevalent in Africa and Latin America. Indeed, in recent years, they have vied for the dubious honor of having the highest rates of homicides per 100,000 people globally. Most analyses conclude that Latin America has edged out Africa to top these charts. As one astute observer notes, "Latin America and the Caribbean are home to 8 of the top 10 most violent countries and 40 of the world's 50 most dangerous cities. Just four countries— Brazil, Colombia, Mexico and Venezuela—account for 1 in 4 violent killings around the world each year."[51]

What is also remarkable about these killings is that perpetrators are rarely identified and brought to justice. In other words, high levels of violence have been both facilitated and perpetuated by a culture of impunity. Such a culture develops when the state lacks either the capacity or the will to implement laws effectively. Indeed, it is closely linked to the ineffectiveness of the criminal justice system, whether this stems from a limited capacity to investigate crimes or a general reluctance to do so because of laziness, ineptness, corruption, or fear within the judicial system. The word "impunity" typically refers to exemption from punishment for certain kinds of crimes; a culture of impunity refers to a situation in which exemption or the lack of punishment has become the norm. In many countries, getting away with murder is not so much an idiom or figure of speech as it is a description of a widespread reality.

The concept of impunity has been most fully articulated in relation to perpetrators of human rights violations. One document submitted to the United Nations Commission on Human Rights (UNCHR) in February 2005, for example, defined "impunity" as "the impossibility, de jure or de facto, of bringing the perpetrators of violations to account— whether in criminal, civil, administrative, or disciplinary proceedings—since they are not subject to any inquiry that might lead to their being accused, arrested, tried and, if found guilty, sentenced to appropriate penalties, and to making reparations to their victims."[52] The same document also notes that:

> Impunity arises from a failure by States to meet their obligations to investigate violations; to take appropriate measures in respect of the perpetrators, particularly in the area of justice, by ensuring that those suspected of criminal responsibility are prosecuted, tried and duly punished; to provide victims with effective remedies and to ensure that they receive reparation for the injuries suffered; to ensure the inalienable right to know the truth about violations; and to take other necessary steps to prevent a recurrence of violations.[53]

If this statement were broadened from the focus on human rights violations to crimes more generally, particularly crimes of violence, it would encapsulate perfectly the notion of a culture of impunity that can develop in particular states. The situation in such states is often so dire that even in rare cases when punishment is imposed, it often fails to have the desired impact. This is especially the case if the criminal justice system in general and the penal system in particular are marred by corruption, weakness, or inefficiency.

Alternative Identities

The crisis of governance in many countries is also evident in a lack of faith in the state and its institutions. The triumphalist mantra of the free market has been accompanied by disappointment and disenchantment with the state, at least among significant sectors of the citizenry. This is hardly surprising; many states have abdicated social and economic responsibilities to accord with neoliberal economic dogma and meet the requirements of global financial institutions, most notably the World Bank and the International Monetary Fund. Ironically, these same institutions emphasize the importance of good governance while simultaneously demanding the pursuit of an economic orthodoxy that requires government stringency, austerity, and the reduction, weakening, or dismantling of social welfare mechanisms. The contradiction either escapes them or is rendered nonexistent by a profound if often implicit tendency to equate good governance with facilitating market primacy.

One result of all this has been that, in some instances, people have not only turned away from the state, but have turned to other entities to meet their needs. It is hardly surprising that alongside disillusionment and disaffection with the state, there has been an upsurge of support for, and loyalty to, these other entities, whether it is ISIL and the Caliphate, the Calabrian 'Ndrangheta (which is based predominantly on familial ties and loyalties), tribes and clans in Afghanistan, Syria, and Iraq, or the gangs in Honduras, El Salvador, and Guatemala—the primary affiliation of these people is not with the state, but with the smaller and, in many respects, more organic and highly functional organizations to which they belong. Whatever the hierarchy of needs, the state no longer seems to be meeting them. The massive significance of alternative loyalties is evident, for example, in the Maras of Central America, where gang membership and allegiance have filled the gaps left by the breakdown of family structures and by states that do little to earn the respect, let alone the loyalty, of young men and women who are marginalized at best and more often are brutally expelled from the formal economy and society. In societies where the chasm between the elites and the poor is unbridgeable, those who are disenfranchised often join gangs, which become a source of identity, support, and status. State authorities then move from indifference to hostility and punishment as gang members are no longer simply disenfranchised but also criminalized, thereby further alienating them and consolidating their identity.

Alternative Governance

Those who are dissatisfied with existing state governance sometimes turn to alternative governance, when it is available. Sometimes, they create their own governance. The gangs in Central America have done this; although they have been predatory, there are also cases where they have provided rudimentary order and justice, and even facilitated conflict resolution among members of the community in urban areas they control. Central America also provides examples of drug trafficking organizations providing governance and engaging in paternalistic behavior. Perhaps the best example of this was in the Guatemalan Department of Zacapa, where the Lorenzana family combined illegal and legal businesses, acted as a major employer, provided services and patronage to the community, and imposed a degree of order. Although the family was heavily involved in drug trafficking, it also owned and operated "15 construction companies (some of which are contracted by the state), transportation fleets, fruit companies and gas stations, which provide employment and therefore popular support as well as opportunities to launder drug trafficking proceeds."[54] The family was very clearly part of the political elite, with very good connections that enabled it to obtain numerous public works contracts.[55] "The family also has large tracts of land where they employ hundreds of people. At Christmas, they give out gifts to kids and bags of food to their parents."[56] Reportedly, in Zacapa, members of the Lorenzana family also "donated land and built 60 houses for families left homeless after the Rio Motagua flooded in 2010."[57] Not surprisingly, this paternalism provided strong social and political capital that enhanced the family's legitimacy and status.[58] It also made the Lorenzanas "hard to capture. On at least two occasions, Guatemalan and U.S. authorities were unable to get past the throngs of protesters who had been called to the streets because of Lorenzana family members' imminent arrests."[59] While such demonstrations were almost certainly orchestrated, at least in part, they also reflected the sentiment that the Lorenzanas were "civic benefactors."[60] In spite of this popular support, in 2011, the patriarch of the family, Waldemar Lorenzana, was arrested and in March 2014, he was extradited to the United States. The family was further weakened with the arrest of two of Waldemar's sons, and a U.S. Treasury designation of another son and a daughter as drug traffickers. While clearly a law enforcement success and a blow to the culture of impunity that had long prevailed in Guatemala, the takedown of the organization eroded, rather than augmented, governance. The state failed to fill the governance vacuum. Indeed, according to the International Crisis Group, the weakening of the Lorenzanas "has brought chaos in its wake. Waldemar and his family maintained a certain order among traffickers in the region that restricted the violence to internal account settling."[61] Moreover, "Zacapa residents say the Lorenzanas have sold off the fruit-export [sic] business that generated local jobs and abandoned much of their charitable work, such as support for a health clinic that reportedly gave the poor free care."[62] One restaurant owner noted that the Lorenzanas had "provided jobs and not just for field workers. They employed engineers and other professionals."[63] Waldemar, in particular, had been regarded with both respect and affection.

Something similar is also evident with ISIL. Although sometimes described as a proto-state, the Caliphate, operating from a territorial base, also has a transnational appeal and seeks to expand its influence and reach far beyond its immediate territory. As one analyst noted, when ISIL declared itself "the Islamic State," and no longer simply "the Islamic State of Iraq and Sham," the implication was that "its sovereignty was to extend across the entire world, not just Iraq and Syria."[64] In addition to this, however, it is worth noting that ISIL emerged as an alternative form of governance in two countries where governance was extremely poor, to say the least. In Syria, the Assad regime had lost legitimacy and was surviving only through repression and military force; in Baghdad, the regime retained a Shi'ite sectarian bent that made many Sunnis receptive to ISIL. In both Iraq and Syria, the state appeared to be self-serving, corrupt, inefficient, and exclusive. As an Egyptian former official notes, "bad governance has led to part of what we see today in the region."[65] For those who were marginalized and excluded, ISIL offered a preferable alternative. It is not that ISIL necessarily has to be good at governance; it simply has to be better, in at least some respects, than the existing states. According to one observer, "the Islamic State's strength in matters of governance consists in doing marginally better than others" and ensuring that the population under its control is "better off than in the hands of criminalized, shifting armed groups, alien militias or a vengeful government determined to punish them."[66]

For all this, there are highly divergent assessments of the capacity of ISIL to provide governance. On one side are those who see ISIL as highly predatory and repressive. Their arguments have considerable credence. One analysis, for example, based on a month's worth of documents obtained from Deir ez-Zor province in eastern Syria, where ISIL has been in control since July 2014, suggests that the bulk of ISIL funding (around 68 percent) comes from taxes and confiscation of property or, in essence, from extortion and predation, while 54 percent of its expenditures are directed to bases and to paying its fighters.[67] According to one report, ISIL has established a "predatory and violent bureaucracy that wrings every last American dollar, Iraqi dinar, and Syrian pound it can from those who live under its control, or pass through its territory."[68] It does this through "exacting tolls and traffic tickets; rent for government buildings; utility bills for water and electricity; taxes on income, crops and cattle; fines for smoking or wearing the wrong clothes," and so on.[69] It is clear from such reports that ISIL has become adept at resource extraction, one of the hallmarks of strong states. Yet, even though it has earned grudging praise for its social provision, its record in this area is far less impressive. Indeed, the report from Deir ez-Zor province noted that only 17.7 percent of ISIL expenditures were made by the Services Department.[70] The implication is that although ISIL has had some short-term success with governance, there is a major imbalance between resource extraction and service provision that could significantly undermine its governance efforts. Some observers have gone even further, and have argued that since ISIL funding—and, therefore, its governance efforts— are based largely on predation, they are not sustainable.[71] There have also been areas where ISIL rigidity has hurt, rather than assisted social provision. According to one journalist, this has been particularly evident in the health sector, where ISIL rules regarding gender and dress have hindered efficiencies, and preferential treatment given to fighters has created significant public resentment.[72]

On the other side are those who contend that "in the midst of the chaos, ISIL is deliberately and methodically establishing clear areas of definable civil governance, breathing new life into the memory of a series of caliphates that united a succession of Muslim empires until 1924."[73] According to this assessment, the group "built a holistic system of governance that includes religious, educational, judicial, security, humanitarian, and infrastructure projects, among others."[74] Although this level of governance has been particularly evident in Raqqa, similar programs have been introduced in towns in Aleppo province and elsewhere. ISIL reportedly offers humanitarian aid and has sought to maintain and repair sewer and electrical infrastructure. According to one report, ISIL has also "treated complaints seriously," arbitrated "old property or financial disputes," and, on occasion, even "punished its own members accused of abuse."[75] Such behavior has clearly enhanced its legitimacy.

Yet even these positive assessments have caveats. As one commentary notes, ISIL has "yet to demonstrate the capacity for the long-term planning of state institutions and processes."[76] Moreover, it has proved difficult to extend the level of governance in Raqqa to Mosul; even in Raqqa, U.S. air attacks have eroded some of the governance initiatives.[77] The decision by the Iraqi government to stop paying civil servants in ISIL-controlled areas has also had an impact on the ISIL economy.[78] Moreover, it bears reiteration that none of the positive assessments of ISIL governance suggests that ISIL provides an ideal form of governance. Much of its behavior is barbaric and reprehensible, and even some of its foreign fighters have become disillusioned, defected, and returned home.[79] ISIL has also driven out non-Muslim sectors of the population in the areas it controls, and its atrocities have contributed significantly to the refugee crisis in Europe in 2014 through the present.

Leaving the State: Refugees and Migrant Flows

Although there is a critically important legal distinction between political refugees and economic migrants, in both cases they are fleeing poor governance. Refugees sometimes seek political asylum, because they live in states that rely excessively on coercion and fail to respect the rights of the individual. In other cases, the state has failed to provide adequate protection from violent groups and individuals within the society. The arrival of tens of thousands of unaccompanied minors at the U.S. border in 2014, for example, was in large part a result of the high levels of violence in the Northern Triangle of Central America.

In 2010, "about fourteen people emigrated from Guatemala every hour" or around 330 people per day.[80] These migrants were seeking "better development opportunities, making an expensive, risky and, above all, difficult journey," in order to improve both their own quality of life as well as that of their relatives', "who remain in the country."[81] For an estimated 97.4 percent, the destination is the United States.[82] In other words, at that time, the pull factors were very important and were strengthened by the well-established communities of Guatemalans in the United States that provided a welcoming network for newcomers. What has changed significantly in recent years, however, is that the push factors have become increasingly salient. It is no accident, for example, that in Guatemala, "49.4 percent of all homicides in 2010 occurred in the five departments with the highest rates of emigration (Guatemala, San Marcos, Huehuetenango, Quetzaltenango and Jutiapa)."[83] Indeed, economic migration has increasingly been accompanied by levels of "forced displacement" that are likely unprecedented outside combat zones.

A United Nations High Commissioner for Refugees (UNHCR) study published in 2012 identified both risk zones and expelling zones in Guatemala, El Salvador, and Honduras.[84] Significantly, Petén in northern Guatemala, which is an important area for drug trafficking into Mexico, contained five expelling zones or hotspots, while the province of Guatemala had three, including the municipality of Guatemala.[85] In El Salvador, expulsion zones were identified in various municipalities in the departments of San Salvador, La Libertad, San Miguel, La Unión, and Usulután.[86] The Maras were the driving force behind forced displacements in most of these municipalities.[87] In Honduras, the pattern of forced displacements was more mixed, with Maras "mainly present in the capital cities (Tegucigalpa, Comayagua) and the country's commercial capital (San Pedro Sula and nearby areas)," while drug trafficking organizations operated "in the east of the countries (Gracias a Dios) and in some areas of the west and northwest (Atlántida, Cortés, Copán, and Ocotepeque)."[88] The zones of expulsion correlate remarkably well with the corridors of narcotic flows and the presence of gangs engaged in forced recruitment of boys and sexual violence against girls.

None of this is meant to suggest that these three states in the Northern Triangle had become failed states. Yet it is clear that the failure of governance, the inadequacies of local control measures, and the inability of the state to develop and maintain a clear monopoly on the use of violence, all contributed to the "pervasive insecurity" for children and young people.

In 2015, the European Union faced a refugee and migrant crisis that dwarfed the flows from Central America to the United States a year earlier. The arrival of hundreds of thousands of refugees and migrants willing to make the hazardous journey across the Mediterranean in small boats, and the many who drowned along the way, gave the issue an unprecedented visibility. The 2015 global report provided by UNHCR identified the top 10 sources of refugees in 2014. The countries of origin and the numbers of people are captured in Table 1.1.[89]

Table 1.1. Top 10 Sources of Refugees in 2014

Country of Origin	Number of Migrants and Refugees
Syria	3,880,000
Afghanistan	2,590,000
Somalia	1,110,000
Sudan	666,000
South Sudan	616,200
Democratic Republic of Congo	516,800
Myanmar/Burma	479,000
Central African Republic	412,000
Iraq	369,900
Eritrea	363,100

The trends continued in 2015, with the number of refugees from Syria increasing to 4.9 million, some changes in placement of countries in the top 10 source countries, and Iraq dropping out and Colombia coming in.[90] It is significant that there is a major crisis of governance and/or a level of conflict and violence that compel people to leave in all of the listed countries. Almost all the countries have suffered from extensive and, to different degrees, protracted conflict. For its part, "Eritrea is among the most closed countries in the world; human rights conditions remain dismal. Indefinite military service, torture, arbitrary detention, and severe restrictions on freedoms of expression, association, and religion provoke thousands of Eritreans to flee the country each month."[91] Indeed, according to Human Rights Watch, "Eritrea has no constitution, functioning legislature, independent judiciary, elections, independent press, or nongovernmental organizations; it does not hold elections," and all power is controlled by the president.[92] Not surprisingly, the rule of law is set aside for arbitrary and inhumane rule.

In other words, the arrival of tens of thousands of refugees and migrants looking for sanctuary as well as economic opportunities in Europe is being driven by political violence, pervasive fear, and fundamental insecurities, buttressed by a sense of deprivation and aspirations for something better. It is not simply that these refugees are leaving states unable or unwilling to guarantee their security and well-being; in many cases, the state is the source of the insecurity. Consequently, what is happening is that significant segments of national populations are literally voting with their feet. It is not coincidental that the 10 countries on the list would also be at, or near, the top of the list of countries in which there is an intense crisis of governance. Nor is it an accident that people are moving from qualified states of one kind or another and seeking refuge among that minority of states that still manages to approximate the Westphalian ideal.

The danger is that the refugee problem has now reached a point at which integrating the refugees itself is a wicked problem. Those who in the past might have been easily assimilated are now more likely to become victims of social, political, and economic exclusion in their destination countries. Consequently, they are more likely to end up alienated and hostile to their adopted state. Moreover, the existing population of states receiving large numbers of refugees is unlikely to be invariably welcoming. In some cases, there will even be a backlash against the state. In other words, the problems of maintaining state legitimacy that have been largely a problem of the developing world or the Global South could all too easily have contagion effects in the Global North, with these states also becoming victims of disappointed expectations. The question arises, however, as to what can be done about this crisis of governance. The final section of this chapter offers at least some preliminary answers.

Conclusion

The erosion of state power and authority seems to be an increasingly global phenomenon, although it is clearly most pervasive and overt in the developing world. Yet states everywhere still claim traditional prerogatives and powers, and more often than not, treat nonstate actors that, for whatever reason, assist in the provision of good governance as upstarts or potential usurpers at best, and existential threats at worst. State claims to exclusive authority, however, run into a whole series of difficulties that makes the claims quixotic and the states themselves appear increasingly hollow. As discussed above, many states in the contemporary international system face increasing capacity, legitimacy, and authority deficits.

One frequent response to such arguments and to the crisis of governance—on those occasions when it is at least recognized or acknowledged—is that the answer lies in strengthening and revitalizing the state. The inclusion in the discussion, however, of countries like Russia and Eritrea, both of which have forms of authoritarianism combined with extensive elite corruption and self-aggrandizement, suggest that this is not invariably a better solution. In all too many cases, the state is not above politics; it is simply a resource to be exploited and looted. It is a prize that easily translates into wealth—and this seems to happen not only in cases of state capitalism such as Russia, but also in states that formally subscribe to the tenets of neoliberalism. It is not coincidental that in spite of the predominance of neoliberalism, there is what appears to be a global pandemic of corruption. This should not really be surprising. If it is all about the market, free enterprise and corporate and personal profit, then it is only a small step to the manipulation and exploitation of the market—or what elsewhere in this book Nils Gilman terms the "plutocratic revolution from above." The state is not out of the equation, but in many cases, it has become primarily a generator of personal wealth for those in control.

The answer, therefore, might lie in the preceding discussion about the need for balance in several critical dimensions of the state. Framing the issues in these terms has several advantages. First, it makes clear that the central problem, except in very few extreme cases, is a long-term imbalance between state and society that creates a systemic crisis of governance. Second, in policy terms, such an approach invariably puts less emphasis on state-building and more on other options that are more appropriate to the conditions that prevail in these states and do not seek replication of an ideal type that in reality is relatively rare. From this perspective, empowering nonstate actors, both good and bad, might be a more realistic and beneficial approach than seeking to empower states, especially where this latter approach might be at the expense of the society and encourage repression of the population. The emphasis should be on good governance, rather than on the state as such, on restoring balance through whatever means are most feasible rather than simply reiterating variations on state-building based on a Westphalian ideal that is increasingly passé.

The implication is that rather than treating governance in zero-sum terms and seeking to maintain a monopoly on governance, states would be well served to accept and encourage alternative forms of governance as part of a shared, mixed, or hybrid approach. Desmond Arias, in a series of brilliant analyses looking at favelas in Brazil and municipalities in Medellin, Colombia, and Kingston, Jamaica, has identified how this can sometimes happen at the local level.[93] There are, of course, dangers in a hybrid approach to governance; one potential problem is the state providing impunity in return for violent armed groups providing governance. This is not acceptable if the armed group is overly repressive or violent. The critical consideration, therefore, should be the quality of governance and the requirement that it be predominantly protective and oriented toward social provision rather than predatory and rent-seeking. The most important task of the state—in at least some parts of the neoliberal world—might be to nudge market actors (even if they operate in the illicit market) and rival governance providers into accepting more of the roles and responsibilities of governance. Rather than a threat to the state, such an approach could prove to be the salvation of the state. In the 21st century, the only forms of governance that are likely to be sustainable in large swaths of the world are those that are, in effect, post-Westphalian. Paradoxically, the adoption of such an approach might be the one thing that enables the state to survive the current crisis of governance. Stephen Krasner once argued that states have adapted successfully because they have been willing to shed functions which they were unable to manage, including relinquishing authority over the way in which their citizens interacted "with the sacred," which was "no small thing."[94] The issue for the future is very similar but is not about the shedding of functions, so much as a willingness of the state to accept its neoliberal shortcomings, acquiesce in the loss of monopoly control over governance, and embrace the sharing of functions and the mutual pursuit of collective goods. It is partly that state predominance is not immutable. Even more important, however, is that the state does not necessarily represent the optimum set of political arrangements for meeting people's needs or for ensuring peace and stability. More organic, bottom-up forms of governance for all their shortcomings, might be the best available in the decades ahead. Difficult as all this might be for the state, the conditions for the citizenry are likely to be improved. And in the final analysis, this is what really matters.

Notes

[1] Sharad Joshi, Gretchen Peters, and Phil Williams, "The Transnational Security Threat from D-Company," in *The Future of Counterinsurgency: Contemporary Debates in Internal Security Strategy,* ed. Lawrence E. Cline and Paul Shemella (Santa Barbara, CA: Praeger, 2015), 259-283.

[2] Frank G. Hoffman, "The Contemporary Spectrum of Conflict: Protracted, Gray Zone, Ambiguous, and Hybrid Modes of War," in *2016 Index of US Military Strength* at *The Heritage Foundation*, available at <http://index.heritage.org/military/2016/essays/contemporary-spectrum-of-conflict/>. For a discussion of the "third offset," see Richard Purcell, "Hagel's 'Third Offset Strategy' Key to Maintaining U.S. Military Supremacy," *World Politics Review*, December 29, 2014, available at <http://www.worldpoliticsreview.com/articles/14744/hagel-s-third-offset-strategy-key-to-maintaining-u-s-military-supremacy>.

[3] Nathan Freier, personal communication with the author.

[4] Weak and fragile are widely used to describe many states. Several of these other terms are quoted in Cameron G. Thies, "Public Violence and State Building in Central America," *Comparative Political Studies* 39, no. 10 (2006): 1263-1282 on page 1267. For the notion of vampire state see Jonathan H. Frimpong-Ansah, *The Vampire State in Africa: The Political Economy of Decline in Ghana* (Trenton, NJ : Africa World Press, 1991); while criminal or mafia states are discussed in Moises Naím, "Mafia States: Organized Crime Takes Office," *Foreign Affairs,* May/June 2012, available at <https://www.foreignaffairs.com/articles/2012-04-20/mafia-states>.

[5] Charles Tilly, "War Making and State Making as Organized Crime," in *Bringing the State Back In*, ed. Peter B. Evans, Dietrich Rueschemeyer, and Theda Skocpol (Cambridge: Cambridge University Press, 1985), 169-191.

[6] Jeffrey Herbst, "War and the State in Africa," *International Security* 14, no. 4 (Spring 1990): 117-139 on page 122.

[7] Jeffrey Herbst, *States and Power in Africa: Comparative Lessons in Authority and Control* (Princeton, NJ: Princeton University Press, 2014).

[8] Robert H. Jackson, *Quasi-states: Sovereignty, International Relations and the Third World* (Cambridge: Cambridge University Press, 1993).

[9] Ibid.

[10] Robert H. Jackson, "Quasi-States, Dual Regimes, and Neoclassical Theory," in *International Law and International Relations: An International Organization Reader,* ed. Beth A. Simmons and Richard H. Steinberg (Cambridge: Cambridge University Press, 2007), 212.

[11] Herbst, *States and Power in Africa: Comparative Lessons in Authority and Control.*

[12] Ibid.

[13] Jackson, "Quasi-States, Dual Regimes, and Neoclassical Theory," 213.

[14] George B.N. Ayittey, *Africa in Chaos* (New York, NY: St. Martin's Griffin, 1999), 151.

[15] Rachel Flanary, "The State in Africa: Implications for Democratic Reform," *Crime, Law and Social Change* 29, nos. 2-3 (1998): 179-196.

[16] William Reno, *Warlord Politics and African States* (Boulder, CO: Lynne Rienner Publishers, 1999).

[17] R.H. Holden uses the term "improvisational," and H.H. Lentner the term "truncated." Both are quoted in Cameron G. Thies, "Public Violence and State Building in Central America," *Comparative Political Studies* 39, no. 10 (2006): 1263-1282 on page 1267. See also Agustin E. Ferraro and Miguel A. Centeno, "Paper Leviathans: Historical Legacies and State Strength in Contemporary Latin America and Spain," in *State and Nation Making in Latin America and Spain: Republics of the Possible,* ed. Agustin E. Ferraro and Miguel A. Centeno (Cambridge: Cambridge University Press, 2014), 399.

[18] Quoted in Tani Marilena Adams, *Chronic Violence and its Reproduction: Perverse Trends in Social Relations, Citizenship, and Democracy in Latin America* (Washington, DC: Woodrow Wilson Center Update on the Americas, September 2011), 16.

[19] Michael Mann, "The Crisis of the Latin American Nation-State" (paper presented at the University of the Andes, Bogotá, Colombia, to the Conference "The Political Crisis and Internal Conflict in Colombia," April

10-13, 2002), 4, available at <https://www.scribd.com/doc/143496955/THE-CRISIS-OF-STATE-IN-LATIN-AMERICA>.

[20] Ibid., 4-5.

[21] Ibid., 6.

[22] Sarah Chayes, "Corruption: The Unrecognized Threat to International Security," *Carnegie Endowment for International Peace,* June 6, 2014, available at <http://carnegieendowment.org/files/corruption_and_security.pdf>.

[23] Azam Ahmed and Elisabeth Malkin, "Otto Perez Molina of Guatemala is Jailed Hours After Resigning Presidency," *New York Times*, September 3, 2015, available at <http://www.nytimes.com/2015/09/04/world/americas/otto-perez-molina-guatemalan-president-resigns-amid-scandal.html?_r=1>.

[24] Sarah Chayes, "Corruption: The Unrecognized Threat to International Security," *Carnegie Endowment for International Peace,* June 6, 2014, available at <http://carnegieendowment.org/files/corruption_and_security.pdf>.

[25] Ibid.

[26] Ibid.

[27] David Lewis, "High Times on the Silk Road: The Central Asian Paradox," *World Policy Journal* 27, no. 1 (Spring 2010): 39-49.

[28] Karen Dawisha, *Putin's Kleptocracy: Who Owns Russia?* (New York, NY: Simon & Schuster, 2014), 1.

[29] Ibid., 2.

[30] Ibid., 317.

[31] Ibid. 350

[32] David Held, ed., *A Globalizing World? Culture, Economics, Politics* (London: Routledge, 2004).

[33] Gene Sperling, "How to Refloat These Boats," *Washington Post*, December 18, 2005, available at <http://www.washingtonpost.com/wp-dyn/content/article/2005/12/17/AR2005121700028.html>.

[34] Thomas Barnett, "The Pentagon's New Map," *Esquire*, March 1, 2003, available at <http://www.esquire.com/features/ESQ0303-MAR_WARPRIMER>.

[35] Thomas L. Friedman, *The Lexus and the Olive Tree* (New York, NY: Farrar, Straus and Giroux, 1999): 7-8.

[36] Robert A. Isaac, *The Globalization Gap: How the Rich Get Richer and the Poor Get Left Further Behind* (New York, NY: Financial Time Prentice Hall Books, 2004).

[37] Ibid., 4.

[38] Saskia Sassen, *Expulsions: Brutality and Complexity in the Global Economy* (Boston, MA: Harvard University Press, 2014).

[39] The notion of social exclusion is discussed in Manuel Castells, *End of Millennium* (Oxford: Blackwell, 1998), 71-72.

[40] Manfred B. Steger and Ravi K. Roy, *Neoliberalism: A Very Short Introduction* (New York, NY: Oxford University Press, 2010), 11.

[41] Ibid., 12.

[42] Ibid., 14.

[43] Jolle Demmers, Alex E. Fernández Jilberto, and Barbara Hogenboom, ed., *Good Governance in the Era of Global Neoliberalism* (London: Routledge, 2004), 9.

[44] Anne Clunan and Harold Trinkunas, ed., *Ungoverned Spaces: Alternatives to State Authority in an Era of Softened Sovereignty* (Palo Alto, CA: Stanford University Press, 2010), 25.

[45] David Harvey, *A Brief History of Neoliberalism* (New York, NY: Oxford University Press, 2005): 2.

[46] Kedron Thomas and Kevin L. O'Neill, ed., *Securing the City: Neoliberalism, Space, and Insecurity in Postwar Guatemala* (Durham, NC: Duke University Press, 2011), 209-220.

[47] See Robert Mandel, *Global Security Upheaval* (Stanford, CA: Stanford University Press, 2013).

[48] Ibid., 29.

[49] This point and the following bullets draw on the standard literature on wicked problems, most of which is identified in Tom Ritchey, "Wicked Problems: Modelling Social Messes with Morphological Analysis," *Acta Morphologica Generalis* 2, no. 1 (2013), available at <http://www.swemorph.com/wp.html>. A good overview can also be found in Nancy Roberts, "Coping with Wicked Problems," accessed at <http://www.csus.edu/ccp/cdn/syllabi/roberts-fall2006-navalpostgradschool-da4302.pdf>.

[50] Philip Mirowski, *Never Let a Serious Crisis go to Waste* (London: Viro, 2013).

[51] Robert Muggah, "Latin America's Poverty Is Down, But Violence Is Up. Why?" *Instituto Igarape,* available at <http://www.igarape.org.br/pt-br/latin-americas-poverty-is-down-but-violence-is-up-why/>.

[52] "Impunity," *Wikipedia*, available at <http://en.wikipedia.org/wiki/Impunity>.

[53] Ibid.

[54] Julie Lopez, "Guatemala: Lorenzana Case Arrest Delay, Land Dispute Matter Addressed," *Plaza Publica*, April 29, 2011; quoted in Adrienna Jones, "Organization Attributes Sheet: Los Lorenzanas" (Pittsburgh,

PA: University of Pittsburgh, 2011), available at <http://research.ridgway.pitt.edu/wp-content/uploads/2012/05/LorenzanaPROFILEFINAL.pdf>.

[55] Ibid.

[56] Steven Dudley, "Guatemala's Underworld 'Patriarch,' Lorenzana, Extradited to US," *Insight Crime,* March 18, 2014, available at <http://www.insightcrime.org/news-briefs/guatemalas-underworld-patriarch-lorenzana-extradited-to-us>.

[57] International Crisis Group, "Guatemala: Drug Trafficking and Violence," in *Latin America Report №39,* October 11, 2011, available at <http://www.crisisgroup.org/en/regions/latin-america-caribbean/guatemala/139-guatemala-drug-trafficking-and-violence.aspx>.

[58] On the importance of social capital, see Vanda Felbab-Brown, *Shooting Up* (Washington, DC: Brookings Institution, 2009).

[59] Dudley, "Guatemala's Underworld 'Patriarch,' Lorenzana, Extradited to US."

[60] International Crisis Group, "Guatemala: Drug Trafficking and Violence," 10.

[61] International Crisis Group, "Corridor of Violence: The Guatemala-Honduras Border," in *Latin America Report №52,* June 4, 2014, 16, available at <http://www.crisisgroup.org/en/publication-type/media-releases/2014/latam/corridor-of-violence-the-guatemala-honduras-border.aspx>.

[62] Ibid.

[63] Ibid.

[64] Cole Bunzel, *From Paper State to Islamic State: The Ideology of the Islamic State*, The Brookings Project on U.S. Relations with the Islamic World Analysis Paper No. 19 (Washington, DC: Brookings Institution, March 2015), 31, available at <http://www.brookings.edu/~/media/research/files/papers/2015/03/ideology-of-islamic-state-bunzel/the-ideology-of-the-islamic-state.pdf>.

[65] Quoted in Hadley Gamble and Jenny Cosgrave, "Has poor governance in the Middle East aided ISIS?," *CNBC,* May 22, 2015, available at <http://www.cnbc.com/2015/05/22/has-poor-governance-in-the-middle-east-aided-isis.html>.

[66] Simon Speakman Cordall, "How ISIS Governs Its Caliphate," *Newsweek,* December 2, 2014, available at <http://europe.newsweek.com/how-isis-governs-its-caliphate-288517>.

[67] Aymenn Jawad Al-Tamimi, "Unseen Islamic State Financial Accounts for Deir az-Zor Province," *Jihadology,* October 5, 2015, available at <http://jihadology.net/2015/10/05/the-archivist-unseen-islamic-state-financial-accounts-for-deir-az-zor-province/>.

[68] Matthew Rosenberg, Nicholas Kulish, and Steven Lee Myers, "Predatory Islamic State Wrings Money from Those It Rules," *The New York Times*, November 29, 2015, available at <http://www.nytimes.com/2015/11/30/world/middleeast/predatory-islamic-state-wrings-money-from-those-it-rules.html>.

[69] Ibid.

[70] Al-Tamimi, "Unseen Islamic State Financial Accounts for Deir az-Zor Province."

[71] See Onur B. Belli, Andrea Böhm, Alexander Bühler, Kerstin Kohlenberg, Stefan Meining, Yassin Musharbash, Mark Schieritz, Ahmet Senyurt, Birgit Svensson, Michael Thumann, Tobias Timm, and Fritz Zimmermann, "The Business of the Caliph," *Die Zeit,* December 4, 2014, available at <http://www.zeit.de/feature/islamic-state-is-caliphate>.

[72] Mona Alami, "SIS's Governance Crisis (Part II): Social Services," *Atlantic Council,* December 24, 2015, available at <http://www.atlanticcouncil.org/blogs/menasource/isis-s-governance-crisis-part-ii-social-services>.

[73] Cordall, "How ISIS Governs Its Caliphate."

[74] Charlie C. Caris and Samuel Reynolds, "ISIS governance in Syria," *Middle East Security Report* 22 (July 2014): 1-41, available at <http://www.understandingwar.org/sites/default/files/ISIS_Governance.pdf>.

[75] Kareem Fahim, "Strikes by U.S. Blunt ISIS but Anger Civilians," *New York Times,* November 13, 2014, available at <http://www.nytimes.com/2014/11/14/world/middleeast/airstrikes-blunt-isis-in-raqqa-but-many-syrians-there-arent-grateful.html>.

[76] Caris and Reynolds, "ISIS governance in Syria," 5.

[77] Sarah Birke, "How ISIS Rules," *New York Review of Books,* December 9, 2014, available at <http://www.nybooks.com/articles/2015/02/05/how-isis-rules/>.

[78] Isabel Coles, "Despair, hardship as Iraq cuts off wages in Islamic State cities," *Reuters*, October 2, 2015, available at <http://www.reuters.com/article/2015/10/02/us-mideast-crisis-iraq-salaries-idUSKCN0RW0V620151002#U5qdyCg2Zcqq6Gdv.99>.

[79] Peter R. Neumann, *Victims, Perpetrators, Assets: The Narratives of Islamic State Defectors* (London: The International Centre for the Study of Radicalisation and Political Violence, 2015), available at <http://icsr.info/wp-content/uploads/2015/09/ICSR-Report-Victims-Perpertrators-Assets-The-Narratives-of-Islamic-State-Defectors.pdf>.

[80] UNICEF, *Going North: Violence, Insecurity and Impunity in the Phenomenon of Migration in Guatemala* (Guatemala: UNICEF in Guatemala, 2011), 5.

[81] Ibid.

[82] Ibid., 6.

[83] Ibid., 48.

[84] CIDEHUM for UNHCR, *Forced Displacement and Protection Needs produced by new forms of Violence and Criminality in Central America*, May 2012.

[85] Ibid., 19. See Table 3.

[86] Ibid., 22. See Table 4.

[87] Ibid., 21.

[88] Ibid., 24.

[89] UNHCR, *Global Trends: Forced Displacement in 2014* (Geneva: UNHCR, 2015), 14, available at <http://unhcr.org/556725e69.pdf>.

[90] UNHCR, *Global Trends: Forced Displacement in 2015* (Geneva: UNHCR, 2016), available at <http://www.unhcr.org/576408cd7.pdf>.

[91] Human Rights Watch, *World Report 2014: Eritrea*, available at <https://www.hrw.org/world-report/2014/country-chapters/eritrea>.

[92] Ibid.

[93] See Desmond Arias, *Criminal Politics: Illicit Activities and Governance in Latin America and the Caribbean* (forthcoming).

[94] Stephen D. Krasner, "Abiding Sovereignty," *International Political Science Review* 22, no. 3, (2001): 229-252.

2

The Twin Insurgencies: Plutocrats and Criminals Challenge the Westphalian State

Nils Gilman

"Everywhere the ceremony of innocence is drowned."
William Butler Yeats, 1919

States within the modern global political economy face twin insurgencies, one from below, and another from above. On the one hand, there is a series of interconnected criminal insurgencies, in which the global disenfranchised resist, co-opt, and route around states as they seek ways to empower and enrich themselves in the shadows of the global economy. Drug cartels, human traffickers, computer hackers, counterfeiters, arms dealers, and others exploit the failures of governance systems to build global commercial empires that, in turn, provide them the resources to corrupt, co-opt, or challenge incumbent political actors. On the other hand, there exists a plutocratic insurgency, in which globalized elites seek to disengage from traditional national obligations and responsibilities. From libertarian activists, to tax haven lawyers, to currency speculators, to mineral extraction magnates, the new global superrich and their hired help are waging a broad-based campaign that aims either to limit the reach and capacity of government tax collectors and regulators, or to manipulate these functions as a tool in their own cutthroat business competition. Unlike classic 20th-century insurgents, who sought control over the state apparatus in order to implement social reforms, criminal and plutocratic insurgents do not seek to take over the state. These modern insurgencies do not wish to destroy the state, since they rely, like parasites, on the state to provide the legacy goods of social welfare: health, education, infrastructure, and so on. Rather, their aim is simpler: to carve out de facto zones of autonomy for themselves by crippling the state's ability to constrain their freedom of (primarily economic) action. The net result: these transnational insurgencies from above and below are challenging the state's control over the domestic economy, and destabilizing many of the conventions and assumptions rooted in the Westphalian model of governance.

The Failures of Social Modernism

Understanding how we arrived at these twin insurgencies requires a brief return to the anterior period. During the social modernist era (1945-1971), virtually all states—whether capitalist or communist, industrialized or developmental, great power or postcolonial— aimed to legitimate themselves by serving the interests of middle classes, whose size they sought to expand.[1] Both capitalist and communist accumulation strategies were based on nurturing industrial laborers, who were expected to work for a living, and who, in turn, were told that the state would not only steadily improve their standard of living, but would also cushion them from outrageous misfortune through various forms of social security.[2] These states were "welfare states" in the sense that they sought to provide for the general welfare, rather than to protect or lift up the poor or defend the prerogatives of the rich. In the noncommunist world, the wealthy were taxed not out of class hostility, but in order to finance public goods for society as a whole.[3] Health care, pensions, schools, and so on were represented less as individual "entitlements" than as collectively enjoyed public goods that are part and parcel of the social contract. While a diversity of social contracts existed during this period, in virtually every country, elites felt a duty to play a "muscular and essential role in steering the economy and underwriting the well-being of the middle class," and income inequality steadily decreased.[4] For Western elites in particular, the fact that the Cold War order made thinkable radical alternatives to capitalism no doubt helped concentrate a certain commitment to larger moral, social, and political purposes.[5]

By the 1970s, however, it was becoming undeniable that social modernist states across each of the "three worlds of development" were failing to deliver on their promises.[6] In the West, the stagflation of the 1970s undermined the technical foundations of the Bretton Woods financial order, as well as the technocratic consensus in favor of Keynesian demand management and the political consensus in favor of sharing productivity gains between labor and capital. In the East, centrally planned economies were revealing themselves as not only politically repressive, but also economically inefficient and environmentally catastrophic. In the Global South, while the commodity boom of the 1970s led to a golden age for primary producers, import substitution industrialization failed to deliver sustained growth and transition to high per capita incomes. Additionally, the commodity price crash of the early 1980s precipitated a debt crisis which put to rest any dreams of global redistribution.[7] From the late 1970s through the early 21st century, a period of reaction to state-centric models of development set in.[8] Levels of economic inequality began to grow again, eventually reaching heights not seen since the 1920s, and prompting some financial analysts to describe the new economy as a "plutonomy."[9] At the same time, states stopped trying to create a more egalitarian society or to provide for the general welfare; instead, they increasingly sought legitimacy by claiming to maximize the opportunities of individuals.[10] From this perspective, the creation of plutocrats counted not as a defeat, but as a success for the new model of governance.

When Communism collapsed in 1989, what died was not just the particular collectivist economic system and authoritarian politics of the Soviet Union and its satellites. Cremated along with the corpse of Communism was the civic-minded conception of development as the central responsibility of the state and allied elites—a conception shared by communists and liberals alike during the Cold War. It was not just that the state "retreated" from the "commanding heights" of the economy, but also that the very ambitions of the state found themselves in eclipse.[11] The best face that the World Bank could put on the new order was to say that, henceforth, the role of the state would be to "steer" rather than to "row."[12] By the turn of the millennium, even the left had come to doubt whether states could be relied on to effectively and disinterestedly promote the public interest.[13]

The nature of the new order was made most explicit in two texts published the year that the Berlin Wall fell: Francis Fukuyama's "The End of History?" and John Williamson's "The Washington Consensus."[14] Fukuyama proposed that big "H" History (in the Hegelian sense of ideological contestation over the proper relationship between state and civil society) had come to an end with a universal agreement that liberal, democratic capitalism was not just the best, but in fact, the only reasonable form of sociopolitical and economic organization. Williamson's text was more pragmatic than metaphysical, filling in the details of this "posthistorical" policy consensus with specific imperatives around fiscal discipline, the redirection of public spending away from subsidies, the rollback of progressive tax codes, the floating of currencies, the liberalization of trade and cross-border investment, the privatization of state enterprises and deregulation of private ones, and above all, the sacrosanctification of private property rights. Taken together, these texts involved not just a dethroning of the state, but a wholesale challenge to the idea that technocratic leadership under the state was the primary way to ensure collective social well-being. Pioneered as domestic policy in Margaret Thatcher's Great Britain and Ronald Reagan's United States, the programs associated with the Washington Consensus—above all, the privatization of national industrial assets (especially of state-owned firms and utilities) and deregulation (especially of financial firms)—soon became a model that London and Washington sought to export to the Global South and the postcommunist world under the rubric of "structural adjustment" and "shock therapy."[15] As Dani Rodrik concluded, "'Stabilize, privatize, and liberalize' became the mantra of a generation of technocrats who cut their teeth in the developing world and of the political leaders they counseled."[16]

This transformation of the role of the state in the wake of the Cold War has led to a very different sort of landscape of political contestation. With the social modernist state in ideological crisis, the middle classes whose interests it was designed to promote find themselves in an increasingly precarious position. From above, they are threatened by a global financial elite in league with ultra-wealthy compradors, who seek to cut the social services that are paid for by taxes that these elites depict as a form of illegitimate expropriation. From below, they find themselves exposed to various forms of criminals, who have reacted to the collapse of hope for inclusion in the middle classes by taking their futures into their own hands. Let us consider each of these phenomena in turn.

Plutocratic Insurgency: The Revolt of Mainstream Globalization's Winners

This ideological retreat of the social modernist state represents the central event that has enabled plutocratic insurgency. During the 1990s, a new class of globe-trotting economic elites emerged, enriched by the opportunities created by globalizing industrial firms, deregulated financial services, and new technology platforms. This new class is an order of magnitude richer in absolute terms than previous generations of the ultra-wealthy.[17] The rise of the new plutocrats reflects an historic shift in the structure of capital accumulation.[18] The accumulation regime that predominated during the heyday of social modernism was predicated on creating a new class of workers who could afford the goods that they were producing.[19] The great fortunes of the late 19th and early 20th century were built on the backs of masses of worker-consumers in primarily inward-looking national contexts. By contrast, today's plutocrats make their fortunes selling their goods and services globally—in real terms, therefore, their ongoing success is less connected to the fortunes of their fellow national citizens than was that of previous generations. Moreover, the two signature types of massive wealth accumulation in the early 21st century have been software and financial services—both industries that do not rely on masses of laborers, and whose productivity is, therefore, detached from the health of any particular national middle class. The result has been a dramatic rise in inequality within countries, even as wealth inequality transnationally has narrowed.

The rise of a new class of plutocrats has been marked by the emergence of new ideological self-conceptions.[20] Many of these contemporary plutocrats see themselves as "the deserving winners of a tough worldwide competition," and regard efforts to make them pay for public goods as little more than organized theft.[21] Whereas the threat of Communism during the Cold War acted as a check on the maximalist ambitions of the ultrarich, the political and ideological collapse of the Soviet Union removed that constraint, enabling an ideological shift in how a significant segment of the new wealthy conceive their relationship with their societies. While some among the wealthy continue to see themselves as owing a debt of obligation to the societies in which they have enriched themselves, there exists a significant subset—particularly among financial elites—who do not see their personal achievements as tied to the success of the national societies in which they reside.[22] Instead of seeing themselves as the ultimate winners of the systems in which they work, they characterize themselves as rebels, outsiders who have made it on their own despite the restraints presented by incumbents, loafers, and parasites in government and society.[23] The popularity of the pseudo-philosophical novels of Ayn Rand—whose ideas George Monbiot refers to as "the Marxism of the new right"—represents the most visible manifestation of this ideology that poses the rich as "makers," as opposed to the mass of shiftless "takers."[24] From Washington to London, plutocrat-funded think tanks are devoted to creating a body of usable ideas and policy proposals geared at dismantling social modernism.[25] This ideological shift heralds the arrival of plutocratic insurgency.[26]

The defining feature of plutocratic insurgency is the effort on the part of holders of this ideology to defund or de-provision public goods, in order to defang a state that they see as a threat to their prerogatives.[27] Practically speaking, plutocratic insurgency takes the form of efforts to lower taxes, which necessitates the cutting of spending on public goods; to reduce regulations that restrict corporate action or that protect workers; and to defund or privatize public institutions, such as schools, health care, infrastructure, and social spaces. The political strategy associated with plutocratic insurgency is to use austerity in the face of economic shocks to rewrite social contracts on the basis of a much narrower set of mutual social obligations, with the ultimate effect of decollectivizing social risks.[28] As a palliative for the loss of public goods and state-backed programs to improve public welfare, plutocratic insurgents typically promote the idea of philanthropy—directed toward ends defined not democratically but, naturally, by themselves.[29] "There's no such thing as society," Margaret Thatcher famously declared, issuing the *cri de cœur* of insurgent plutocrats everywhere—since, if there's no such thing as society, then the very category of social services collapses, along with any responsibility on the part of the rich to contribute to them. From this perspective, plutocratic insurgency signifies the reimportation back into the industrial core of the aforementioned policies of structural adjustment that were applied across the Global South during the 1980s and 1990s.

For plutocratic insurgents, this strategy is dictated at bottom by a raw cost-benefit analysis: the price the social modernist state asks them to pay in taxes and the regulatory burdens it imposes on them outweighs the benefit they believe they personally receive from living in such a state. Plutocratic insurgents believe they can afford (and, therefore, everyone should be required) to buy for themselves the sorts of goods that before required a state to provide. The need for state-provided security is reduced, as they live in gated communities; public transport is unnecessary for those who travel via personal jets and private bus fleets; public education seems an unnecessary expenditure for the class that already sends their children to exclusive (and expensive) schools.[30] While each of these decisions may, at first, be motivated by lifestyle choices or a desire for social differentiation, the result is a progressive moral disinvestment and civic disengagement from the quality of these traditionally public services, especially as the habit of opting out of public services trickles down from the oligarchs to the upper middle classes.[31] Leaving aside the matter of the undemocratic nature of such private services, or the adverse selection problems that arise from partial privatizations, what marks the arrival of plutocratic insurgency is when the rich begin to revolt against paying taxes for public services they never plan to use. The result is a reinforcing cycle, whereby plutocratic insurgents increasingly see no reason to contribute anything to their host societies, and indeed actively make war on the idea that citizenship imbues them with economic or social responsibilities.

Criminal Insurgency: The Revolt of Deviant Globalization's Winners

Many of the same processes that are driving plutocratic insurgency also underpin the process of criminal insurgency: the globalization of economic flows, growing wealth inequality, and a collapse of state provisioning of public goods and services. From Latin America to Africa to the former Eastern Bloc, the 1980s and 1990s structural adjustment and shock therapy programs led to the "hollowing out" of the state: the physical buildings and institutions of "adjusted" states remained in place, but their ambitions and capacities shriveled.[32] The states in these countries dramatically decreased their spending on social services—ranging from subsidies for food and fuel, to broader social services like public health and pensions. State-owned industries were either shut down or privatized, with wages and employment slashed. The state, in other words, further decreased its capacity to deliver a decent life to its citizens, leading to a collapse in the popular expectation that the state should serve as a guarantor of progress.[33] At the same time, however, the economies of these countries opened rapidly to cross-border financial and trade flows. This combination of the failure of the public goods-providing state and a dramatic increase in the openness of national economies created both an opportunity for enterprising individuals to make money in new ways and an imperative to do so as a matter of survival. These effects were in fact the explicit intention of the structural adjustment and shock therapy programs: rolling back the dirigiste state and opening up the economy was meant to unleash a flood of pent-up entrepreneurial energy and, indeed, it did.

Alas, structural adjustment- and shock therapy-driven globalization of the formerly closed economies of the Eastern Bloc and the Global South turned out to have an unfortunate bug.[34] While the mainstream globalization celebrated by the likes of Thomas Friedman grabbed the headlines, what most distinguished the post-Cold War global economy from the earlier era was the parallel development of a shadowy "deviant" globalization in industries like narcotics, immigration, wildlife harvesting, and antiquities.[35] Though the weakness of the postcommunist and postdevelopmental state represented a dire problem for mainstream businesses and for imploding middle classes in these countries, it offered certain comparative advantages for illicit commerce. Deviant entrepreneurs realized that arbitraging the moral and regulatory differences that existed in different jurisdictions worldwide presented fantastic business opportunities. While big multinational corporations were able to sew up the licit opportunities afforded by the integration of the global economy, they were unable to play in arenas of goods and services banned for moral reasons.[36] The great unsung globalizers of the 1990s and 2000s, therefore, were the criminals who rapidly scaled up their local mom-and-pop criminal organizations to become globe-spanning deviant commercial empires.[37]

These avatars of deviant globalization are also the leaders of the second of our twin insurgencies—the criminal sort. What distinguishes criminal insurgents from classic social revolutionaries is that rather than seeking to build or capture institutionalized state power, they seek merely to protect their rents in various (usually deviant) markets that they control. Organizations such as the First Command of the Capital in Brazil, the 'Ndrangheta in Italy, or the Zetas in Mexico have no interest in taking over the states in which they operate. Instead, like plutocratic insurgents, what criminal insurgents seek is to cripple the state— that is, to establish a zone of economic autonomy while continuing to rely on the state to supply vestigial social services.[38] These actors thrive in (and, indeed, prefer and try to foster) weak state environments, and their activities reinforce the conditions of this weakness. As deviant globalization takes root in a particular locale, however, it soon begins to generate a positive feedback loop; in much the same way as many successful animal and plant species, as they invade a natural ecosystem, deviant globalizers reshape their ecosystem in ways that improve their ability to exclude competitors.[39] The state weakness that, at first, was merely a permissive enabling condition for their business becomes something that the now-empowered criminal insurgents seek to perpetuate and even exacerbate. They siphon off money, loyalty, and sometimes territory; they increase corruption; and they undermine the rule of law. They also force well-functioning states in the global system to spend an inordinate amount of time, energy, and attention trying to control what comes in and out of their borders.

In building their business empires, deviant globalizers inevitably come into conflict with host states in three distinct ways that render them de facto political actors. First, they control huge, growing swaths of the global economy, operating most prominently in places where the state is hollowed or hollowing out. Corruption fueled by drug money on both sides of the U.S.-Mexico border exemplifies this point.[40] Second, many deviant entrepreneurs control and deploy a significant quota of violence—an occupational hazard for people working in extralegal industries, who cannot count on the state to adjudicate their contractual disputes. This use of violence brings deviant entrepreneurs into primal conflict with one of the state's central sources of legitimacy, namely its monopoly (in principle) over the socially sanctioned use of force, transforming them from merely deviant businessmen into criminal insurgents. Third, these criminal insurgents, in some cases, are beginning to emerge as private providers of justice, health care, and infrastructure—that is, precisely the kind of goods that functional states are supposed to provide to their citizens. (However, since they are provided privately, to the deviant entrepreneurs' personal constituents, they are not public goods in the sense of goods equally accessible to all citizens.) Criminal syndicates in Brazil, the Movement for the Emancipation of the Niger Delta (MEND) in Nigeria, narco-traffickers like the Sinaloa Cartel in Mexico—all are criminal insurgents who not only have demonstrated that they can shut down areas of their host states' basic functional capacity, thereby upsetting global markets half a world away, but who are also providing social services to local constituencies.[41]

Thus, criminal insurgency is the form that deviant globalization takes as it scales and reaches political self-consciousness. On the one hand, the more deviant industries grow, the more damage they do to the political legitimacy of the states within which the criminal insurgents operate; therefore, undermining the capacity of the state to provide the infrastructure and services that the criminal insurgents want to free ride on. On the other hand, the people living in the semiautonomous zones controlled by criminal insurgents increasingly recognize the insurgents rather than the hollowed-out state as the real source of local power and authority.[42] Of course, just because these deviant providers of alternative governance functions end up seeming "legitimate" in the eyes of local stakeholders, this type of governance is usually poorly institutionalized and untransparent about both ends and means. Nonetheless, as these groups take over functions that would have been expected of the state, their stakeholders increasingly lose interest in the hollowed-out formal state institutions.[43] Thus, even though criminal insurgents have no desire to kill their host state, they may end up precipitating a process whereby the state implodes catastrophically.

The Enclavization of Microsovereignties and the End of the Middle Classes

During the 1990s, it became fashionable to declare that in the new post-Cold War era, the state was destined to wither away. In fact, something more subtle was taking place:

the double collapse of the social modernist state's capacity and legitimacy was giving birth not to the posthistorical utopia of universal consensus in favor of liberal democratic capitalism, but rather to a conjoined monster in the form of plutocratic secession and deviant globalization. Instead of projects of collective emancipation, what both plutocratic and criminal insurgents desire is for the social modernist state to remain intact except insofar as it impinges on them personally. Neither criminal nor plutocratic insurgents are revolutionaries in the classic modernist sense of political actors who seek to take over the state.[44] As the social modernist state failed to realize its promise, the very notion of a revolution that aspires to a project of national-scale collective social reform has come to seem quaint.[45] Neither category of insurgent is interested in taking control over the state to enact a process of national (or international) social reform. Nor do they seek a political revolution in the Arendtian or Burkean sense of a contest for direct operational and ideological control over the organs of the state.[46] Instead of being in revolt against a particular political regime, with the goal of building a better government, they aim instead to cripple their hosts states in order to gain de facto zones of private autonomy that can enable individual, tribal, or interest-group enrichment.[47] Thus, they are parasitic in a very specific sense: they wish to free ride on the institutional legacy of social modernism so as to avoid costs to their businesses.

Seen from a spatial perspective, what both insurgencies represent is the replacement of the Westphalian ideal of uniform authority and rights within national spaces by a kaleidoscopic array of de facto and de jure microsovereignties. Rather than a single national space in which power is exercised and rights are enjoyed in a consistent and homogeneous way by all residents, the cartography of the dual insurgency represents diverse enclaves of political authority and of social service provisioning arrangements.[48] As these unique arrangements emerge, national and local authorities proliferate a variety of increasingly one-off exceptions to the general rules, incrementally traducing the liberal notion of equality before the law. Just as the 1930s saw a multiplication of conditions poised between war and peace, so our present conjuncture witnesses the multiplication of various forms of authority between the full-blown modern state and outright anarchy, symbolized by the blurring lines between police, military, and private security contractors, in terms of both kinetic capabilities and legal authorities.[49] The process itself is, of course, self-reinforcing: the proliferation of exceptional and unique microsovereignties only increases the scope for the insurgents to engage in jurisdictional arbitrage, and further demands by other insurgents for their own personalized sovereign exceptions. In the space of the dual insurgency, citizenship no longer signifies the liberal ideal of an identical package of rights for all, but instead means very different things depending on where individuals are in the physical and the social space.[50]

Within plutocratic enclaves, the source of authority and loyalty is, at the bottom, money. From a geographic perspective, plutocratic insurgents seek to create zones of private authority and legal autonomy where they can privately command goods once considered public, including not just security, but also increasingly, schooling, transportation, health care, shopping, contract enforcement, and so on.[51] The paradigmatic case for plutocratic spatial segregation and secession are so-called gated communities, which are themselves the subject of a minor academic subfield.[52] These spaces are much more than simple residential enclaves, but increasingly offer full-service operations that contain virtually everything their denizens need, so that residents only need to leave in order to travel to other such enclaves.[53] Rights within such spaces, it goes without saying, accrue to dollars rather than to citizenship. The vision of the future here is of a global archipelago of "privatopias," linked by air and internet to other such spaces, protected by high ramparts from the roiling dystopian ocean of the hoi polloi.[54] Moreover, in addition to these zones of physical separation, plutocratic insurgents also seek out (or seek to create) virtual zones of legal exception, in the form of offshore tax havens, which allows them to avoid income taxes; and special economic zones, which allows them to avoid tariffs as well as laws designed to protect labor or the environment.[55] Plutocratic insurgents are adept at playing off one jurisdiction against another, threatening to take their capital elsewhere if the local authorities do not grant them the exceptions that they seek.

The enclaves of the criminal insurgents are more precarious, as one would expect. Unlike the visible separation that the plutocratic insurgents enjoy in the form of high walls and armed guards, the autonomous zones of the underclasses are more temporary and, naturally, less secure for their masters. From the favelas of Sao Paolo, the slums of Karachi, the waterfront of Kingston, and the suburbs of Beirut or Naples, to the remotest corners of Afghanistan, Honduras, or Sudan, such autonomous spaces take the form of feral "no-go zones"—no-go, that is, to the rich—in which some notionally social modernist state may claim authority, but in which true power is wielded by warlords, gangsters, or other kinds of organized criminals, who take de facto control over local security and whatever meager social service provisioning may be on offer.[56] In these zones, sources of authority and loyalty and the application of raw power tends toward what might be called "neotribalism"—"neo" in the sense that primal loyalties adhere not just to those who share (perceived) bonds of ancient kinship, but rather in accordance to all manner of intense and ritualized personal connections among young male specialists in the use of violence.[57] In short, while globalization is, indeed, undermining national political institutions and thus national identities and loyalties, what appears to be replacing the national is not a "global" political identity—as "cosmopolitical" dreamers have long aspired to—but rather a return to localized identities rooted in clan, sect, ethnicity, corporation, gang, and control over financial resources.[58] It may be that analysis of social relations in such spaces of social fracture is best approached via narratological as opposed to social scientific methods.[59]

The central difficulty that both plutocratic and criminal insurgents face is that it is unclear whether the political objective they seek can produce the sort of stable equilibria of governance that older, Westphalian modalities once promised. There are least two separate reasons to question the ability of these arrangements to produce stability. First, the fracturing of sovereign homogeneity increases transaction costs for people traversing them—it requires a constant expenditure of time and effort to determine exactly what zone of governance one is in and who, therefore, is due respect and obeisance. This is equally true whether one considers the spaces of the plutocratic or the criminal insurgency: in the former case, the price is paid to lawyers; in the second, to gangsters. Second, the kaleidoscope proliferates opportunities for arbitrage and defection of customers and foot soldiers to other governance spaces.[60] The ultimate losers in all of this, of course, are the middle classes—the sorts of people who try to "play by the rules" by going to school and getting traditional middle-class jobs, whose chief virtue is stability. These sorts of people—who lack the ruthlessness to act as criminal insurgents and the resources to act as plutocratic insurgents—can only watch with a certain passivity as the institutions which were built over the course of the 20th century to ensure a high quality of life for a broad majority of citizens are progressively eroded. As the social bases of solidaristic collective action crumble, individuals within the middle classes increasingly face the choice between accepting a progressive loss of social security and de facto social degradation, or attempting to join one of the two insurgencies.

Notes

[1] Anthony B. Woodiwiss, *Postmodernity USA: The Crisis of Social Modernism in Postwar America* (New York, NY: Sage Publications Ltd., 1993); Nils Gilman, *Mandarins of the Future: Modernization Theory in Cold War America* (Baltimore, MD: Johns Hopkins University Press, 2007).

[2] Odd Arne Westad, *The Global Cold War: Third World Interventions and the Making of Our Times* (Cambridge: Cambridge University Press, 2005).

[3] Peter Lambert, *The Distribution and Redistribution of Income, Third Edition* (Manchester: Manchester University Press, 2002).

[4] Peter B. Evans, Dietrich Rueschemeyer, and Theda Skocpol, ed., *Bringing the State Back In* (Cambridge: Cambridge University Press, 1985); Gøsta Esping-Anderson, *The Three Worlds of Welfare Capitalism* (Princeton, NJ: Princeton University Press, 1990); Mark S. Mizruchi, *The Fracturing of the American Corporate Elite* (Cambridge, MA: Harvard University Press, 2013).

[5] The ideal of the modernist welfare state may have been mainly honored in the breach but the point is that it was, in fact, honored despite contestation of the liberal-welfarist model by various actors, whether by leftists who sought a more explicit policy of class leveling, or by rightists who sought to uphold or enforce various forms of racial-, national-, or class-based exclusions. The Westphalian welfare state remained firmly ensconced as the hegemonic model during this period—that is, as the baseline against which other political discourses and proposed political-economic models had to define themselves. With that said, the relations between labor and management in the West (and particularly in the United States) were conflictual even during the postwar heyday of social modernism. Plutocratic pushback against both organized labor and the regulatory and tax reach of the liberal state was present from the beginning of the New Deal and became a formal political strategy by mid-1940s. As Nelson Lichtenstein has observed, "There was no 'labor-management accord,' although labor's strength did generate a kind of armed truce in key oligopolistic sectors of the economy." Despite this pre-history of the plutocratic insurgency, however, it is clear that the end of the Cold War represented a watershed. One cannot help but contrast Tony Judt's (2005) descriptions of Europe's public-minded postwar statesmen to the shameless way that former presidents (GHW Bush, Clinton) and chancellors (Schroeder) and prime ministers (Blair) are happy to receive $100+ million payouts from hedge funds and foreign governments upon leaving office. See Kim Phillips-Fein, *Invisible Hands: The Making of the Conservative Movement from the New Deal to Reagan* (Jakarta: Yayasan Pustaka Obor Indonesia, 2009); Kim Phillips-Fein, "Top-Down Revolution: Businessmen, Intellectuals, and Politicians Against the New Deal, 1945–1964," *Enterprise and Society* 7, no. 4 (2006); Angus Burgin, *The Great Persuasion: Reinventing Free Markets since the Depression* (Cambridge, MA: Harvard University Press, 2012); Nelson Lichtenstein, "Class Politics and the State during World War Two," *International Labor and Working-Class History* 58 (2000); Tony Judt and Denis Lacorne, *With Us or Against Us: Studies in Global Anti-Americanism* (New York, NY: Springer, 2005).

[6] Carl E. Pletsch, "The Three Worlds, or the Division of Social Scientific Labor, Circa 1950-1975," *Comparative Studies in Society and History* 23, no. 4 (1981): 565-590.

[7] Carmen M. Reinhart, "Capital Flow Bonanzas: An Encompassing View of the Past and Present," *The National Bureau of Economic Research* (working paper, 2008), available at <http://www.nber.org/papers/w14321>; Nils Gilman, "The New International Economic Order: A Reintroduction," *Humanity: An International Journal of Human Rights, Humanitarianism, and Development*, March 19, 2015, available at <http://humanityjournal.org/issue6-1/the-new-international-economic-order-a-reintroduction/>.

[8] David Harvey, *A Brief History of Neoliberalism* (Oxford: Oxford University Press, 2007); Christian Caryl, *Strange Rebels: 1979 and the Birth of the 21st Century* (New York, NY: Basic Books, 2013); Daniel J. Sargent, *A Superpower Transformed: The Remaking of American Foreign Relations in the 1970s* (Oxford: Oxford University Press, 2015).

[9] Ajay Kapur, Niall Macleod, and Narendra Singh, "Plutonomy: Buying Luxury, Explaining Global Imbalances," *CitiGroup*, October 16, 2005, available at <https://docs.google.com/file/d/0B-5-JeCa2Z7h-NWQyN2I1YjYtZTJjNy00ZWU3LWEwNDEtMGVhZDVjNzEwZDZm/edit?hl=en_US>.

[10] Philip Bobbitt, *The Shield of Achilles: War, Peace, and the Course of History* (New York, NY: Anchor, 2003).

[11] Susan Strange, *The Retreat of the State: The Diffusion of Power in the World Economy* (Cambridge: Cambridge University Press, 1996); Daniel Yergin, *The Commanding Heights: The Battle for the World Economy* (New York, NY: Free Press, 2002).

[12] World Bank, *World Development Report 1997* (New York, NY: Oxford University Press, 1997).

[13] James C. Scott, *Seeing Like a State: How Certain Schemes to Improve the Human Condition Have Failed* (New Haven, CT: Yale University Press, 1999).

[14] Francis Fukuyama, *The End of History and the Last Man* (New York: Free Press, 2006), 3-18.

[15] Marion Fourcade-Gourinchas and Sarah L. Babb, "The Rebirth of the Liberal Creed: Paths to Neo-liberalism in Four Countries," *The American Journal of Sociology* 108, no. 3 (2002); Raghuram Rajan and Luigi Zingales, "The great reversals: the politics of financial development in the twentieth century," *Journal of Financial Economics* 69 (2003).

[16] Dani Rodrik, "Goodbye Washington Consensus, Hello Washington Confusion?" *Journal of Economic Literature* 44, no. 4 (2006).

[17] Just a few statistics give a sense of the scale: When *Forbes* magazine first started tracking the ultrarich in 1982, there were 12 billionaires in the United States; by 2012, there were 425. In 1982, there were fewer than 200,000 millionaires in the United States; by 2012, there were over 3.7 million. In 2013, there were also 98,700 "ultra-high net worth individuals" (with assets > \$50 million), of which 45 percent were American. To speak of the habits, ideological or otherwise, of the very rich is thus largely to speak of Americans. See Luisa Kroll, "Forbes World's Billionaires in 2012," *Forbes,* March 7, 2012, available at <http://www.forbes.com/sites/luisakroll/2012/03/07/forbes-worlds-billionaires-2012/#6978a7909b13>; Capgemini, *World Wealth Report 2013*, available at <https://www.capgemini.com/resource-file-access/resource/pdf/wwr_2013_0.pdf >; Credit Suisse, *Global Wealth Report 2013*, available at < https://publications.credit-suisse.com/tasks/render/file/?fileID=BCDB1364-A105-0560-1332EC9100FF5C83>.

[18] George Irvin, "Growing Inequality in the Neo-liberal Heartland," *Post-Autistic Economics Review* 43 (2007).

[19] David Harvey, *Spaces of Capital: Towards a Critical Geography* (New York, NY: Routledge, 2001).

[20] Chrystia Freeland, "An Elite Deserving of the Name," *Democracy: A Journal of Ideas*, Summer 2013, available at <http://democracyjournal.org/magazine/29/an-elite-deserving-of-the-name/>.

[21] Chrystia Freeland, "The Rise of the New Global Elite," *The Atlantic*, January/February 2011, available at <http://www.theatlantic.com/magazine/archive/2011/01/the-rise-of-the-new-global-elite/308343/>.

[22] Jörg Huffschmid, "Finance as a Driver of Privatization," *Transfer: European Review of Labour and Research* 14, no. 2 (2008); Jim Taylor, Doug Harrison, and Stephen Kraus, *The New Elite: Inside the Minds of the Truly Wealthy* (New York, NY: AMACOM, 2008).

[23] G. William Domhoff, *Who Rules America? The Triumph of the Corporate Rich* (New York, NY: McGraw-Hill Education, 2013); Robert Frank, *Richistan: A Journey Through the American Wealth Boom and the Lives of the New Rich* (New York, NY: Crown Business, 2008).

[24] Jennifer Burns, *Goddess of the Market: Ayn Rand and the American Right* (New York, NY: Oxford University Press, 2011); Gary Weiss, *Ayn Rand Nation: The Hidden Struggle for America's Soul* (New York, NY: St. Martin's Press, 2013).

[25] James A. Smith, *The Idea Brokers: Think Tanks and the Rise of the New Policy Elite* (New York, NY: Free Press, 1993); Thomas Medvetz, *Think Tanks in America* (Chicago, IL: University of Chicago Press, 2012).

[26] The locus of the plutocratic insurgency today lies in the West—in particular, the world headquarters for the global plutocratic insurgency is London, the world's largest "offshore" financial center that is home to (or at any rate has the homes of) more plutocrats than any other city. Elsewhere, the evidence is less clear: Russia experienced a huge plutocratic insurgency in the 1990s, but the arrival of Putin and the defenestration of the first-generation oligarchs represented the reassertion of the prerogatives of the state—that is, a successful *plutocratic counterinsurgency*. In China, the rise of the super-rich has happened mainly through state-sponsored (though not state-owned) enterprises, which means that plutocrats there remain dependent on the state and the Communist Party and, as such, relatively insecure politically. There, and elsewhere in East Asia, rent-seeking rather than insurgent remains the norm among plutocrats. See Chris Vellacott, "London impoverished by rise of the plutocrats," *Reuters,* March 20, 2012, available at <http://www.reuters.com/article/london-incomedisparity-

idUSL5E8EE3K320120320>; Nicholas Shaxson, "A Tale of Two Londons," *Vanity Fair*, April 1, 2013, available at <www.vanityfair.com/society/2013/04/mysterious-residents-one-hyde-park-london>.

[27] Conceptually, plutocratic insurgencies differ from kleptocracies—the latter involve the using the institutions of state to loot the population, whereas the former wish to neutralize those institutions in order to facilitate private sector looting. In practice, these may overlap or comingle. See Janet Rothenberg Pack, "Privatization of public-sector services in theory and practice," *Journal of Policy Analysis and Management* 6, no. 4 (1987).

[28] Naomi Klein, *The Shock Doctrine: The Rise of Disaster Capitalism* (New York, NY: Picador, 2008); Ulrich Beck, *World at Risk* (Cambridge: Polity, 2008).

[29] Joanne Barkan, "Plutocrats at Work: How Big Philanthropy Undermines Democracy," *Dissent*, Fall 2013, available at <https://www.dissentmagazine.org/article/plutocrats-at-work-how-big-philanthropy-undermines-democracy>.

[30] Chrystia Freeland, *Plutocrats: The Rise of the New Global Super-Rich and the Fall of Everyone Else* (New York, NY: Penguin Books, 2013).

[31] Susan Bickford, "Constructing Inequality: City Spaces and the Architecture of Citizenship," *Political Theory* 28, no. 3 (2000); Somini Sengupta, "Inside Gate, India's Good Life; Outside, the Servants' Slums," *The New York Times*, June 9, 2008, available at <http://www.nytimes.com/2008/06/09/world/asia/09gated.html?_r=0>; Edward J. Blakely and Mary Gail Snyder, "Separate Places: Crime and Security in Gated Communities," in *Reducing Crime through Real Estate Development and Management* (Washington, DC: Urban Land Institute, 1998); Teresa P. R. Caldeira, *City of Walls: Crime, Segregation, and Citizenship in São Paulo* (Berkeley, CA: University of California Press, 2001).

[32] H. Brinton Milward and Keith G. Provan, "Governing the Hollow State," *Journal of Public Administration Research and Theory* 10, no. 2 (2000).

[33] Mark Duffield, "Post-modern conflict: Warlords, post-adjustment states and private protection," *Civil Wars* 1, no. 1 (1998); James Ferguson, *Expectations of Modernity: Myths and Meanings of Urban Life on the Zambian Copperbelt* (Berkeley, CA: University of California Press, 1999); Christian Parenti, *Tropic of Chaos: Climate Change and the New Geography of Violence* (New York, NY: Nation Books, 2012).

[34] Maria Los, "Crime in transition: The post-Communist staterkets and crime," *Crime, Law and Social Change* 40, no. 2 (2003); Misha Glenny, *McMafia: A Journey Through the Global Criminal Underworld* (New York, NY: Vintage, 2009).

[35] Thomas Friedman, "It's a Flat World, After All," *The New York Times*, April 3, 2005, available at <http://www.nytimes.com/2005/04/03/magazine/its-a-flat-world-after-all.html?_r=0>.

[36] Nils Gilman, Jesse Goldhammer, and Steven Weber, ed., *Deviant Globalization: Black Market Economy in the 21st Century* (London: Bloomsbury Academic, 2011).

[37] Roberto Saviano, *Gomorrah: A Personal Journey into the Violent International Empire of Naples' Organized Crime System,* trans. Virginia Jewiss (New York, NY: Picador, 2008); Patrick Radden Keefe, "Cocaine Imported," *The New York Times*, June 15, 2012, available at <http://www.nytimes.com/2012/06/17/magazine/how-a-mexican-drug-cartel-makes-its-billions.html>.

[38] Liberal enthusiasts of globalization assert poverty, insecurity, and state fragility are the result of "disconnectedness" from the world economy. This is false: even paradigmatically "failed" states—Congo, Somalia, Afghanistan—are deeply connected to the global economy. While it is true that they remain weakly connected to the *formal* and *legal* parts of the global economy, such places are *deviantly* connected—via the illicit trade in minerals, via piracy, via the global drug trade, and so on. The crucial issue, in other words, is not connectedness or disconnectedness, but rather *what kind* of connectedness. See Thomas Barnett, "Global Transaction Strategy," *Review – Institute of Public Affairs* 57, no. 1 (2005); Thomas Friedman, *The World is Flat: A Brief History of the Twenty-first Century* (New York, NY: Macmillan, 2005).

[39] Robert J. Bunker and John P. Sullivan, "Integrating Feral Cities and Third Phase Cartels/Third Generation Gangs Research: The Rise of Criminal (Narco) City Networks and Blackfor," *Small Wars and Insurgencies* 22, no. 5 (2011); Max. G. Manwaring, *Street Gangs: The New Urban Insurgency* (Carlisle, PA: U.S. Army War College Press, 2005); Enrique Desmond Arias, "The Dynamics of Criminal Governance: Networks and Social Order in Rio de Janeiro," *Journal of Latin American Studies* 38, no. 2 (2006).

[40] Judith Miller, "The Mexicanization of American Law Enforcement," *City Journal*, Autumn 2009, available at <http://www.city-journal.org/html/mexicanization-american-law-enforcement-13231.html>.

[41] William Langewiesche, "City of Fear," *Vanity Fair News*, April 2007, available at <http://www.vanityfair.com/news/2007/04/langewiesche200704>; Sebastian Junger, "Blood Oil," Vanity Fair News, February 2007, available at <http://www.vanityfair.com/news/2007/02/junger200702>; Keefe, "Cocaine Imported;" John Robb, *Brave New War: The Next Stage of Terrorism and the End of Globalization* (Hoboken, NJ: Wiley, 2007).

[42] William Finnegan, "Silver or Lead," *The New Yorker*, May 31, 2010, available at <http://www.newyorker.com/magazine/2010/05/31/silver-or-lead>.

[43] Diane E. Davis and Anthony W. Pereira, ed., *Irregular Armed Forces and their Role in Politics and State Formation* (Cambridge: Cambridge University Press, 2008).

[44] Rebels who seek to take over or direct the state toward projects of social reform do continue to exist of course—from Marx-inspired movements like the Zapatistas in Mexico or the Naxalites in India to Allah-inspired movements like al Shabaab in Somalia or the Moro insurgency in the Philippines. These sorts of movements, as well as the so-called "color revolutions" that have befallen various post-Soviet states represent a different phenomenon than either described in this essay.

[45] Theda Skocpol, *States and Social Revolutions: A Comparative Analysis of France, Russia and China* (Cambridge: Cambridge University Press, 1979).

[46] Hannah Arendt, *On Revolution* (New York, NY: Penguin Classics, 2006); Edmund Burke, *Reflections on the Revolution in France* (New York, NY: Dover Publications, 2006).

[47] The ideological collapse of the labor-centric, social welfare-providing nationalist state helps to explain why the post-2007 crisis has failed to produce organized opposition movements geared at reining in the secessionist impulses of plutocrats or at addressing the abjections that drive deviant globalization. See Nancy Fraser, "A Triple Movement?" *New Left Review* 81 (2013).

[48] Aihwa Ong, "Graduated Sovereignty in South-East Asia," *Theory, Culture & Society* 17, no. 4 (2000); James D. Sidaway, "Enclave space: a new metageography of development?" *Area* 39, no. 3 (2007).

[49] P. W. Singer, *Corporate Warriors: The Rise of Privatized Military Industry* (Ithaca, NY: Cornell University Press, 2011); Leopold Lambert, *Weaponized Architecture: The Impossibility of Innocence* (Barcelona: dpr-barcelona, 2013); Michael Shank and Elizabeth Beavers, "America's police are looking more and more like the military," *The Guardian*, October 7, 2013, available at <http://www.theguardian.com/commentisfree/2013/oct/07/militarization-local-police-america>.

[50] Marieke Krijnen and Mona Fawaz, "Exception as the Rule: High-End Developments in Neoliberal Beirut," *Built Environment* 36, no. 2 (2010).

[51] Caldeira, *City of Walls*; Tim Hope, "Inequality and the Clubbing of Private Security," in *Crime, Risk and Insecurity*, Tim Hope and Richard Sparks, ed. (New York, NY: Routledge, 2012); Rita Abrahamsen and Michael Williams, *Security Beyond the State: Private Security in International Politics* (Cambridge: Cambridge University Press, 2010); Yves Dezalay and Bryant G. Garth, *Dealing in Virtue: International Commercial Arbitration and the Contruction of a Transnational Legal Order* (Chicago, IL: University of Chicago Press, 1996); Eric Rodenbeck, "Mapping Silicon Valley's Gentrification Problem through Corporate Shuttle Routes," *Wired*, September 6, 2013, available at <http://www.wired.com/2013/09/mapping-silicon-valleys-corporate-shuttle-problem/>.

[52] Sarah Blandy and Diane Lister, "Gated Communities: (Ne)Gating Community Development?" *Housing Studies* 20, no. 2 (2005).

[53] John Connell, "Beyond Manila: Wallslls, and Private Spaces," *Environment and Planning* 31, no. 3 (1999); Chris Webster, "Gated Cities of Tomorrow," *The Town Planning Review* 72, no. 2 (2001); Sengupta, "Inside Gate;" Werner Breitung, "Enclave Urbanism in China: Attitudes Towards Gated Communities in Guangzhou," *Urban Geography* 33, no. 2 (2012).

[54] Evan McKenzie, *Privatopia: Homeowner Associations and the Rise of Residential Private Government* (New Haven, CT: Yale University Press, 1996); Steve Graham and Simon Marvin, *Splintering Urbanism: Networked Infrastructures, Technological Mobilities and the Urban Condition* (New York, NY: Routledge, 2001); Mike Davis, *Evil Paradises: Dreamworlds of Neoliberalism* (New York, NY: The New Press, 2008).

[55] Godfrey Baldacchino, *Island Enclaves: Offshoring Strategies, Creative Governance, and Subnational Island Jurisdictions* (Montreal: McGill-Queen's University Press, 2010); Nicholas Shaxson, *Treasure Islands: Uncovering the Damage of Offshore Banking and Tax Havens* (New York, NY: St. Martin's Press, 2011); Jonathan Bach, "Modernity and the Urban Imagination in Economic Zones," *Theory, Culture & Society* 28, no. 5 (2011).

[56] Langewiesche, "City of Fear;" Sobia Ahmad Kaker, "Enclaves, insecurity and violence in Karachi," *South Asian History and Culture* 5, no. 1 (2014); David Kilcullen, *Out of the Mountains: The Coming Age of the Urban Guerrilla* (New York, NY: Oxford University Press, 2015); Mona Fawaz, Mona Harb, and Ahmad Gharbieh, "Living Beirut's Security Zones: An Investigation of the Modalities and Practice of Urban Security," *City & Society* 24, no. 2 (2012); Saviano, *Gomorrah: A Personal Journey*; Tim Hetherington, "Into the Korengal," *World Policy Journal* 28, no. 1 (2011); Mattathias Schwartz, "A Mission Gone Wrong," *The New Yorker*, January 6, 2014, available at <http://www.newyorker.com/magazine/2014/01/06/a-mission-gone-wrong>; Kenneth Omeje, "Markets or Oligopolies of Violence? The Case of Sudan," *African Security* 3, no. 3 (2010); Richard J. Norton, "Feral Cities," *Naval War College Review* 56, no. 4 (2003); Bunker and Sullivan, "Integrating Feral Cities;" Kimberly Marten, *Warlords: Strong arm Brokers in Weak States* (Ithaca, NY: Cornell University Press, 2012).

[57] David Ronfeldt, "Tribes: The Once and Forever Form," *RAND Corporation* (working paper, 2007), available at <http://www.rand.org/pubs/working_papers/WR433.html>; John Robb, *Brave New War: The Next Stage of Terrorism and the End of Globalization* (Hoboken, NJ: Wiley, 2007).

[58] Pheng Cheah and Bruce Robbins, *Cosmopolitics: Thinking and Feeling beyond the Nation (Studies in Classical Philology)* (Minneapolis, MN: University of Minnesota Press, 1998).

[59] Kenneth DiMaggio, "Seceding from the Narrative: How the Criminal Underworlds in William Burroughs' Naked Lunch Map out a Non-Linear Narrative through the Creation of 'Temporary Autonomous Zones,'" *The International Journal of the Book* 8, no. 1 (2011).

[60] Albert O. Hirschman, *Exit, Voice, and Loyalty: Responses to Decline in Firms, Organizations, and States* (Cambridge, MA: Harvard University Press, 1970).

3

The Islamic State Revolution

Scott Atran

"Virtue, without which terror is destructive; terror, without which virtue is impotent.
Terror is only justice prompt, severe and inflexible; it is then an emanation of virtue."
Maximilien Robespierre, *On the Principles of Political Morality*, 1794

In response to yet more slaughters perpetrated by the Islamic State (ISIL), security services deployed across Europe, Africa, and America.[1] U.S. and Russian forces ratcheted up air attacks in Iraq and Syria, while politicians and pundits hammered their publics into existential dread. Perhaps never in history have so few, with such meager means, caused such fear in so many. But it is easy amid the bullets, bombs, and bluster, to lose sight of a central fact in the fight against the violent forces of radical Islam: not only are we not stopping its spread, but our efforts to contain the contagion appear to contribute to its strength, while further constraining our own freedoms.

What accounts for the failure of "The War on Terror" and efforts to counter the spread and growth of "violent extremism?" Apart from the heedless reactions in anger and revenge that consistently engender more savagery than security is the failure to understand the revolutionary character of radical Arab Sunni revivalism, which ISIL now spearheads. For it is a dynamic countercultural movement of world historic proportions, with the largest and most diverse volunteer fighting force since World War II, and which, in less than two years, has created a dominion over thousands of square kilometers and millions of people.[2] What is more, though ISIL is the focus of this chapter and the most dynamic, it is not the only manifestation of the countercultural revolution, which has possible counterparts in other regions, embracing other ideologies and motivations.

What the United Nations and most of the international community regard as senseless acts of horrific violence are, to ISIL's acolytes, part of an exalted campaign of purification through sacrificial killing and self-immolation: "Know that Paradise lies under the shade of swords," says a hadith, now a motto of ISIL fighters, from the *Sahih al-Bukhari*, a collection of the Prophet's sayings considered second only to the Quran in authenticity.

This is the purposeful plan of violence that Abu Bakr al-Baghdadi, ISIL's self-anointed Caliph, outlined in his call for "volcanoes of jihad:" to create a globe-spanning jihadi archipelago that will eventually unite to destroy the present world and create a new-old world of universal justice and peace under the Prophet's banner. A key tactic in this strategy is to inspire sympathizers abroad to violence—do what you can, with whatever you have, wherever you are, whenever possible.

Dozens of structured interviews and behavioral experiments with youths in Paris, London, and Barcelona, as well as with captured ISIL fighters in Iraq and members of Jabhat al-Nusra (al-Qaeda's affiliate in Syria), have demonstrated clear lines of commonality among fighters. These interviews focused on youths from distressed neighborhoods previously associated with violence or jihadi support—for example, the Paris suburbs of Clichy-sous-Bois and Épinay-sur-Seine, the Moroccan neighborhoods of Sidi Moumen in Casablanca, and Jamaa Mezuak in Tetuán.[3]

Because many foreign volunteers—especially from Europe—are marginalized in their host countries, a pervasive belief in governments and nongovernmental organizations (NGOs) is that offering would-be enlistees jobs, education, or spouses could be the best way to reduce violence and counter the Caliphate's pull. But a yet unpublished report by the World Bank shows no reliable relationship between job production and violence reduction.[4] (When a World Bank representative was asked why this was not published, he responded, "Our clients [governments] wouldn't like it because they've got too much invested in the idea.") If people are ready to sacrifice their lives, then it is not likely that offers of greater material advantages will stop them. In fact, research shows that material incentives, or disincentives, often backfire and instead raises the commitment of devoted actors.

Research also shows that most of those who originally joined al-Qaeda were married, and prior marriage does not seem to be a deterrent to those now volunteering for ISIL.[5] And among the senior ranks of such groups, there are many who have had access to considerable education—especially in scientific fields, such as engineering and medicine, which require great discipline and willingness to delay gratification. Ever since the anarchists, this sort of specialized preparation holds for much of the leadership of insurgent and revolutionary groups.

Many in the West dismiss radical Islam as simply nihilistic. According to U.S. Secretary of State John Kerry, "ISIL is offering nothing to anyone except chaos, nihilism, and ruthless thuggery."[6] As we shall see, ISIL does deal in chaos but works with a script and a purpose; however, nihilist it is not. Research suggests something far more menacing: a profoundly alluring mission to change and save the world. Indeed, jihadi volunteers believe they are combating the "nihilism" of the West—that is, a certain way of life that ends up destroying all moral constructs, religions, and metaphysical convictions (by relativizing everything, assigning it monetary value, etc.).

Terror's Sublime Virtue

In the West, the seriousness of this mission is denied. Olivier Roy, usually a deep and subtle thinker, writes in *Foreign Policy* that the Paris plotters represent most of those who flock to ISIL. They are marginal misfits largely ignorant of religion and geopolitics, and bereft of real historical grievances.[7] They ride the wave of radical Islam as an outlet for their nihilism, because it is the biggest and baddest countercultural movement around.

However, the worldwide ISIL revolution is hardly just a bandwagon for losers. Although attacked on all sides by internal and external foes, ISIL has thus far only been contained and somewhat degraded in Iraq and Syria, while continuing to take root in ISIL-controlled areas and expanding its influence in deepening pockets throughout Eurasia and Africa. Repeated claims that ISIL was on the way to inevitable defeat ring hollow for almost anyone who has had direct experience in the field. Only Kurdish frontline combatants and some Iranian-led forces have managed to fight ISIL to a standstill on the ground, and only with significant French and U.S. air support. As of this writing, the first phase of the Iraqi army offensive to retake Mosul, aided by U.S. Marines and coalition air forces, was bogged down despite overwhelming superiority in manpower and firepower.[8]

Despite our relentless propaganda campaign against ISIL as vicious, predatory, and cruel, there is little recognition of its genuine appeal, and even less of the joy it engenders. The many young people who volunteer to fight for it unto death feel a joy that comes from joining with comrades in a glorious cause, as well as a joy that comes from satiation of anger and the gratification of revenge (whose sweetness, says science, can be experienced by brain and body much like other forms of happiness).[9] As Osama bin Laden wrote in an elegy for the 9/11 hijackers, "embracing death, the knights of glory found their rest. They gripped the towers with the hands of rage and ripped through them like a torrent." One young man from the Balkans, who is now fighting in Syria, expressed his joy as the "happiness of martyrdom," sending us the following image:

Figure 3.1.

But there is also a subliminal joy felt across the region for those who reject ISIL's murderous violence, yet yearn for the revival of a Muslim caliphate and the end to a nation-state order that the Great Powers invented and imposed. It is an order that has failed, in their view, and that the United States, Russia, and their respective allies are trying willy-nilly to resurrect, and it is an order that many in the region believe to be the root of their misery. What the ISIL revolution is *not,* is a simple desire to return to the ancient past. The idea that ISIL seeks a return to medieval times makes no more sense than the idea that the U.S. Tea Party wants to return to 1776. "We are not sending people back to the time of the carrier pigeon…" Abu Mousa, ISIL's press officer in Raqqa, has said. "On the contrary, we will benefit from development. But in a way that doesn't contradict the religion."

ISIL's Caliphate seeks a new order based on a culture of today. Unless we recognize these passions and aspirations, joining with comrades in a glorious cause, the joy that comes from satiation of anger, and the gratification of revenge, and deal with them using more than just military means, we will likely fan those passions and lose another generation to war and worse.

Treating ISIL as merely a form of terrorism or violent extremism masks the menace. All novel developments are "extremist" compared with what was the norm before. What matters for history is whether these movements survive and thrive against the competition. Throughout history, success has depended on willingness to shed blood, including the sacrifice of one's own, not merely for family and tribe, wealth, or status, but for some greater cause. This has been especially true since the start of the Axial Age more than two millennia ago. At that time, large-scale civilizations arose under the watchful gaze of powerful divinities, who mercilessly punished moral transgressors—thus, ensuring that even strangers in multiethnic empires would work and fight as one.

Call it "God," or whatever secular ideology one prefers, including any of the great modern salvational -*isms*: colonialism, socialism, anarchism, communism, fascism, and liberalism. In *Leviathan*, Thomas Hobbes deemed sacrifice for a transcendent ideal "the privilege of absurdity to which no creature but man is subject."[10] Humans make their greatest commitments and exertions, for ill or good, for the sake of ideas that give a sense of significance. In an inherently chaotic universe, where humans alone recognize that death is unavoidable, there is an overwhelming psychological impetus to overcome this tragedy of cognition: to realize "why I am" and "who we are."

In *The Descent of Man*, Darwin cast this devotion as the virtue of "morality…the spirit of patriotism, fidelity, obedience, courage, and sympathy" with which winning groups are better endowed in history's spiraling competition for survival and dominance.[11] Across cultures, the strongest forms of primary group identity are bounded by sacred values that are immune to material tradeoffs, carrots, or sticks—like unwillingness to sell one's children or sell out one's religion or country. Devotion to such values, as when land or law become holy or hallowed, leads some groups to prevail because of nonrational commitment from at least some members to actions that drive success, independent or out of proportion, from expected rational outlays and outcomes, risks and rewards, and costs and consequences.

Often such values are attributed to Providence or Nature, and embedded in notions whose meanings one can never quite pin down, and which cannot ever be definitively verified nor falsified by logic or empirical evidence, such as, "God is great; bodiless but omnipotent," or "free markets are always wise." Thus, while "sacred values" intuitively denote religious belief, as when land becomes holy, it can also include the "secularized sacred," such as the hallowed ground of Gettysburg, or the site of the 9/11 attacks at New York City's "Ground Zero." For example, the foundational doctrines and beliefs of the great ideological *–isms*; the quasi-religious notion of the Nation itself, ritualized in song, ceremony, and sacrifice; and those "self-evident" aspects of "human nature" that humankind is supposedly endowed with, such as "inalienable rights of life, liberty, and the pursuit of happiness," which are anything but inherently self-evident and natural in the life of our species (cannibalism, infanticide, slavery, oppression of minorities, and male domination of women were more standard fare). It was not inevitable or even reasonable that conceptions of individual freedom and equality concocted by 18th-century European intellectuals should emerge, much less prevail. They did, only through revolution, intensive social engineering, economic competition, and belief in "just war."

"Nothing human is alien to me," said Terence, the Roman slave who became a playwright and gave the field of anthropology an enduring credo: to empathize with those most different from one's own moral culture, without necessarily sympathizing. This is our call to comprehend. If we can only grasp why otherwise normal humans would want to die amidst killing masses of other humans who have harmed no one, we ourselves might better avoid killing and being killed.

In our preferred world of liberal democracy and human rights, violence—especially extreme forms of mass bloodshed—is generally considered pathological or an evil expression of human nature gone awry, or collateral damage as the unintended consequence of righteous intentions. But across most of human history and across cultures, violence against other groups is universally claimed by the perpetrators to be a sublime matter of moral virtue. For without a claim to virtue it is difficult, if not inconceivable, to kill large numbers of people innocent of direct harm to others.

What many in the international community do not understand is that these apparently senseless acts of horrific violence are, to ISIL's followers, part of an exalted campaign of purification through sacrificial killing and self-immolation, to destroy what is presently corrupt in order to save what was pure in some past "Golden Age," and to serve as a basis for the creation of a brave new world.

Besides the emotional appeal, brutal terror scares the hell out of enemies and fence-sitters. According to interviews with Kurdish leaders, when 350 to 400 ISIL fighters came in a convoy of some 80 trucks (each truck carrying about 4 or 5 fighters) to free Sunni captives (and massacre more than 600 Shia inmates) from Badoush prison in Mosul, Iraq's second largest city, a relatively well-equipped Iraqi army of some 18,000 troops under American-trained leaders immediately melted into the city or ran away. When one Arab Sunni soldier embedded with a Kurdish Peshmerga force on the Mosul-Erbil front was asked why fellow soldiers fled, he simply said, "They wanted to keep their heads."

The shutdown of Brussels in the wake of the Paris attacks, like that of Boston in the aftermath of the Boston Marathon bombings, speaks of comparable fear, and perhaps an underlying lack of faith in the solidity of our own societies and values. During World War II, not even the full might of the German Luftwaffe at the height of the Blitz could compel the British government and the people of London to cower so. Now, the mere mention of an attack on New York in an ISIL video has American officials scurrying to calm the public. Media exposure, which is the oxygen of terror in our age, not only greatly amplifies the perception of danger; but, in generating such hysteria, makes the bloated threat to society real. Because nowadays media is mostly designed to titillate rather to inform, it is has become child's play for ISIL and its ilk to turn our own propaganda machine and the world's mightiest into theirs—a novel, highly potent jujitsu style of asymmetric warfare that we could counter with responsible restraint, but which we do not.

The outcome is dangerous and preposterous. The U.S. Justice Department now considers the common kitchen pressure cooker to be a weapon of mass destruction if used for terrorism.[12] This ludicrously levels a cooking pot with a thermonuclear bomb, which has a destructive power that is a billion times greater. It trivializes true weapons of mass destruction, making their acceptance more palatable and their use more conceivable. In this present hyperreality, messaging is war by other means. ISIL's manipulation of our media creates a sense of foreboding of mass destruction where not really possible, and at the same time obscures any real future threat.

Asymmetric operations involving spectacular killings to destabilize the social order is a tactic that has been around as long as recorded history. Violent political and religious groups routinely provoke their enemies into overreacting, preferably by committing atrocities to get the others to drive in the sheep and collect the wool.

The violence of ISIL, like the revolutionary violence of many who came before, is perhaps best characterized by what Edmund Burke referred to as "the sublime:" willingness—indeed, need and passion—for the "delightful terror" of a sense of power, destiny, giving over to the infinite, ineffable, and unknown.[13] "No passion so effectually robs the mind of all its powers of acting and reasoning as *fear*," notes Burke, "For fear being an apprehension of pain or death, it operates in a manner that resembles actual pain. Whatever therefore is terrible, with regard to sight, is sublime."

But for terror to succeed in the service of the sacred and sublime, "obscurity seems in general to be necessary," Burke goes on, "Those despotic governments which are founded on the passions of men, and principally upon the passion of fear, keep their chief as much as may be from the public eye."[14] Al-Baghdadi, Prince of the Faithful, surely fits that bill. More generally, notes France's Charles De Gaulle in 1932, "there can be no prestige without mystery, for familiarity breeds contempt." And so, too, "great leaders have always carefully stage-managed their effects" to "concentrate all efforts on captivating men's minds," so that they may transcend themselves to act on behalf of a glorious, group-defining cause.[15]

The sublime is also intensely physical and visceral, steeped in emotion and identity, and not a core part of our recent and current ideologies that would favor reason and "the mind" as the driver rather than a slave of the passions. There is no brainwashing, which is a leftover canard about Allied soldiers during the Korean War being broken like Pavlov's dogs by Red China's psychological manipulation wizards. In *Mein Kampf*, Adolf Hitler declared that, "All great movements are popular movements, volcanic eruptions of human passions and emotional sentiments, stirred either by the cruel Goddess of Distress or by the firebrand of word hurled among the masses."[16] But the word must be framed within the spectacular theater of the sublime. When both Charlie Chaplin and French filmmaker René Clair viewed Leni Riefenstahl's visual paean to National Socialism, *Triumph of the Will*, at a showing at the New York Museum of Modern Art, Chaplin laughed but Clair was terror-stricken, fearing that if it were shown in the West all might be lost.[17]

The Revolutionary Vanguard

"O soldiers of the Islamic State, continue to harvest soldiers," Baghdadi intones, "erupt volcanoes of jihad everywhere," and "dismember [enemies] as groups and individuals" to liberate mankind from the "satanic usury-based global system" leached by "the Jews and crusaders"[18]—an appeal that resonates with many and stirs at least some to atrocity. Although there has yet to be replication, a recent poll suggests that a quarter of France's young adults of all creeds, from ages 18 to 24, have at least a "somewhat favorable" attitude towards ISIL. Other research with young people in the hovels and grim housing projects of the Paris *banlieues* found fairly wide tolerance or support for ISIL even among the non-Muslim underclasses.[19]

It matters little that, as J.M. Berger wrote in *The Atlantic*, "the Islamic State's ideological sympathizers make up less than one percent of the world's population…and the fact that active, voluntary participants in its caliphate project certainly make up less than a tenth of a percent."[20] Few, if any, revolutionary vanguards in history achieved success by first capturing a significant portion of the world's population, or even the people in their home regions. During the surge of American troops in Iraq, up to three-fourths of the fighters were neutralized in al-Qaeda's Iraqi affiliate, which would become ISIL, and an average of about a dozen high-value targets were eliminated monthly for 15 consecutive months, including its top leader, Abu Musab al-Zarqawi. Yet, the organization survived and the group went on to thrive beyond all expectations amidst the chaos of Syria's civil war and Iraq's factional decomposition.

Just since World War II, revolutionary movements have, on average, emerged victorious with as little as one-tenth of the firepower and manpower of the state forces against them.[21] Behavioral research in conflict zones indicates that sacred values (e.g., national liberation, God, and Caliphate) mobilized for collective action by devoted actors enables outsized commitment in initially low-power groups (e.g., Viet Cong, ISIL) to resist and often prevail against materially more powerful foes who depend on standard incentives, such as police and armies that rely on pay, promotion, and punishment (e.g., South Vietnamese Army, Iraqi Army).[22]

As history and empirical studies show, what has mattered in revolutionary success is commitment to cause and comrades that, even in the face of initial failures and often devastating defeats, can trump overwhelming material disadvantages.[23] In 1776, American colonists were primarily frustrated not over economics, but over perceived denial of truths "sacred and undeniable"—Thomas Jefferson's original words for the Declaration of Independence.[24] They were willing to sacrifice "our lives, our fortunes, and our sacred honor" against the world's mightiest military empire. Britain sent the largest naval expeditionary force in the 18th century (30,000 men) against the fledgling American Revolution in New York (20,000 inhabitants), and initially beat Washington's army to a pulp. At year's end, revolutionary forces were starving, although it was a bumper crop year. Enlistments in the highly fractious revolutionary army were coming to an end, and its remnants were beginning to return to their homes. Eyewitness reports indicate that Washington saved the incipient republic with an evidently sincere appeal to a higher moral calling: "You will render that service to the cause of liberty which you can probably never do under any other circumstances."[25] And so the army fused together in the harsh winter at Valley Forge, henceforth able to withstand any adversity.

But the sort of liberal democracy initiated by the American Revolution has never been very good at adjudicating across religious and ethnic boundaries, especially when, as in much of the Middle East and Central Asia, such boundaries are tribally based. Democracy took root in Britain's American colonies, which had the world's highest standard of living at the time and unprecedented opportunities for people other than Native Americans and African slaves to strike out on their own into virtually limitless territory, relatively free to realize their aspirations.[26]

In Western Europe, democracy gradually developed during the 19th century under the tutelage of authoritarian rule. France's Napoleon III not only continued Napoleon Bonaparte's promotion of cultural secularism and tolerance of religious plurality, but also went on to introduce legislative elections, permit organized political opposition, and legalize the right to strike. In Europe, people were torn from their ancestral lands (under laws closing the commons) to work mostly in urban centers of the industrial revolution, bound in toil and war to a novel, overarching notion of national identity.

In this landscape, liberal institutions began to develop, enabling hitherto anonymous strangers to work with one another and, if necessary, to fight together. These institutions included free and universal education, a press accessible to a wide range of information and argument, equality of all citizens before the law (at least in principle), and a culture of growing tolerance towards minorities and others. Without an overarching national identity and the liberal values and institutions to sustain it, popular choice and elections lead only to a tyranny of the majority, as both ancient Athens and post-Saddam Iraq confirm.

The chasm between the values of the West and those of ISIL and its sympathizers is compounded by alternate historical arcs. The West and the Arab and Muslim worlds have long lived mostly separate and parallel histories. In the West, people generally believe history began with Ancient Sumeria around the 26th century BCE. Centered in the southern part of modern-day Iraq, Sumeria was the birthplace to written law and literature, and to Abraham and his monotheistic creed. Civilization then moved west to Greece and Rome.

After the fall of Rome, came the Middle Ages, the Renaissance, the Industrial Revolution and the Enlightenment, the first political revolutions, the World Wars, and the Cold War. By the end of the 20th century, human rights and democracy became triumphant and seemingly inevitable.

The Arab and Muslim worlds also begin with Sumeria, but until the World Wars, Rome, Greece, and the rest were peripheral. Christian Europe was the "dark continent." Muslim heroes, myths, legends, and references were all basically different. Indeed, there are Moses, Alexander the Great, and Jesus, but their profiles in Islam are distinct. Musa's (Moses') life paralleled Mohammed's and foretold the Prophet's coming. Iskandar (Alexander), or Dhul-Qarnayn (Arabic for "The Two-Horned One"), was a religious figure to whom Allah gave great power and the ability to build a wall of civilization to provisionally keep out the forces of chaos and evil. And Isa (Jesus) was Allah's righteous messenger, not his son, who did not die on the cross but, like Mohammed, was raised to heaven.

All of the European political imports (and even nationalism itself, except maybe for Turkey, Egypt, and Iran, which are still more built around ethnicity and confession than national identity per se) have failed, and miserably so. People are yearning for something in their history, in their traditions, with their heroes and their morals. And ISIL, however brutal and repugnant to us and even most in the Arab and Muslim worlds, is speaking directly to that.

Yet, there is little apparent in the response of the U.S. and Western powers that even recognizes that revival. The hackneyed solutions amount to a tired call to shore up the broken nation-state system imposed in the aftermath of World War I by the European victors, Great Britain and France, and a reaffirmation of "moderate Islam," which appeals to young people's longings for adventure, glory, ideals, and significance even less than does the promise of eternal shopping malls.

Still, the popular notion of a "clash of civilizations" between Islam and the West—current to many of our own politicos and the public as well as to ISIL and al-Qaeda—is woefully misleading. Violent extremism represents not the resurgence of traditional cultures, but their collapse, as young people unmoored from millennial traditions flail about in search of a social identity that gives personal significance and glory. This is the dark side of globalization. The young radicalize to find a firm identity in a flattened world where vertical lines of communication between the generations are replaced by horizontal peer-to-peer attachments that can span the globe, albeit in informationally narrow and tight ways.

As I testified to the U.S. Senate Armed Service Committee, and before the United Nations Security Council, what inspires the most lethal assailants in the world today is not so much the Quran or religious teachings (although for leadership this is important) as a thrilling cause and call to action that promises glory and esteem in the eyes of friends; through friends, eternal respect and remembrance in the wider world that many will never live to enjoy.[27] Foreign volunteers for ISIL are often youth in transitional stages in their lives—immigrants, students, between jobs and before finding their mates, having left their homes, and looking for new families of friends and fellow travelers to find purpose and significance.

France's Centre for the Prevention of Sectarian Drift Related to Islam estimates that 80 percent come from nonreligious families; West Point's Center for Combating Terrorism finds that their average age is 25.[28] For the most part, they have no traditional religious education and are "born again" to religion through the jihad. About one in four, often the fiercest followers, are converts. Research suggests that French converts from families of Christian origin are often the most vociferous of ISIL's defenders. There is something about joining someone else's fight that makes one fierce. A former body builder from Épinay-sur-Seine, a northern suburb of Paris, when asked why he converted to Islam, said that he had been in and out of jail, constantly getting into trouble. "I was a mess, with nothing to me, until the idea of following the *mujahid's* way gave me rules to live by"—to channel his energy into jihad and defend his Muslim brethren under attack from infidels in France and everywhere, "from Palestine to Burma."

Self-seekers who have found their way to jihad reach out through private gatherings or the internet. They might be people who feel uncomfortable with binge drinking or casual sex, or have seen their parents humiliated by employers or the government, or their sisters insulted for wearing a headscarf. Most do not follow through to join the jihad, but some do. More than 80 percent who join ISIL do so through peer-to-peer relationships, mostly with friends and sometimes family.[29] Very few join in mosques or through recruitment by anonymous strangers.

What we know about the 2015 Paris attackers, for example, fits this pattern. As with the perpetrators of the 2004 Madrid train bombings and the 2005 London Underground bombings, several of the principal plotters in the January and November Paris attacks lived for a time in the same neighborhood, several enlisted friends and family members, and some moved in the same criminal networks and spent time together in jail.

In France, as elsewhere in Europe, many of these young people identify neither with their country of origin nor their country of domicile. Other identities are weak and nonmotivating. One woman in the Paris suburb of Clichy-sous-Bois told of feeling like a transgender person who opts out of the gender they were assigned to at birth. "I was like a Muslim trapped in a Christian body," believing herself only able to live fully as a Muslim with dignity in the Islamic State.[30]

Unlike the United States, Europe was not built to absorb immigrants. In America, Muslim immigrants attain parity or surpass the average American in wealth and education in the first generation.[31] In Europe, they are much more likely to be poorer than the average citizen and poorer still after the second generation, a legacy of decolonization left largely to fester unattended.[32]

France and Germany have the largest Muslim populations in Europe. In France, seven to eight percent of the total population is Muslim. At the same time, up to two-thirds of the prison population is Muslim, contributing significantly to an underclass ripe for radicalization.[33] One 24-year-old who joined Jabhat al-Nusra in Syria, described his experience in Germany:

They teach us to work hard to buy a nice car and nice clothes but that isn't happiness. I was a third-class human because I wasn't integrated into a corrupted system. But I didn't want to be a street gangster. So, [my friends and I] decided to go around and invite people to join Islam. The other Muslim groups in the city just talk. They think a true Muslim state will just rain down from heaven on them without fighting.

Most European volunteers join ISIL, rather than Jabhat al-Nusra, because they believe the Caliphate is here today and there is no need to wait for tomorrow. Yet, many ISIL volunteers are far from marginal in their home countries. As one family physician wrote to me earlier this year:

> During 2015, two groups of medical students [17 in all] from the University [of Medical Sciences and Technology in Khartoum, Sudan] fled to the Levant in order to join IS. The families of those students have had difficulties coping with their loss. It was almost grievousness of death. The students who left from our university…are well-funded by their parents (higher middle class with multi-background). I find difficulty identifying the factors that led those smart, straight-A students, to [IS]. Could it be lack of identity? Could it be the universities' fault? Could it be…the family's lack of influence?

A banker from Mosul recounted:

> Daesh [ISIL] fighters came into the bank and our staff was terrified. They offered to help in any way. An Algerian, about 25, polite, asked only to be led to our computers. In a short time he downloaded all of our bank's transactions. He said that he came to the Islamic State to put his education in computer engineering to good use.

The Caliphate is an attractor to all of these young people, providing purpose and freedom from what they have come to see as the vice of a meaningless, material world. ISIL is supposed to conform to the pure, Salafi vision of the Prophet's initial followers (of the *salaf*, or "forebears"). It is an imperial enterprise that demands offensive jihad, or holy war, against the infidel (*kafir*), as an "individual obligation" (*fard al-'ayn*) of everyone who belongs to the "House of Islam" (*Dar al-Islam*).

Adherents of this pure Caliphate are violently opposed to the idea of greater jihad as an inner spiritual struggle. They consider this bogus notion of jihad to be the heart of the Sufi heresy introduced in the later Abbasid Caliphate, which corrupted the pure Arab-led form of the Caliphate and led to its decay and downfall.

Reviving the Muslim Caliphate, under its original Arab cast, is a powerful attractor to these young people, providing purpose and freedom from what they have come to see as the vice of a material world based on a specious freedom to make only false and meaningless choices. Some speaking for Western governments at the East Asia Summit in Singapore last April argued that the Caliphate is mythology covering traditional power politics. Research with those drawn to the cause show that this is a dangerous misconception. The Caliphate has reemerged as a mobilizing cause in the minds of many Muslims, and even has some appeal to Muslims who favor interfaith cooperation. "I am against the violence of [al-Qaeda] and ISIL," an imam, who helps to run an interfaith dialogue initiative with Christians and Jews, in Barcelona told us, "But they have put our predicament in Europe and elsewhere on the map. Before, we were just ignored. And the Caliphate…. We dream of it like the Jews long dreamed of Zion. Maybe it can be a federation, like the European Union, of Muslim peoples. The Caliphate is here, in our hearts, even if we don't know what real form it will finally take."

Whatever form it assumes, we can be sure it will be rooted in the history and culture of the Arab states, not the West. That perspective includes the reality of Muslim dominance of middle Eurasia until the European industrial revolution and a rejection of the Western world order, be it liberal democracy or socialism, imposed after the Ottoman Empire's collapse in the early 20th century.

Perhaps above all else, ISIL aims to put an end to Sykes-Picot, the neocolonial order that Britain and France imposed on the Arab provinces of the Ottoman Empire after the First World War—an order solidified in borders drawn by Churchill, T.E. Lawrence, Gertrude Bell, and others at the Cairo Conference in March 1921, to ensure British control of unfettered lines of communication, resources (especially oil), and transport from Suez to India. In the spring of 2014, when ISIL bulldozed the border markers between Iraq and Syria, it generated shudders of liberation and joy for many across the region and beyond. Unlike the United States and other great powers, including Russia and China, many people in the region do not consider the current mayhem to result from failed states that now must be revived and reinforced at whatever cost, but from the expedient fictions that created those states in the first place.

Revolutions Past and Present

Revolutions past and present are moral events. Deteriorating or rapidly changing economic and social conditions can initiate a cascading series of events that produces a political crisis. However, this will lead to a "revolutionary" challenge to the prevailing order, and the costly commitment to basic political and social change, only when action becomes morally motivated by a shift in core cultural norms, or "sacred values," and the seizure of state power to enforce those values. Thus, despite the fact that the influence of the Islamic clergy and canon had declined precipitously within Iran's civil institutions and government under the shah's regime, the failure of secular forces (from liberal to Marxist-Leninist) to cohere around a new political morality left the way open for Islamic forces to seize the moral high ground. In Egypt, the Muslim Brotherhood's influence had been on the rise well before the Arab Spring, and although the Brotherhood initially refused to participate, the disunity of secular forces allowed it to rush in and fill the moral void. But unlike the founders of the Islamic Republic of Iran who purged the army, controlled the *bazaari* (the urban commercial class), and took root in the rural religious population, Egypt's Muslim Brotherhood leadership believed (as Safwat Hegazi, head of the Egyptian Revolutionary Council, stated) that the economy and army would fall into line if the Islamic leadership first managed to control the messaging and Ministry of Information.

By contrast, ISIL has moved swiftly and ruthlessly to impose a new-old ethos among Arab Sunnis in the war-torn wastelands of the Middle East. It promises total war against the "satanic" morality of Iran and the Shia and their helpers (including America, its allies, and Russia) in a mortal struggle for the Muslim soul and ultimately for the salvation of all humankind.

Historical analogies are always of limited usefulness, but they are also one of the only means by which we can make sense of what is new, or at least recognize where true novelty begins. There are striking historical parallels in the history of modern revolutions ever since the Jacobin faction of French revolutionaries, led by Maximilien Robespierre, introduced the political concept of terror and decapitation by guillotine as an extreme measure for the defense of democracy and Republican virtue. These were a divine form of violence "supported by the most sound [sic] and wholesome of all laws, the salvation of the people." For a decade, at the end of the 18th century, the French Revolution consumed its own like bloodied sharks, all the while fighting a fractious coalition of great powers that sought to destroy it.[34] Yet, it thrived. United and transformed into an imperial mission to reform and save humankind—as all revolutions since have endeavored to do—revolutionary forces conquered nearly all of Europe before the Empire's fall. And ever after, revolutionary commitment to "total war" in the service of some indomitable moral and spiritual force has continued to inspire nearly all revolutions.

The current rivalry between al-Qaeda and ISIL echoes that between the anarchists and social revolutionaries versus the Bolsheviks in the early 20th century. Beginning in Russia in the 1870s as a countercultural agitation against the power of the state and capital, the anarchist and social revolutionary movements soon spread throughout Europe and on to the Americas. Between 1881 and 1900, assassins closely linked to the anarchist and social revolutionary movements had killed the czar of Russia, the president of France, the prime minister of Spain, the king of Italy, and the empress of Austria. In September 1901, the anarchist Leon Czołgosz assassinated the U.S. President William McKinley.

The Great Powers considered anarchism to pose the greatest threat to the internal political and economic order, and to international stability. America beefed up the Secret Service and created the Federal Bureau of Investigation (FBI). Britain's Scotland Yard, Russia's Okhrana (forerunner of the NKVD and KGB), and France's *La Brigade spéciale des Renseignements* généraux were all formed largely to meet the anarchist threat. In the face of repeated anarchist attacks randomly targeting Parisians in "bourgeois" cafés, theaters, and the like, French leaders and the popular press repeatedly demanded that the French people "awaken" and "unify" to fight a scourge that threatened civilization itself (while confounding the many currents of anarchism, including the many peaceful and communitarian strands of the multidimensional movement).[35] The political (and to some extent, social and economic) consequences from this first wave of modern terror were similar in many respects to those of the 9/11 attacks. Teddy Roosevelt made the defeat of anarchism an overriding mission of his administration. "When compared with the suppression of anarchy, every other question sinks into insignificance. The anarchist is the enemy of humanity, the enemy of all mankind; and his is a deeper degree of criminality than any other."[36]

But Roosevelt did not restrict the fight against terrorism to anarchists alone. He expanded the war on anarchy into an imperial mission to intervene in any country around the world if necessary to protect it from foreign evil and preserve it from chaos. "Chronic wrongdoing," he said, "or an impotence which results in a general loosening of the ties of civilized society, may in America, as elsewhere, ultimately require intervention by some civilized nation, and may lead the United States, however reluctantly, in flagrant cases of such wrongdoing or impotence, to the exercise of an international police power."[37] Most tellingly, the war against anarchy and terror helped to justify the brutal repression of an ethnic Muslim (Moro) insurgency against U.S. rule in the Philippines.

Despite political and popular belief in the existence of an "Anarchist Central," there never really was anything of the sort. As with al-Qaeda, the anarchist movement was largely a decentralized movement of volunteers led by fairly well-off and well-educated folk. What ultimately killed off the anarchist movement as a geopolitical force were not the armies and police of the Great Powers, but the Bolsheviks. They knew much better how to manage a somewhat shared political ambition through military and territorial management. They were also, on the whole, much more ruthless.

In a series of interviews with Jabhat al-Nusra fighters from the Aleppo and Dara regions of Syria, it has become increasingly evident that, in the words of a former ISIL imam whom we interviewed in Jordan, "Daesh (ISIL) is eating Qaeda" in much the same way that the Bolsheviks co-opted and practically annihilated the anarchist movement. Even some Jabhat al-Nusra fighters echoed this imam's sentiment, conceding that ISIL is better led, organized, supplied, rooted in territory, more uncompromising, and brutal in action. "Daesh [ISIL] has taken our power and financial resources from us, their media is more powerful, their military commanders are more efficient, and so we are like a fish out of water (*tatakhet*)."

Opponents of Germany's National Socialist Workers (Nazi) Party argued that the Nazis were neither a party of workers nor socialists. Today, we are told again and again that ISIL is "neither a State nor Islamic" (at least I am, nearly every time I talk to political or religious leaders), and that using the term "Islamic State" only "feeds into its hands." In fact, the contrary is true: believing that refusing to call the Islamic State by its own name can somehow delegitimize it is only self-deluding (a rose, or a National Socialist, by any other name is still what or who it is).

In fact, there is a deeper connection between the Nazi movement and ISIL, an association that I noted some time ago.[38] George Orwell, in his review of *Mein Kampf* in 1940, describes the essence of the problem:

> Hitler knows…that human beings *don't* only want comfort, safety, short working-hours, hygiene…and, in general, common sense; they also, at least intermittently, want struggle

and self-sacrifice…. Whereas Socialism, and even capitalism in a more grudging way, have said to people 'I offer you a good time,' Hitler has said to them 'I offer you struggle, danger and death,' and as a result a whole nation flings itself at his feet.[39]

Man for man, the German army outfought all Allied armies by any measure. In classical military doctrine about a 30 percent loss in a fighting unit usually leads to entropy, so when that degree of destruction is confirmed, the victorious army moves on to the next task (this was basically how the Israeli Army fought the Six-Day War). But German forces often suffered in excess of 50 percent loss and still held fast, fought bravely—and sometimes knowingly—to the death, in defense of a devoutly believed cause, however horrible it may seem (as for example, in the Waffen-SS volunteer "death squads" that fought to the end against the Soviets in Budapest).

Postwar social psychological studies reveal that the German soldier believed in what he was doing, and fought for a cause as much as for comrades, whereas there is little evidence that the Allies fought for democracy or communism, despite Hollywood and Soviet propaganda.[40] The German armies were destroyed only by the massive superiority of American firepower and by the massive manpower of more than 20 million Russians given over to slaughter. Perhaps it will come to something like that with ISIL, when and if ISIL is ever perceived to be a true existential threat. But for now, the means arrayed against this dynamic revolutionary movement look feeble and what the U.S. government grandly dubs the "global ISIL coalition" of 65 nations seems a very tenuous, if not fatuous, thing (with several of its members ever-ready to stick knives into one another's backs).[41]

Over the course of the 20th century, America and its allies used three different strategies to meet the great international threats of the day:

1. First came general policing, at home and abroad, to meet the anarchist menace. This had only very modest and intermittent success until it was overtaken and subdued by a more potent revolutionary movement, Bolshevism.

2. Then, "total war" was waged against the Axis powers. That succeeded because of America's massive productive capacity, the Allies' overwhelming manpower, the fact that the Axis had clearly targetable industrial infrastructures and political hierarchies that could be destroyed, and strong national identities that could be mobilized to rapidly rebuild under the victors' different yet familiar value systems.

3. Lastly, a two-fold strategy of "containment" was employed against the military and political challenge posed by the Soviet Union and its confederates. For Paul Nitze, head of President Truman's Policy Planning Staff, containment principally involved measures placing greater emphasis on strengthening our own military capabilities, rather than relying on extensive economic assistance and military aid to our allies.[42] Although for Nitze's successor, George Kennan, measures involving "adroit and vigilant application of counter-force at a series of constantly shifting geographical and political points, corresponding to the shifts and maneuvers of Soviet policy" were critical, political countermeasures were paramount. These involved both economic assistance (e.g., the Marshall Plan) and "psychological warfare" (overt propaganda and covert operations) to counter the spread of Soviet influence until the

"internal contradictions" of the Soviet political economy compelled collapse.[43] In the end, containment seemed to work, although considerable debate remains over the relative importance of the different countermeasures.

Recent calls to counter al-Qaeda and now ISIL first focused on general policing, then moved to containment. Now there are calls for total war (at least among some of the leading presidential candidates). But total war is hardly more likely to succeed than general policing or containment against a global jihadi archipelago because of its lack of the very conditions that fostered Allied success against the Axis powers (i.e., strong industrial base, national identity, familiarity with victors' values, etc.). What we need, it appears, is a new military, political, and psychological strategy that targets the peculiarly novel features of the ISIL Revolution (i.e., dispersed infrastructure, confessional and tribal allegiances, wholly different values, etc.).

The United States and its allies may yet opt for force of arms, with all of the unforeseen and unintended consequences that are likely to result from all-out war. But even if ISIL is destroyed in its core lands—and even if we were to do something serious about ISIL's growth in Africa across areas totaling millions of square miles—its message could still captivate many in coming generations and in disparate regions. Empowering and ennobling the legions of Muslims opposed to ISIL, including Islamists who reject democracy but who can coexist with democracies, is likely a better bet. Unfortunately, nothing today in the Muslim world competes with ISIL's voice and strength. Nearly everyone is either for it or against, and though overwhelmingly against, as advertising wisdom has it, a lot of bad publicity for one side still beats little or none for another.

"Will to Fight": Sacred Values, Identity Fusion, and Spiritual Formidability

One 25-year-old Jabhat al-Nusra fighter who originally joined ISIL but tired of "blowing up innocent civilians" describes a fairly general path to "the Syrian Revolution which has turned to jihad" as a desire for struggle and self-sacrifice more than anything in life:

As a teen I just wanted to play football and video games. I used to love reading fiction books. Looking back on my thoughts it seems that my mind was too focused and distracted by the mundane: studying, getting a good job, socializing, having fun and being a family man. The concept of Jihad was something scary at the time, something of sacrifice and hardship and impossible to pull off. It wasn't long before I was informed about the concept of martyrdom (*shohada*).... Immediately my mind would conjure images of two armies fighting each other on an open plane. Warriors wielding their swords and riding along on beautiful horses, my mind in overdrive with thoughts of fighting in the way of Allah and attaining martyrdom. I never really watched much jihadi propaganda online and I was so eager to get to Syria I walked in Blind with two brothers I was with, who were locals from the UK...[to] rid society of its many filths and return the earth to a state of purity where the law of God is supreme and surpasses everything else, jealous about brothers who had been killed fighting in the way of Allah.

Of course, wars are won in the material world, but a spiritual commitment to cause and comrades conveys great advantage, all things being equal. As 14th-century Arab historian Ibn Khaldûn first noted, comparing Muslim dynasties in North Africa with similar military might, long-term differences in success "have their origin in religion...group feeling (*asabiyah*) [wherein] individual desires come together in agreement [and] mutual cooperation and support flourish."

In September 2014, President Obama endorsed the judgment of National Intelligence Director James Clapper: "We underestimated the Viet Cong...we underestimated ISIL and overestimated the fighting capability of the Iraqi [A]rmy.... It boils down to predicting the will to fight, which is an imponderable."[44] In fact, predicting who is willing to fight and who is not, and why, is ponderable and amenable to scientific study.

Recent interviews and psychological experiments on the frontlines with Kurdish fighters of the Peshmerga and the Kurdistan Workers' Party, with captured ISIL fighters, and with Jabhat al-Nusra fighters in Syria provide a good initial indication of willingness to fight. Two principal factors interact to predict readiness to make costly sacrifices (e.g., going to prison, fight, die, have one's family suffer, etc.).

The first factor is perception of relative commitment of one's own group versus those of the enemy to a sacred cause. This can be measured through behavioral experiments and tracked via neural imaging to show four elements.[45]

1. Disregard for material incentives or disincentives: attempts to buy people off from their cause ("carrots") or punish them for embracing it through sanctions ("sticks") do not work, and even tend to backfire.

2. Blindness to exit strategies: people cannot even conceive of the possibility of abandoning their sacred values or relaxing their commitment to the cause. This fosters unconditional cooperation and intractable conflict in ways that social contracts born of shared convenience and utility do not.

3. Immunity to social pressure: it matters not how many people oppose your sacred values, or how close to you they are in other matters. Such values are not social or cultural norms but defining and circumscribing features of culture itself. They provide the moral frame for which social interactions and material exchanges are permissible or taboo.

4. Insensitivity to discounting: according to most economic and political theory, and usual in most everyday affairs, distant events and objects have less significance for people than things in the here and now. But matters associated with sacred values, regardless of how far removed in time or space, are more important and motivating than mundane concerns, however immediate.

The second factor in predicting willingness to fight is the degree of fusion with one's comrades. Consider, by way of illustration, a pair of circles where one circle represents "me" and a larger circle represents "the group" (see Figure 3.2). In one set of experiments, participants were asked to consider five possible pairings: in the first pairing, the "me" circle and "the group" circle do not touch; in the second pairing, the circles touch; in the third, they slightly overlap; in the fourth, they half overlap; and in the fifth pairing, the "me" circle is entirely contained within "the group" circle. People who choose the last pairing think and behave in ways entirely different from those who choose any of the other pairings. They experience what social psychologists call "identity fusion," wedding their personal identity ("who I am") to a unique collective identity ("who we are"). Such total fusion demonstrably leads to a sense of group invincibility and a willingness of each and every individual in the group to sacrifice for each and every other.[46]

The following diagram consists of two circles measuring identify fusion. The small circle represents you (I) and the big circle represents your close circle of friends/religion/country (here, ISIL). Those individuals that are fused (far right) indicate that the group and the individual become one and measures of willingness to commit costly sacrifices are dichotomous with all other fusion pairings.

Figure 3.2. Fusion Measure

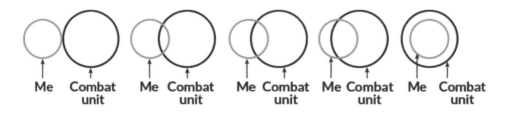

| Me | Combat unit | Me | Combat unit | Me | Combat unit | Me | Combat unit | Me | Combat unit |

(Example: Individual/IS)

Only among the Kurds do we find commitment to the sacred cause of "Kurdeity" (their own term) and fusion with fellow Kurdish fighters comparable to commitment to cause and comrade among ISIL fighters.[47]

Willingness to fight and make costly sacrifices is also strongly associated with perceptions of physical formidability on the battlefield and, even more importantly, with spiritual strength (see Figure 3.3). Research indicates that Jabhat al-Nusra fighters consider Iran (by which, they also mean Hezbollah) to be the most formidable foe in Syria, both in terms of physical and spiritual strength, but they consider ISIL growing to parity on both scores. These al-Qaeda combatants consider the United States to be of middling formidability, and the Syrian and Iraqi Armies to be relatively weak physically, and spiritually worthless; and thus, an inconsequential enemy in the long run (see Figure 3.4). Such perceptions appear to correspond to performance and results on the battlefield.

Figure 3.3. Formidability Measure
(Example: Individual/ISIL)

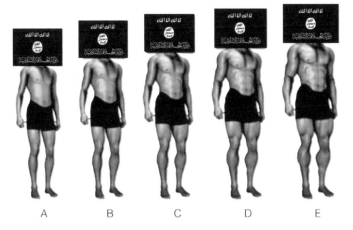

Here is a series of human bodies that represent the strength of one group (e.g., ISIL). You can choose one representative body to indicate the size and strength of the group as a whole. This holds constant for measures of physical strength and spiritual strength.

Figure 3.4. Perception of Physical vs. Spiritual Formidability by Jabhat al-Nusra Fighters

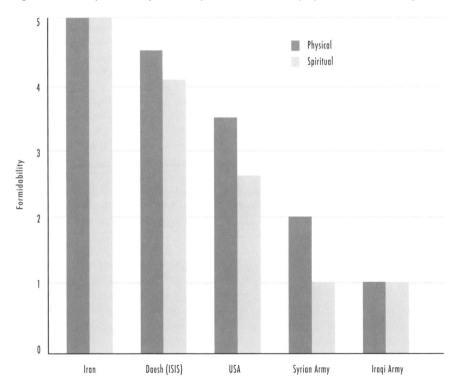

To be sure, not all who fight with ISIL are committed zealots. Captured ISIL fighters recounted growing up in the failed Iraqi state during the last decade: a hellish world of guerrilla war, disrupted families, constant fear, and utter lack of hope. They see Iran and the Shi'ites as their greatest enemies, but they also believe that America allowed them to oppress the Arab Sunni minority for the sake of majority rule. When prisoners were asked, "What is Islam?" they answered, "My life." Yet, it was clear that they knew little about the Quran, or Islamic history, other than what they had heard from al-Qaeda and ISIL propaganda. They could neither cite passages from the Quran relevant to their actions nor even name the first four Caliphs and companions of the Prophet who founded Islam's first Empire. For them, the cause of religion was fused with the vision of a caliphate—a joining of political and religious rule—that kills or subjugates any nonbeliever (but which in the face of almost sure execution by the Kurds, they were ready to recant).

In one conversation picked up by a Kurdish walkie-talkie, a fighter with a local accent asked for help: "My brother has been killed. I am surrounded. Help me take his body away." The reply: "Perfect, you will join him soon in Paradise." The fighter retorted: "Come for me. This Paradise, I don't want." The Islamic State will say to a local sheikh: "Give us 20 young men or we loot your village." To a father with three sons, they will say: "Give us one or we take your daughter as a bride for our men." (One young girl we were told of, who came from a village near Mosul Dam, was "wed" for this reason 15 times in in a single night.)

In the face of such brutality, wavering ISIL supporters could well rally to an Arab Sunni force, possibly allied with the Kurds who fight with remarkable strength of spirit—although this was not initially the case—but with the barest of means. Despite suffering almost nightly grenade attacks and suicide assaults via steel-plated vehicles, few Kurdish frontline units had night vision goggles (or even binoculars) or armor-piercing weapons.

Nevertheless, it is foreign fighters that the Kurds most fear. As the chief of the Kirkuk police station housing the prisoners puts it, "the foreign fighters are the most dangerous and fearless. They fight to win and they fight to die. They believe in what they are doing and will not surrender."

Revolutionary Strategy and the Headless Tiger

ISIL's core strategy is not a mystery, although surprisingly few people engaged in policy and decisionmaking with regard to ISIL pay heed, preferring more familiar paradigms, of power politics and war as simply politics by other means. Think of reactions to the horrors of Paris, Ankara, Beirut, or Bamako, and then consider the following axioms drawn from *The Management of Chaos-Savagery* (Idarat at-Tawahoush, required reading for every ISIL political, religious, military leader, or amir), and from the February 2015 editorial in *Dabiq* (the online ISIL publication), on "The Extinction of the Gray Zone."[48] ISIL's actions have been, and likely will continue to be, consistent with these axioms:

- Work to expose the weakness of the so-called Great Powers by pushing them to abandon the media psychological war and war by proxy until they fight directly.
- Draw these powers into military conflict. Seek the confrontations that will bring them to fight in our regions on our terms.
- Diversify the strikes and attack soft targets—tourist areas, eating places, places of entertainment, sports events, and so forth—that cannot possibly be defended everywhere. Disperse the infidels' resources and drain them to the greatest extent possible, and so undermine people's faith in the ability of their governments to provide security, most basic of all state functions.
- Target the young, and especially the disaffected, who tend to rebel against authority, are eager for self-sacrifice, and are filled with idealism; and let inert organizations and their leaders foolishly preach moderation.
- Motivate the masses to fly to regions that we manage, by eliminating the "Gray Zone" between the true believer and the infidel, which most people, including most Muslims, currently inhabit. Use so-called "terror attacks" to help Muslims realize that non-Muslims hate Islam and want to harm all who practice it, to show that peacefulness gains Muslims nothing but pain.
- Use social media to inspire sympathizers abroad to violence. Communicate the message: Do what you can, with whatever you have, wherever you are, whenever possible.
- Pay attention to what works to hold the interest of people, especially youth, in the lands of the Infidel (e.g., television ratings, box office receipts, music and video charts), and use what works as templates to carry our righteous messages and calls to action under the black banner.

Thus, the 2015 Paris and 2016 Brussels attacks, for example, did not represent a "game change" in ISIL's strategy, or even tactics, contrary to statements by U.S. leaders, senior intelligence officials, and the *New York Times*.[49] In reality, the attacks were just an ever more effective installment for fomenting chaos in Europe, just as attacks in Turkey and Lebanon sought to instigate more savagery and chaos in the Middle East. A welcome to refugees would clearly represent a winning response to this strategy, whereas wholesale rejection of refugees just as clearly represents a losing response to ISIL. We may wish to celebrate diversity and tolerance in the gray zone, but the general trend in Europe and the majority segment of America's political establishment and population is to collude in erasing it.

There is a disheartening dynamic between the rise of radical Islamism and the revival of the xenophobic ethno-nationalist movements that are beginning to seriously undermine the middle class—the mainstay of stability and democracy—in Europe, in ways reminiscent of the hatchet job that the communists and fascists did on European democracy in the 1920s and 1930s. The fact is Europe's replacement rate is less than 1.6 children per couple and so needs considerable immigration to maintain a productive workforce that can sustain the middle class standard of living.[50] This is at a time where there has never been less tolerance for immigration, creating a situation of chaos that ISIL is effectively exploiting.[51]

In areas under ISIL control, or adjacent to it, the general populations likely do not support either ISIL or the Western- (and now also Russian-) dominated forces arrayed against it. They are not zealots nor samurai, and do not want to die as martyrs. ISIL knows this and entices its enemies to attack the population centers that it controls, even though the ability of ISIL to diffuse its highly mobile military assets and personnel in a regime without borders means that there is little infrastructure available to target. Mostly, the local populations suffer. Although many would flee from both ISIL and the bombs of its enemies if given half a chance, they cannot move and must exclusively depend for protection on the black banner, where evidence of gray can be punished with death. And history shows that aerial bombing campaigns generally harden populations against the bombers, whatever the regime.

In the West, the imminent death of ISIL has been greatly oversold. ISIL is destined to fail on its own, in part because it is a "desperately poor nation trying to fight a three-front war," in part because of a noxious ideology of governance, as two professors recently argued in *Politico*.[52] The authors invoke the doomed destiny of the current Zimbabwe state and the collapse of the Soviet Union to bolster their argument.

However, historical precedent and present evidence do not support their point of view. Poverty, multifront wars, and extreme or exclusive ideologies can also end in revolutionary triumph or lasting influence, as with Republican France and possibly the Islamic Republic of Iran. The authors' contention that, "as the Soviet Union was to communism, so ISIL is to jihadism" might be on the mark.[53] However, before ISIL's inherent contradictions confine it to the dustbin of history, there are likely miles and miles of grief to go. Before the revolutionary flame burns itself out, it can also burn away much in its path, and profoundly reshape the region and beyond.

The 9/11 attacks cost between $400,000 and $500,000. According to Brown University's "Cost of War Project," the response by the United States alone is 10 to 100 million times that figure, including related security arrangements and military actions that make up the vast bulk of that spending.[54] On a strictly cost-benefit basis, this violent movement has been wildly successful, beyond even bin Laden's original imagination, and is increasingly so. Herein lies the full measure of jujitsu-style asymmetric warfare. After all, who could claim that we are better off than before or that the overall danger declines rather than rises?

This alone should inspire a radical change in our own counterstrategies. Yet, in keeping with the proverbial notion of insanity as repeating the same mistakes and expecting different results, the West continues to focus almost exclusively on security and military responses to the violent consequences of other's actions, with all of the unforeseen, unintended, and uncontrollable consequences that can result from war. Some of these repeated responses have proven almost hopelessly ineffective from the get-go, such as relying on the Iraqi, Afghan, or Free Syrian Armies. By contrast, there is precious little attention to the social and psychological causes that are likely to reassert themselves ever more vehemently unless we address them in serious, concrete ways. In brief, we are wastefully reactive, and incompetently proactive.

In contrast with, say, the off-target tweets of the U.S. State Department's "Think Again Turn Away" campaign, ISIL may spend hundreds of hours trying to enlist single individuals, to learn how their personal frustrations and grievances can fit into a universal theme of persecution against all Muslims, and thus translate anger and unrealized aspiration into moral outrage. To pass their message, ISIL employs some 50,000 Twitter accounts, with about 1,000 followers each. ISIL also pays close attention to the pop songs, video clips, action movies, and television shows that garner high ratings among youth, and uses them as templates to tailor their own messages.

Any serious engagement must be attuned to individuals and their networks, not to mass marketing of repetitive messages. Young people empathize with each other; they generally do not lecture at one another. From Syria, a young woman messages another:

> I know how hard it is to leave behind the mother and father you love, and not tell them until you are here, that you will always love them but that you were put on this earth to do more than be with or honor your parents. I know this will probably be the hardest thing you may ever have to do, but let me help you explain it to yourself and to them.

Yet, the U.S. government has few operatives who personally engage with youth before they become a problem. The FBI is pressing to get out of the messy business of prevention to focus on criminal investigation. "No one wants to own any of this," one group from the U.S. National Counterterrorism Center recently confided to us. And public diplomacy efforts do not quite get that hackneyed appeals to "moderation" fall flat on restless and idealistic youth seeking adventure, glory, and significance. As the imam and former ISIL recruiter in Jordan states:

> The young who came to us were not to be lectured at like witless children; they are for the most part understanding and compassionate, but misguided. We have to give them a better message, but a positive one to compete. Otherwise, they will be lost to Daesh [ISIL].

Without universal appeal, and quality individual time, little progress can be made beyond what is achievable by force of arms. Local grassroots approaches have had better luck in pulling people away. The United Network of Young Peacebuilders has had remarkable results in convincing young Taliban in Pakistan that enemies can be friends, and then encouraging those so convinced to convince others.[55] But this will not challenge the broad attraction of ISIL for young people from nearly 100 nations and every walk of life. The lessons of local successes must be shared with governments, and ideas allowed to bubble up before they boil over.

To date, no such platform exists. Young people with good ideas have no really good institutional channels to develop them: their often naive demands such as "governments must do this or that"—so apparent at the summer 2015 UN-sponsored Global Forum on Youth in Amman—are dismissed out of hand by people in government, who have to deal with real world constraints on power and its exercise, and the youth are left in the lurch with their ideas unrealized and unrealizable for lack of practical guidance and refinement.

Even if good ideas find ways to emerge from youths and obtain institutional support for their development to application, they still need intellectual help to persuade the public to adopt them. But where are the public intellectuals to do this? In the Muslim world, we see PowerPoint presentations intoning on "dimensions of ideology, grievance, and group dynamics," notions that originate exclusively with Western "terrorism experts" and think tanks. When asked, "What ideas come from your own people?" we are told in moments of candor, as I was most recently informed by a Muslim leadership council in Singapore, that, "We don't have many new ideas and we can't agree on those we have."

And where among our own current or coming generation are the intellectuals who might influence the moral principles, motivations and actions of society towards a just and reasonable way through the morass? In academia, you will find few willing to engage with power. Thus, they render themselves irrelevant and morally irresponsible by leaving the field of power entirely to those they censure. Accordingly, politicians pay them little heed, and the public could not care less, often with good reason. For example, in the immediate aftermath of the 9/11 attacks, many in the field of anthropology principally occupied themselves with the critique of empire: is the United States a classic empire or "empire light?" This was arguably a justifiable academic exercise, and perhaps a useful reflection in the long run, but hardly helpful in the context of a country moving fast to open-ended war, with all the agony and suffering that extended wars inevitably bring.

Responsible intellectual endeavor in the public sphere was once a vibrant part of our public life: not to promote "certain, clear, and strong" action, as Martin Heidegger writes in support of Hitler, but to generate just and reasonable possibilities and pathways for consideration. Now this sphere is largely abandoned to the Manichean preaching of blogging pundits, radio talk show hosts, product-pushing podcasters, and television evangelicals. These people rarely do what responsible intellectuals ought to do. "The intellectual," explained France's Raymond Aron 60 years ago, "must try never to forget the arguments of the adversary, or the uncertainty of the future, or the faults of one's own side, or the underlying fraternity of ordinary men everywhere."[56]

Awe of God and its myriad representations in art and ritual was once the West's sublime, followed by the violent struggle for liberty and equality. Civilizations rise and fall on the vitality of their cultural ideals, not their material assets alone. History shows that most societies have sacred values for which their people would passionately fight, as "devoted (rather than principally rational) actors," risking serious loss and even death rather than compromise.[57] Research suggests this is for many who join ISIL, and for many Kurds who oppose them on the frontlines.[58] But, so far, we find no comparable willingness among the majority of youths that we sample in Western democracies. With the defeat of fascism and communism, have their lives defaulted to the quest for comfort and safety? Is this enough to ensure the survival, much less triumph, of values we have come to take for granted, on which we believe our world is based? More than the threat from violent jihadis, this might be the key existential issue for open societies today.

Notes

[1] The Islamic State has evolved over recent years, adopting at various times, or being given, different names and acronyms. Throughout this chapter, I will refer to the Islamic State, IS, and ISIS, as ISIL.

[2] Estimates vary widely with regard to both territory and population under ISIL control, as do views on the degree of control. See Harleen Gambhir, "ISIS Sanctuary: January 29, 2016," *Institute for the Study of War*, January 29, 2016, available at <http://www.understandingwar.org/backgrounder/isis-sanctuary-january-29-2016>.

[3] These interviews were conducted in 2015 by a team of researchers from the Centre for the Study of Intractable Conflict at the University of Oxford. Except where otherwise attributed, quotations throughout this chapter are from these interviews.

[4] Carlos Lozada, "Does Poverty Cause Terrorism," *The National Bureau of Economy*, available at <http://www.nber.org/digest/may05/w10859.html>; Darcy Noricks, Todd C. Helmus, Christopher Paul, Claude Berrebi, Brian A. Jackson, Gaga Gvineria, Michael Egner, and Benjamin Bahney, "Social Science for Counterterrorism: Putting the Pieces Together," ed. Paul K. Davis and Kim Cragin, *RAND Corporation*, 2009, available at <http://www.rand.org/content/dam/rand/pubs/monographs/2009/RAND_MG849.pdf>; "Youth & Consequences: Unemployment, Injustice and Violence," *Mercy Corps*, 2015, available at <https://www.mercycorps.org/research-resources/youth-consequences-unemployment-injustice-and-violence?source=WOW00088&utm_source=release&utm_medium=media%20relations&utm_campaign=youth%20conflict%20report>.

[5] Marc Sageman, *Understanding Terror Networks* (Philadelphia, PA: University of Pennsylvania Press, 2011).

[6] "Kerry: ISIL Fights to Divide, Destroy Iraq," *DoD News*, August 7, 2014, available at <http://www.defense.gov/News-Article-View/Article/603018/kerry-isil-fights-to-divide-destroy-iraq>.

Olivier Roy, "France's Oedipal Islamist Complex," *Foreign Policy*, January 27, 2016, available at <http://foreignpolicy.com/2016/01/07/frances-oedipal-islamist-complex-charlie-hebdo-islamic-state-isis>.

[7] Scott Atran, "On the Front Line against ISIS: Who Fights, Who Doesn't, and Why," *Daily Beast*, April 19, 2016, available at <http://www.thedailybeast.com/articles/2016/04/19/on-the-front-line-against-isis-who-fights-who-doesn-t-and-why.html>.

[8] Johnross12, "ISIS Video 6 6 2015 The Joy of Muslims with the Victories in Anbar," YouTube video, June 10, 2015, available at <https://www.youtube.com/watch?v=lVkIptIAl0I>; Kellan Howell, "ISIS Hands Out Celebratory Sweets in Syria after Brussels Attack," *Washington Times*, March 23, 2016, available at <http://washingtontimes.com/news/2016/mar/23/isis-hands-out-sweets-syria-after-brussels-attacks>; Dominique J.-F. de Quervain, Urs Fischbacher, Valerie Treyer, Melanie Schellhammer, Ulrich Schnyder, Alfred Buck, and Ernst Fehr, "The Neural Basis of Altruistic Punishment," *Science* 305, No. 5688 (August 2004), available at <http://science.sciencemag.org/content/305/5688/1254>.

[9] Thomas Hobbes, "Of Reason and Science," in *Of Man Being the First Part of Leviathan* (New York, NY: P.F. Collier & Son, 1909), available at <http://bartleby.com>.

[10] Charles Darwin, *The Descent of Man* (London: John Murray, 1871), accessed at <http://darwin-online.org.uk/content/frameset?pageseq=1&itemID=F937.1&viewtype=text>.

[11] Michael Crowley, "Did the Boston Bombers Really Use WMD?" *Time*, April 22, 2013, available at <http://swampland.time.com/2013/04/22/dont-panic-if-the-boston-bomber-is-charged-with-wmd-use>; Susannah Cullinane, "WMD: From A-Bombs to Pressure Cookers," *CNN*, December 4, 2014, available at <http://edition.cnn.com/2013/04/26/world/weapons-of-mass-destruction-explainer>.

[12] Edmund Burke, *The Sublime and Beautiful* (Adelaide: University of Adelaide, 1756), accessed at <https://ebooks.adelaide.edu.au/b/burke/edmund/sublime>.

[13] Ibid.

[14] Alden Whitman, "De Gaulle Rallied France in War and Strove to Lead Her to Greatness," *New York Times*, November 11, 1970, available at <http://www.nytimes.com/learning/general/onthisday/bday/1122.html>.

[15] Adolf Hitler, *Mein Kampf*, trans. Ralph Manheim (Boston, MA: Houghton Mifflin, 1971).

[16] Nuclear Vault, "Triumph des Willens (1935) - Triumph of the Will," YouTube video, September 22, 2011, available at <https://www.youtube.com/watch?v=GHs2coAzLJ8>.

[17] David D. Kirkpatrick and Rick Gladstone, "ISIS Chief Emerges, Urging 'Volcanoes of Jihad,'" *New York Times*, November 13, 2014, available at <http://www.nytimes.com/2014/11/14/world/middleeast/abu-bakr-baghdadi-islamic-state-leader-calls-for-new-fight-against-west.html?_r=0>.

[18] Kathryn Chamberlain, "ISIS poll for Rossiya Segodnya," *ICM Unlimited*, August 19, 2014, available at <http://www.icmunlimited.com/media-centre/press/isis-poll-for-rossiya-segodnya>.

[19] J.M. Berger, "ISIS Is Not Winning the War of Ideas," *The Atlantic*, November 11, 2015, available at <http://www.theatlantic.com/international/archive/2015/11/isis-war-of-ideas-propaganda/415335/>; Daveed Gartenstein-Ross, "How Many Fighters Does the Islamic State Really Have?" *War on the Rocks*, February 9, 2015, available at <http://warontherocks.com/2015/02/how-many-fighters-does-the-islamic-state-really-have>.

[20] Ivan Arreguin-Toft, "How the Weak Win Wars: A Theory of Asymmetric Conflict," *International Security* 26, no. 1 (2001).

[21] Scott Atran and Jeremy Ginges, "Religious and Scared Imperatives in Human Conflict," *Science* 336, no. 6083 (2012).

[22] Scott Atran, Hammad Sheikh, and Angel Gomez, "Devoted Actors Sacrifice for Close Comrades and Sacred Cause," *National Academy of Sciences* 111, no. 50 (2014).

[23] Thomas Jefferson, "The Papers of Thomas Jefferson, Vol. 1, 1760-1776," in *Declaring Independence: Drafting the Documents*, ed. Julian P. Boyd (Princeton, NJ: Princeton University Press, 1950), 243-247, accessed at <http://www.loc.gov/exhibits/declara/ruffdrft.html>.

[24] David McCullough, *1776* (New York, NY: Simon & Schuster, 2007).

[25] Bureau of International Information Programs, *Outline of U.S. History,* 2011, available at <http://photos.state.gov/libraries/amgov/30145/publications-english/history_outline.pdf>.

[26] Statement of Scott Atran, "U.S. Government Efforts to Counter Violent Extremism," hearing before the Subcommittee on Emerging Threats and Capabilities, Committee on Armed Services, U.S. Senate, March 10, 2010, available at <https://www.gpo.gov/fdsys/pkg/CHRG-111shrg63687/html/CHRG-111shrg63687.htm>; Greg Downey, "Scott Atran on Youth, Violent Extremism and Promoting Peace,"*Artis*, April 25, 2015, available at <http://blogs.plos.org/neuroanthropology/2015/04/25/scott-atran-on-youth-violent-extremism-and-promoting-peace/>; Statement of Scott Atran, "Pathways To And From Violent Extremism: The Case For Science-Based Field Research," testimony before the Subcommittee on Emerging Threats & Capabilities, Committee on Armed Services, U.S. Senate, March 9, 2010.

[27] Dounia Bouzar, Christophe Caupenne, and Sulayman Valsan, "Metamorphose opérée chez le jeune par les nouveau discours terroristes," *CPDSI*, November 2014, available at <http://www.bouzar-expertises.fr/metamorphose>; Daniel Milton, "The French Foreign Fighter Threat in Context*," Combating Terrorism Center at West Point*, November 14, 2015, available at <https://www.ctc.usma.edu/posts/ctc-perspectives-the-french-foreign-fighter-threat-in-context>.

[28] *Final Report of the Task Force on Combating Terrorist and Foreign Fighter Travel*, Homeland Security Committee, 2015, available at <https://homeland.house.gov/wp-content/uploads/2015/09/TaskForceFinalReport.pdf>.

[29] Scott Atran and Nafees Hamid, "Paris: The War ISIS Wants," *New York Review of Books*, November 16, 2015, available at <http://www.nybooks.com/daily/2015/11/16/paris-attacks-isis-strategy-chaos>.

[30] "Muslim Americans: Middle Class and Mostly Mainstream," *Pew Research Center,* May 22, 2007.

[31] Yann Algan, Christian Dustmann, Albrecht Glitz, and Alan Manning, "The Economic Situation of First and Second-Generation Immigrants in France, Germany and the United Kingdom," *The Economic Journal* 120, no. 572 (February 2010), available at <http://www.ucl.ac.uk/~uctpb21/Cpapers/AlganDustmannGlitzManning2010.pdf>.

[32] Lucas Martin, "Jack Lang: 2/3 Des Prisonniers Sont Musulmans," *MediaPart,* February 19, 2015, available at <https://blogs.mediapart.fr/lucas-martin/blog/190215/jack-lang-23-des-prisonniers-sont-musulmans>.

[33] Maximilien Robespierre, "Justification of the Use of Terror," in *On the Moral and Political Principles of Domestic Policy*, February 1794, available at <https://www.marxists.org/history/france/revolution/robespierre/1794/terror.htm>.

[33] "Attaques Anarchistes," in *Le Proces des Anarchistes* (Lyon: 1883).

[34] Theodore Roosevelt, "First Annual Message," in *The American Presidency Project,* December 3, 1901, available at <http://www.presidency.ucsb.edu/ws/?pid=29542>.

[35] Ibid.

[36] Scott Atran, "État islamique: l'illusion du sublime," *Anthropologie,* No. 66 (Novembre – Decembre 2014), available at <http://artisresearch.com/wp-content/uploads/2014/10/Satran-Cerveau-Psycho-oct-nov-2014.pdf>.

[37] Cory Doctorow, "Orwell's review of Mein Kampf," *Boing Boing,* August 17, 2014, available at <http://boingboing.net/2014/08/17/orwells-review-of-mein-kampf.html>.

[38] Scott Atran, Hammad Sheikh, and Angel Gomez, "For Cause and Comrade: Devoted Actors and Willingness to Fight," *Cliodynamics* 5, no. 1 (December 2014): 41-57.

[39] "Senior Administration Officials on Counter-ISIL Coalition Efforts," *U.S. Department of State,* July 28, 2015, available at <http://www.state.gov/r/pa/prs/ps/2015/07/245403.htm>.

[40] S. Nelson Drew, ed., *NSC-68: Forging the strategy of containment, with analyses by Paul Nitze* (Washington, DC: National Defense University Press, 1994), 15, accessed at http://www.au.af.mil/au/awc/awcgate/whitehouse/nsc68/nsc68.pdf>.

[41] George Kennan, "The Sources of Soviet Conduct," *Foreign Affairs*, July 1947, available at <https://www.foreignaffairs.com/articles/russian-federation/1947-07-01/sources-soviet-conduct>.

[42] Sebastian Payne, "Obama: U.S. misjudged the rise of the Islamic State, ability of Iraqi army," *The

Washington Post, September 28, 2014.

[43] Hammad Sheikh, Jeremy Ginges, and Scott Atran, "Sacred values in the Israeli-Palestinian conflict: resistance to social influence, temporal discounting, and exit strategies," *Annals of the New York Academy of Sciences* 1299, (September 2013): 11-24.

[44] Atran et al., "For Cause and Comrade."
 Scott Atran and Douglas Stone, "The Kurds' Heroic Stand against ISIS," *The New York Times,* March 16, 2015.

[45] Abu Bakr Naji, trans. William McCants, *The Management of Savagery: The Most Critical Stage Through Which the Umma Will Pass* (Cambridge, MA: John M. Olin Institute for Strategic Studies at Harvard University, May 2006), available at <https://azelin.files.wordpress.com/2010/08/abu-bakr-naji-the-management-of-savagery-the-most-critical-stage-through-which-the-umma-will-pass.pdf>; "Dabiq VII Feature Article: The World Includes Only Two Camps – That Of ISIS And That Of Its Enemies," *Memri,* February 18, 2015, available at <http://www.memrijttm.org/dabiq-vii-feature-article-there-is-no-longer-any-gray-zone-the-world-includes-only-two-camps-that-of-isis-and-that-of-its-enemies.html>.

[46] Michael D. Shear and Peter Baker, "Supporting France, Obama Loath to Add Troops to ISIS Fight," *The New York Times,* November 15, 2015.

[47] "Eurostate – Statistics Explained: Fertility Statistics," March 2016, accessed at <http://ec.europa.eu/eurostat/statistics-explained/index.php/Fertility_statistics>.

[48] Kurdish security services and police have given my research team information on attempts to infiltrate Europe, even with young Yazidis and Kurds who ISIL captured to cultivate for such a mission. Nevertheless, KRG leadership and police pretty much do manage to control threat in a very threatening environment, while also providing shelter and basic aid to an IDP and refugee population of nearly 2 million, more than a third as large as their own. And this, despite a serious economic crisis owing to falling oil prices and the high cost of war against ISIL. This contrasts with Turkey, a population of 75 million, which threatens to unleash its 2 to 3 million refugees on Europe unless it receives billions of dollars in aid; or France, as reflected in its expulsion of refugees from Calais; or America, whose majority agrees with Republican leaders who oppose accepting Muslim refugees from the region and whose leading presidential candidate would ban all Muslims from entering America for a time. Of course, there are security risks from the refugee problem; however, attitudes of European and American host populations towards Middle East refugees are also evidently based more on perceived cultural threats than the controlled evaluation of probable risks versus likely benefits from this refugee population based on previous experience and the present experience of other nations with similar refugee populations. On the risks of the refugee crisis for Europe, see George Soros and Gregor Schmitz," "Europe is on the Verge of Collapse – An Interview," *New York Review of Books*, February 11, 2016, available at <http://www.nybooks.com/articles/2016/02/11/europe-verge-collapse-interview>.

[49] Eli Berman and Jacob N. Shapiro, "Why ISIL will fail on Its Own," *Politico,* November 29, 2015.

[50] Ibid.

[51] "Costs of War," *Watson Institute International & Public Affairs at Brown University*, available at <http://watson.brown.edu/costsofwar/>.

[52] "United Network of Young Peace Builders," available at <http://unoy.org/>.

[53] Raymond Aron, *The Opium of the Intellectuals* (Piscataway, NJ: Transaction Publishers, 2001).

[54] Scott Atran, "The Devoted Actor: Unconditional Cooperation and Intractable Conflict across Cultures," *Current Anthropology* 57, no. S13, June 2016, available at <http://www.journals.uchicago.edu/doi/full/10.1086/685495>.

[55] It is worth examining whether a similar disenchantment with the Westphalian order of the West, and the same passions propel youth to join the Maras of Central America, Boko Haram in Nigeria, or even the lone-wolf perpetrators of San Bernardino.

4

The March Is Not Linear: Big Party Politics and the Decline of Democracy Worldwide

Francis Fukuyama and Hilary Matfess

Alternative models of governance have gained credibility in recent years, as the expectations set by post-Cold War democracy promotion programs initiated following the collapse of the Soviet Union were not met. In particular, China has increasingly become a source of inspiration for a number of developing countries seeking to replicate the country's remarkable economic growth. As a part of this process of examination and emulation, the Chinese Communist Party (CCP) is increasingly seen as instrumental to China's national development and the party's centrality is now considered a long-term feature of their political economy, rather than being a transient phase. This revelation confronts the assumption of modernization theory, that economic growth prompts democratization; instead of fading into the background amid the growth of multiparty democracy, the CCP seems capable of maintaining hegemony within China—in fact, in a number of ways, the growth that was supposed to undermine the party seems to have strengthened its claims to legitimacy. The means by which the CCP has woven itself into a secure position within China's political economy has not gone unnoticed by other political parties globally.

Perhaps no place is the emulation of the Chinese development model more evident than in Ethiopia and Rwanda under their respective ruling parties, the Ethiopian People's Revolutionary Democratic Front (EPRDF) and the Rwandan Patriotic Front (RPF). Both the EPRDF and RPF have adopted a form of governance, dubbed "developmental authoritarianism," that seems to be an isomorphic mimicry of China's model of state growth that reinforces party dominance.[1] The attempt by these parties to imitate the CCP, lacking the bureaucratic capacity that China enjoys, has given rise to a specific, anti-democratic form of governance that relies on heavy-handed repression of dissent. The rise of powerful political parties worldwide, capable of subsuming and displacing the state, represents a serious challenge to post-Cold War liberal democratic norms.

The CCP model, a political economy centered on a hegemonic ruling party that subsumes the state through the manipulation of investment, suppression of dissent, and delivery of impressive economic growth records, has both endogenous appeal to leaders attempting to retain power in their respective countries, and exogenous momentum, as China gains international clout and becomes an increasingly important source of aid and investment.

Undoubtedly, the CCP, the EPRDF, and the RPF have overseen robust economic growth by intervening in the economy and guiding investment towards certain "high priority" sectors. In China, this was done through a variety of State-Owned Enterprises (SOEs); whereas in Ethiopia and Rwanda, this has been accomplished through party-owned investment companies (party-statals).[2] It is estimated that 40 percent of China's nonagricultural gross domestic product (GDP) is state-owned or controlled. If urban collectives and township and village enterprises (TVEs) are included, the estimate rises to 50 percent.[3] In Ethiopia and Rwanda, the party-led investment companies have been a way to expand and cement patronage networks, fortifying their political positions while promoting growth. While it is difficult to estimate what share of the economy they control, it is clear that the seemingly robust private sectors in Ethiopia and Rwanda are dominated by party-held investments and companies that act as party appendages. A rising tide may lift all ships, however, as the EPRDF and RPF have ensured that their countries' growth is directed by, and confers benefits to, the ruling party. This is a marked difference from the use of SOEs in other contexts which, while conferring political benefits and generating rents, also bolstered state capacity.

Across all three examples, the parties' strength has also been bolstered by limitations on civil rights and political freedoms, enforced through an impressive surveillance apparatus and politicized security sector. This surveillance capacity has kept pace with technological change. The so-called "Great Firewall of China" has been imitated by the EPRDF and the RPF through their vast internet and telecom surveillance programs, made all the more potent by the parties' central roles in the expansion of such services within their respective countries. China's legal restrictions on free speech have also been mimicked by Ethiopia and Rwanda, both of which have passed draconian legislation curtailing the freedom of speech and association under the guise of national security concerns.

Low-tech methods have been continuously updated by these parties, as well. The CCP has used "neighborhood committees" for a variety of purposes, not the least of which was "snooping on ordinary citizens."[4] In Ethiopia, the *kebele* system of community organization has been used to expand party rolls, and in Rwanda, special training camps (*ingando)* have been used to encourage the RPF's message. In all three countries, the military is politicized, acting as a party appendage rather than a government body. As Richard McGregor notes, "Unlike in the West, where controversies often arise about the potential politicization of the military, in China the party is on constant guard for the opposite phenomenon, the *depoliticization* of the military;" the military's legacy, reputation, and political affiliation play a similarly critical role in quelling dissent and furthering the party's agenda.[5]

In short, the CCP and its emulators worldwide have embarked on a process of "party development" in the name of "national development." However, emulation is an imperfect process; the imperfect implementation of the China model and the lack of bureaucratic accountability, turnover, and capacity in these countries has created vulnerabilities within the regimes that could catalyze domestic resistance. While the role of ideas and perceptions is important in motivating the adoption of this model, the undeniable differences in the social profiles and bureaucratic capacities of East Asian and African countries have stark implications for the potential effects of mimicking such a development strategy. As Phil Williams writes in his contribution to this book, the shortcoming of the international state system stems from the weakness of its components. A proliferation of states emulating the China model, lacking bureaucratic capacity, will further erode the stability of the global system.

The sources of legitimacy that the CCP has cultivated for itself, in particular through the imposition of term limits for bureaucrats and credible anticorruption efforts, are absent from those mimicking the CCP's success. While Ethiopia managed a smooth transition of executive power to another party member following Meles Zenawi's death, no such transition seems possible in Rwanda, as constitutional reform has been undertaken to validate Paul Kagame's ambitions for a third term. Further, the accountability mechanisms that have been adopted in these countries, including the *kebele* system and *imihigo* (both of which will be addressed below) have been criticized as oppressive expansions of party influence. The narrow base of support enjoyed by ruling parties in Ethiopia and Rwanda prompts these regimes to rely even more heavily on crudely repressive measures and their military credentials to retain power.

This chapter will address the growth record of autocratic and democratic regimes in sub-Saharan Africa, the global democratic recession, and existing Western and Chinese aid paradigms. The success of Ethiopian and Rwandan developmental authoritarianism under the EPRDF and the RPF will then be explored, paying particular attention to the mechanisms of party dominance within the economic, social, and political apparatuses, in order to construct an image of the emerging alternatives to liberal democracy. This discussion will highlight the role of party-owned investment companies, attempts at bureaucratic institutionalization, and the robust surveillance regimes of these two countries. The Ethiopian and Rwandan model will then be compared to governance patterns in China, in order to examine areas of commonality and difference between the regimes. Whether or not the model is being promoted by China or emulated by developing states will also be discussed. Finally, the implication of the spread of alternative governance models will be discussed; if "emulation can also be understood as norm diffusion," the rise of these alternative models inspired by Chinese development represents a challenge to the norms that underpin the international community and to global stability and security.[6]

Throughout this chapter, we hope to demonstrate that the rise of this model presents three threats to the liberal democratic system of norms and interstate relations: the first is the legitimization of undemocratic rule by virtue of economic performance; the second threat is the normalization of the repressive measures employed by the parties when the legitimacy garnered from economic growth wanes; and the third threat is to the system of interstate relations. In these regimes, the removal of the party (through force or credible elections) would involve the removal of the most critical aspects of the state's economic sector and leave the political sphere lacking institutions. The international system of order is dependent upon the stability of states; the longevity of developmental authoritarian regimes is subject to speculation. Thus, the implications and characteristics of the spread of this model must be considered.

While the CCP model may not represent a wholesale challenge to the liberal democratic paradigm in the short term, it has contributed to an international political *zeitgeist* that prioritizes economic results over human rights and governmental accountability, and may facilitate the continuation of the global democratic recession. The ending of the Cold War seemed to represent the ascendency of the liberal democratic model the United States represented; however, the march towards the end of history is not linear and the rise of alternative governance models is certainly a stumbling block in the process of global democratization. As one of the authors has written elsewhere, the durability of the Chinese economic model in the long run is completely irrelevant to the challenges American foreign policy faces today—one of which is certainly the rise and emulation of the CCP in developing countries.[7]

Even in the absence of Asian efforts to promote such governance or even the recognition of their role as an inspiration to developing countries, if political leaders in developing countries adopt authoritarian, party-elevating policies "based on their understanding of the Chinese model, then it [really] does exist."[8] The result of this model, if left unchecked, will be a more violent, unstable world in which the promotion of democracy will be challenged by the institutionalization of unaccountable governance regimes.

Democratization and Development

More than 25 years after the fall of the Berlin Wall, global democratization efforts appear to have stalled, confounding those who assumed that the end of the Cold War would usher in an era of liberal democratic governance. According to a count by Larry Diamond in 2011, one in five of the democracies that existed during the "third wave" of democracy has subsequently experienced a democratic reversal.[9] In addition to a declining "head count" of democracies, "the level of freedom in the world, as measured annually by Freedom House, is now in decline—for several consecutive years."[10]

The failure of democratization to proceed as predicted, as well as the narrative surrounding the rise of the autocratic East Asian Tigers, notably China, has dampened global enthusiasm for democracy. In sub-Saharan Africa, freedom across the continent has declined as measured by annual Freedom House reports over the past decade.[11] At the same time, economic growth across the continent hovered just below five percent. However, overall there is little relationship between governance and growth in sub-Saharan Africa. Figure 4.1 shows a scatterplot demonstrating the relationship between Freedom House's 2015 rankings and average growth rate since 2000. Clearly, there is little relationship between current levels of freedom and economic growth for the past 15 years.

Figure 4.1. Freedom House Score (2015) and Economic Growth

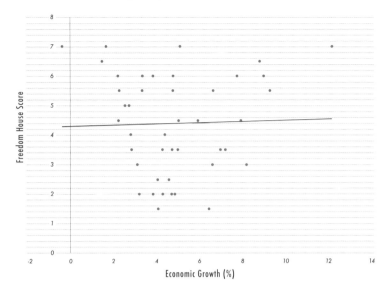

The relationship between changes in governance and economic growth is slightly more comforting to democracy advocates. Figure 4.2 illustrates the relationship between changes in Freedom House scores between 2001 and 2015 and economic growth since 2000. The slight positive trend line would suggest that growth and good governance have, at least, a weak correlation in sub-Saharan Africa. However, the highest-performing economies (including Equatorial Guinea, Chad, and Angola) are categorized as "Unfree" by Freedom House.

Figure 4.2. Difference in Freedom House Scores Between 2001 and 2015 (positive = democratization; negative = closing of political space)

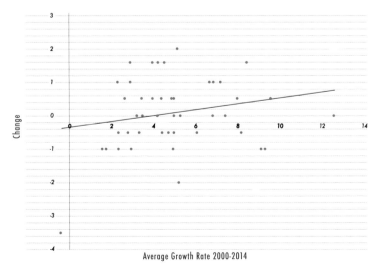

Though the relationship between governance and growth across the continent is muddled and unclear, Rwandan and Ethiopian growth is exceptional. Once diamond-, gold-, and oil-exporting countries are removed from the top 10 fastest-growing economies, Ethiopia and Rwanda are the top 2 performers, boasting average growth rates of 10 percent and 8 percent respectively since 2005. Since assuming power in the mid-1990s, both of these parties have pursued and delivered economic growth and human development. Between 1990 and 2012, Rwandan GDP per capita nearly doubled, from $353 to $620 and its Human Development Index (HDI) score rose from 0.238 to 0.506 between 1990 and 2013. Ethiopia boasted similarly impressive growth; over the same time periods, Ethiopia's GDP per capita rose from $251 to $470 and its HDI score rose from 0.284 to 0.435. These gains in human development scores come not just from their economic growth, but also from the investments in education and public health that both countries have prioritized. Based on forecasts from the World Bank's Global Economic Prospects, *Business Insider* projected that Ethiopia would be the fastest-growing economy and Rwanda the 12th fastest between 2014 and 2017.[12] The question of how these countries achieved growth rates nearly double the continental average requires an understanding of their tightly controlled political economies and their ruling parties.

The strong economic performances of these countries are not incidental to their models of governance; both countries have undertaken massive and well-publicized developmental agendas since coming to power in the 1990s, with clear targets for domestic production in a handful of sectors that are tightly controlled by the parties, and relatively robust provisions of state services. While data concerning the revenues of their party-owned investment vehicles is difficult to come by, estimates suggest that these two parties are among the wealthiest on the continent—despite the lack of mineral resources and petroleum within their borders. These parties have promoted growth, and lined party coffers, through investments in export-oriented industries (such as coffee in Rwanda and cut flowers in Ethiopia), as well as through domestically-oriented investments (notably cement in both countries). These countries' ability to maintain high growth rates in the absence of political liberalization suggests that the worldwide threat to liberal democracy stems not just from autocracy in East Asia and China, but also from emulation and adaptation of such models in the Global South.

The failure of democratization and the fading enthusiasm for the model worldwide stems, in part, from the inability of states to keep pace with citizens' expectations—for both economic development and political accountability. Though the empirical relationship between governance and democracy in sub-Saharan Africa is inconclusive, a handful of autocratic success stories may spur emulation. Further, the model of democracy promoted by the United States and other traditional donor countries rests upon certain assumptions concerning state capacity that are not universal.[13] Democracy promotion efforts and their advocates, particularly those from the United States, have overlooked that only a handful of countries have historically institutionalized bureaucratic traditions, and have demanded, at times, an unreasonable level of capacity. Frustration with the "Western model" and expectations adds to the allure of alternative governance paradigms.

In contrast, Chinese involvement with African economies has taken a "hands-off" approach to governance issues. While recent studies have demonstrated that Chinese companies are no more likely to invest in natural resources than other countries and are just as likely as other companies to submit transparency reports, a review by Dreher et al. found that Chinese aid has been disproportionately allocated to the home district of the incumbent executive.[14] Further, China and the CCP have facilitated the purchase of surveillance technology for a number of African countries, enabling these governments to crack down on dissent and restrict the freedom of association.[15] Unsurprisingly, "Africans who rank human rights as high in importance were more likely to have an unfavorable opinion" of Chinese activities within their countries.[16] Unlike the political leaders and government officials that have seemed receptive to Chinese aid, "civil society, trade unions, and some sectors of local business have been more wary."[17] Local capacity to raise concerns about the impacts of Chinese partnership on labor practices and the domestic political economy, however, is stymied by oppressive political atmospheres and low organizational capacity.

Both Ethiopia and Rwanda have made use of the CCP's example, and increasingly, Chinese assistance and investment, in order to bolster the power of their respective parties; however, both of these parties lack the bureaucratic capacity and institutional accountability of their Chinese counterpart, required to be stable in the long run. The spread of big-party rule, seeking to emulate East Asian and specific instances of African economic success, threatens to light the long fuse of resistance to unaccountable governments. The nuances of party rule in Ethiopia and Rwanda have made these countries "African success stories," but have also created regime vulnerabilities that will provoke violent resistance and court instability.

The CCP Model

The Chinese Communist Party has presided over impressive economic growth, while maintaining tight social and political control over its citizens. The country is among the least free in the world. Freedom House rated the country 6.5 in 2015, on a scale where 7 means the most autocratic; this is the same score that it received in 1998.[18] While levels of freedom have been stagnant, growth rates in China have regularly been in the double digits in recent decades, and human development in the country (as measured by the Human Development Index) has skyrocketed from 0.389 in 1975 to 0.718 in 2010. Across the three countries discussed in this paper, these parties' ability to promote economic growth has granted them a measure of legitimacy and public support; when that is insufficient to legitimize their rule, they resort to explicitly repressive policies. In China, as in Ethiopia and Rwanda, this is facilitated through a vast surveillance network and finely-tuned oppression tactics centered on curbing the freedoms of speech, information, and association.

The oppressive tactics adopted in China to foster self-censorship (including "a number of restrictive regulations issued since 2005…requiring publishers not to reprint politically sensitive books, restricting popular access to foreign films and television programs," and blocking certain websites the CCP considers "politically threatening"), and the vast surveillance network (which includes the monitoring of "personal communications, including cellular telephone text-messaging"), are not wildly different from other authoritarian governments.[19] The centrality of the ruling parties and the subversion of the state to the party in all three of these countries are their distinctive features.

While recent reforms in China have slightly altered the balance in favor of the state, it is not out of the question to suggest that the CCP exercises social, political, and economic control throughout the country. The CCP's political dominance is obvious from its position as the only political party in the country; however, according to Xiao Ma, the party's dominance has leaked into the social sphere such that "the influence of the CCP is so encompassing to the extent that there is little to no independent political space beyond the Party itself...being a member of the Party is the only way to powerful positions."[20]

This growth has been achieved in the context of the Party adopting a multipronged strategy, in which the party "substitutes itself" for the state in the economic, social, and military spheres.[21] Throughout this process, "the organization line between the Party and government blurred, and the delegation relationship disappeared almost entirely."[22]

Economically, the CCP has taken an active role in organizing and owning the commanding heights of the economy. As previously discussed, it is estimated that the CCP controls roughly half of the Chinese nonagricultural economy, largely through SOEs staffed by Party members. Despite the process of gradually liberalizing the economy, the Party has designated "a number of industries that are important to China's economic and national security and indicated that these strategic industries will remain wholly or largely under the government's control. In other important so-called pillar industries, the state will remain a major player, with significant, though not majority, ownership."[23] The country's success with the "national champion" model of development, in which certain sectors and companies are given preferential treatment, is reflected in its inclusion in subsequent five-year development plans.

This strategy has not only led to economic growth; it has facilitated the sophistication of the Party structure and control. Zhao Ziyang stated in 1987 that Party involvement with factory management acted as a "yardstick for supporting or opposing party leadership.... Every time we undertook a campaign, this setup was strengthened, to the extent that the Party committees monopolized many administrative matters."[24]

With the sophistication of the Party structure came the ability to penetrate more deeply into Chinese society. The CCP "dictates all senior personnel appointments in ministries and companies, universities, and the media, through a shadowy and little known [sic] body called the Organization Department."[25] The result is that the CCP has the final say in who fills "just about every significant position in every field in the country."[26] Party membership is thus a necessity for those wishing to take part in, and benefit from, the country's upward trajectory. Maria Edin describes the "cadre responsibility system," which manages this patronage network, as such:

Political rewards are linked to the result of the annual evaluation and the subsequent ranking of leaders. Top-ranking township leading cadres will be awarded with the political title of advanced leader (*xianjin lingdao*) or declared to be a model leader. As shown above, if a township has failed to attain the priority targets with veto power, it disqualifies the township government from becoming an advanced unit and the responsible cadre from becoming an advanced leader. In one county, leading cadres of the first three ranked townships in the annual evaluation were entitled advanced leaders in accordance with local regulations. The results are officially announced, thereby putting pressure on those involved, during a large meeting to mark the end of the year and the beginning of the next working year. In the county above, the top 15 percent and bottom 5 percent of cadres on the list were respectively praised and disgraced at this meeting. In another county, a list of the first 100 cadres was both published in the local media and circulated as a government document to all relevant government department.[27]

The accountability mechanisms that the Party has adopted, in order to increase transparency and efficiency, have thus strengthened the Party's control over subnational political units.

In addition to this control over China's political and economic institutions, the CCP has undertaken robust efforts to censor the internet and the country's press. China has a robust legal framework supporting such suppression; nearly all media in the country are subject to "an extensive and burdensome licensing scheme over all media that bars those without money and political connections from establishing publishing enterprises."[28] The internet, promised to be an equalizing and democratizing force, has been tightly monitored and regulated by the Chinese government. Not only must users cope with the "Great Firewall of China," "anyone wishing to operate a commercial website must acquire a Telecommunications Business Operating License and an Internet Content Provider License;" even noncommercial sites must obtain a license and all internet publishers must receive government authorization to publish content.[29]

Content itself is heavily regulated. Article 105 of the Criminal Law prohibits "rumor mongering [sic]" or "defamation" to "incite subversion," without defining any of these terms.[30] Human Rights Watch (HRW) notes that Article 19 of the Criminal Law prohibits content:

1. violating the basic principles as they are confirmed in the Constitution;
2. jeopardizing the security of the nation, divulging state secrets, subverting of the national regime or jeopardizing the integrity of the nation's unity;
3. harming the honor or the interests of the nation;
4. inciting hatred against peoples, racism against peoples, or disrupting the solidarity of peoples;
5. disrupting national policies on religion, propagating evil cults and feudal superstitions;
6. spreading rumors, disturbing social order, or disrupting social stability;
7. spreading obscenity, pornography, gambling, violence, terror, or abetting the commission of a crime;
8. insulting or defaming third parties [sic], infringing on the legal rights and interests of third parties [sic];

9. inciting illegal assemblies, associations, marches, demonstrations, or gatherings that disturb social order;
10. conducting activities in the name of an illegal civil organization; and
11. any other content prohibited by law or rules.

HRW also commented that these regulations have been applied liberally, extending "well beyond the narrow band of information that might truly incite hatred or disturb social order," to include "information that the government deems too embarrassing, or is too candid in its discussion of particularly entrenched social problems."[31]

One of the means by which the CCP has been able to carry out such measures has been through its control over the People's Liberation Army (PLA). Put simply: "the PLA is the party's military, not the country's."[32] As such, the CCP seeks to preserve the politicization of the armed forces, even within the context of increasing accountability and bureaucratic institutionalization in other aspects of the country. The Army crackdown on protestors challenging the CCP's reign in 1989 was a critical juncture in the nation's history; the CCP's leaders recall both the importance of the military in quelling dissent and their vulnerability when one general refused to clear protesting students from Tiananmen Square. Even more directly, the politicization of the PLA reduces the likelihood of a challenge to CCP power from the military.

As a result of the party's historical memory, CCP elites have "worked hard to keep the generals on their side, should they be needed to put down protests again."[33] According to David Shambaugh, "one way this is done is to co-opt the military elite into the party elite;" the creation of an "interlocking directorate," where the upper echelons of the party's organs are staffed by senior military officers or those that have previously served in the armed forces.[34]

Further, China and its African counterparts have appealed to national security concerns in order to legitimate their positions. According to Dan Slater, "authoritarianism is at its strongest when it is widely perceived as a necessary stabilizer."[35] While the argument for party-induced stability is easier to make for the EPRDF and the RPF who led their countries through civil conflicts in the 1990s, the CCP has also implicitly and explicitly tied the survival and stability of China to its tenure.

Additionally, among the three countries, and over time, there has been a measure of adaptation and learning. For example, Singaporean opposition leaders were able to use the internet to challenge claims made by the People's Action Party and the party-controlled press; no such recourse is available in Ethiopia or Rwanda and China is strictly limited by "the Great Firewall of China."[36] The specifics of this model will vary from country to country; however, there is an undeniable rise in the logic of dominant-party politics, in which national strength is funneled into the ruling party to promote stability. While these projects have been described as "state-building," given their economic dividends, they are perhaps more accurately described as "party-building."

Economic Growth, Accountability Mechanisms, and Party Development: Where the Differences Lie

Perhaps the most prominent distinction between the party politics of East Asia and their African counterparts are the differences in bureaucratic capabilities and their nominal adoption of multiparty democracy; however, these two characteristics are not unrelated. The lack of bureaucratic capacity in states seeking to emulate the CCP will have serious ramifications for stability globally.

Critical to Chinese development has been the "successive administrative reforms…to create a slimmer and more efficient state," which have entailed rigorous efforts to improve state efficiency, not merely to reduce state involvement in the economy.[37] Chinese SOEs, while staffed by Party members and subject to special treatment from the government (not unlike the party-statals in Rwanda and Ethiopia), are ultimately arms of the states and are regulated by the State-owned Assets Supervision and Administration Commission (SASAC). Though this regulation is weak, it still represents a sort of institutionalized oversight lacking in Ethiopia and Rwanda. Further, the appointment of Party members to positions of authority within these SOEs is managed according to the *nomenklatura* system. According to Kjeld Erik Brødsgaard, "the *nomenklatura* system prevents business leaders from successfully challenging Party rule and helps to ensure that the achievements and the continued growth of the business sector will benefit the Party and contribute to regime stability."[38] Using the *nomenklatura* system was a specific policy choice, as staffing the SOEs with people outside of the system might have "preserved" the separation between the Party and the government.[39]

Not only has China endeavored to create a leaner state; it has instituted a robust bureaucratic system to promote the circulation of Party members and institute regularity into the one-party system. The refinement and institutionalization of bureaucratic accountability mechanisms began under Deng Xiaoping in the 1980s. Under these reforms, the Party Congress was reinstitutionalized and both term and age limits for elite government cadres were introduced. According to Zhengxu Wang and Anastas Vangeli, these reforms were tested at the 12th Party Congress held in 1982; within 20 years, "a number of implicit and explicit rules seemed to have been established," regulating the behavior of Party members.[40] Most notably, a two-term limit for major political posts was instituted in the 1982 Party Constitution and has been abided by since. As Ma notes, "term limit[s] that effectively [bind] top leaders however [are] rarely seen among dictatorships." The Chinese model of constraining and regulating power within the Party is, thus, a significant development and an innovative governing characteristic.[41]

Accompanying the term limits are mechanisms for promoting Party members with some measure of transparency and accountability. The Alternative Central Committee (ACC), for example, is seen to be a sort of antechamber to the powerful Central Committee (CC), which oversees the *nomenklatura* system. The predictability and stability introduced by this visible cadre of potential future leaders reduces uncertainty in political contracts and facilitates continuity. Further, the ACC has acted as a representative pressure gauge for aggrieved minorities. A study conducted by Ma, regarding which members of the ACC eventually are promoted to the CC, found that "[w]omen and ethnic minorities are appointed as ACC members to showcase the Party's commitment in promoting equality along the gender and ethnicity lines. Yet once it comes to the competition into more powerful positions like CC, these members are hugely disadvantaged or are even not considered for promotion at all."[42] Ethiopia and Rwanda lack such a mechanism for channeling identity politics and grievances through their parties. Indeed, Rwanda under Paul Kagame has emerged as a Tutsi dictatorship, where Hutus are largely excluded from positions of power. As minority groups have frequently organized in opposition political parties to advocate for change, the result has been harsh political oppression from the state and mounting risk of political violence.

While China and East Asian countries have adopted provisions for political turnover within the party bureaucracy and increased accountability, it is unclear how African developmental authoritarian states will manage succession. In Rwanda and Ethiopia, *imihigo* and *kebele* programs introduce measures of political accountability and a broad base of party members (if not deep party support); however, the parties remain clientelistic and opaque. Following Zenawi's death in 2012, the EPRDF was able to maintain its hegemonic position within the country's political sphere, suggesting the durability of the model; however, Kagame's intention to run for a third term as president of Rwanda suggests that the RPF may not have similar plans in place for succession.

Additionally, and perhaps obviously, Rwanda and Ethiopia also differ from China by holding regular elections, however manipulated they may be. Nevertheless, the adoption of multiparty democracy by these countries may be a reflection of their dependence upon the international community for foreign aid. Further, these reforms were adopted in an era of unquestioned American hegemony in the international system; the China model was not yet a competitor to Western liberal democracy promotion. Timing and the aid dependence of these two countries can partially explain their adoption of multiparty elections.

Developmental Authoritarianism in Ethiopia and Rwanda

While Ethiopia and Rwanda lack the long history of bureaucratization of East Asian countries, in both of these countries, the ruling parties have been able to leverage traumatic events in their national histories to consolidate their power and engage in sweeping economic and political reforms. Both EPRDF and the RPF began as militias and transitioned to political parties as part of the resolution of dramatic civil conflicts. The EPRDF assumed power in 1991, when the coalition of four ethnic militias overthrew the oppressive, communist Derg and vaulted Zenawi to the country's top political post, where he remained until his death in 2012. Similarly, the RPF invaded Rwanda in 1990 and came to power in 1994, after a civil war and the horrors of the Rwandan Genocide; since then, Kagame has been the dominant political figure within Rwanda.

While overall levels of development in these countries remain significantly below China's, their growth rates, coupled with tangible improvements in health and education indicators, make them veritable "African miracles" in many circles. These improvements were not made through laissez-faire economic liberalization; Ethiopian and Rwandan growth is the result of concerted party efforts. A 2012 paper, written by Ethiopia's Zenawi, outlines his party's philosophy—demonstrably shared by the RPF—regarding the role of the state at both the "ideological" and the "structural" level:

> At the ideological level, accelerated development is the mission, its source of legitimacy. Moreover, the development project is a hegemonic project in the Gramscian sense— the key actors voluntarily adhere to its objectives and principles. Structurally it has the capacity to implement policy effectively, which is the result of various political, institutional, and technical factors, which in turn are based on the autonomy of the state. This autonomy enables the state to pursue its development project without succumbing to myopic interests.[43]

At the inaugural Meles Zenawi Foundation Symposium on Development, held in August 2015, Rwandan President Paul Kagame highlighted the continued relevance of the conference's namesake in championing of guided national development. In his opening remarks, Kagame asserted that "every developed economy, without exception, is the fruit of a free market, and a strong developmental state, working in tandem. The orthodoxy of shrinking the state to the bare minimum, and replacing it with externally-funded nonstate actors (here you can say NGOs), left Africa with no viable path out of poverty."[44]

Recalling the continent's disastrous experiences with "one-size-fits-all" political and economic prescriptions under Structural Adjustment Programs, he portrayed Ethiopia and Rwanda as vanguards of a new governance model, stating "ours is the true democracy of citizens, not the false one of institutionalized corruption and division (or "rent-seeking," as Comrade Meles usually said). We cannot be bullied into accepting policies that misrepresent us and do us harm in the end, as we have seen over many years," he said.[45]

While Kagame's characterization of the two countries as "democratic" is, at best, euphemistic, he is certainly accurate in portraying Ethiopia and Rwanda as innovators in governance models. Jones et al. identify an "emerging mode of illiberal state building [sic]," of which, Ethiopia and Rwanda are at the forefront. In addition to achieving growth through domestic intervention using party-statals, these countries have demonstrated how foreign aid can "bolster, rather than undermine undemocratic leaders."[46] While Zenawi and Kagame frequently reference the importance of a strong state, their policies generally bolstered the capacity of their parties. State power and capacity in these two countries are entirely dependent upon their respective ruling party, both of which have embarked on endeavors to incorporate all citizens into their parties and to permeate the economic and social spheres through their political positions.

Developmental Authoritarianism and Party-Statals

In attempting to accelerate and manage their economies, ruling parties in Ethiopia and Rwanda have relied heavily upon party-owned investment vehicles, described as "party-statals" by a number of observers drawing comparisons between such investments and parastatals. The use of party-statals is not unlike the CCP's use of SOEs in China to prioritize certain sectors and reward party members; the most critical difference is that party-statal investment in Rwanda and Ethiopia is recorded as being "private sector" activity. In Rwanda, Tri-Star/Crystal Ventures Limited (CVL), controlled by the RPF, has invested in critical sectors such as building materials, road construction, and mobile communications technology. Academics David Booth and Frederick Golooba-Mutebi assert that the "willingness of Tri-Star/CVL to use its financial clout to fund investments with high expected social benefits and/or positive economic externalities, including those associated with venture capitalism, is a significant aspect of Rwandan political economy."[47] The company has monopolized production in a number of the sectors it has invested in, in large part because of its role as an extension of the RPF.

In Ethiopia, the government-investment vehicles of choice are endowment-funded companies; most notably, the Endowment Fund for the Rehabilitation of Tigray (EFFORT). EFFORT is managed by the Tigray People's Liberation Front, a crucial ethnic political organization in the EPRDF alliance. EFFORT's investments in critical sectors like logistics and construction, and the "high degree of vertical and horizontal integration of its companies" has allowed it to develop a formidable presence as a patron, linking the private sector, the government, ethnic advocacy groups, and citizens writ large.[48] There is not sufficient space to discuss all of the party-statals owned by the RPF and the EPRDF, so our discussion will be limited to CVL and EFFORT, despite both parties owning additional investment vehicles.

Table 4.1. Selected Party-Statals and Their Subsidiaries

CVL's Subsidiaries	EFFORT's Subsidiaries
Crystal Telecom	Beruh Tesfa (Manufacturing)
Inyage Industries (Beverages)	Almeda Textiles
NOD-COTRACO (Civil Engineering and Construction)	Addis Pharmaceutical Factory
Mutara Enterprises (Furniture and Office Furnishings)	Messebo Building Material Production
Bourbon Coffee	Mesfin Industrial Engineering
Intersec Security	Bruh Tesfa Irrigation and Water Technology
Ruliba Clays	Sheba Leather Industry
Real Contractors Limited	Guna Trading House
East African Granite Industries	Ezana Mining Development PLC
	Hiwot Agricultural Mechanization PLC
	Saba Dimension Stones PLC
	SUR Construction
	Enterprise Ethiopia Travel
	Express Transit Service PLC
	TransEthiopia (Transportation)

These investment vehicles do more than promote economic growth; they are significant sources of economic clout for the owning parties and their profits are funneled into party coffers, rather than into the state. Unlike SOEs, which previously were important features of developing countries' economies—including Ethiopia and Rwanda under different ruling parties—the party-owned companies register as private investment in national accounting. Thus, they do not create an accountable bureaucratic class. Instead, these companies are entirely party-owned and operated by party-appointed heads "in the guise of the new business elite."[49]

When Rwanda was pressured to sell off its state-owned companies by the International Financial Institutions (IFIs), many of the SOEs were replaced by these party-statals.[50] While Booth et al. have praised Rwandan party-statals for their developmental dividends, others have noted that these institutions have "become extractive economic institutions, concentrating power and opportunity in the hands of only a few."[51]

The RPF's party-owned investment holdings have a range of investments, detailed in Table 4.1. An estimate of their worth suggests that, in addition to controlling the country's most critical sectors, these companies have the highest combined worth in terms of "fixed assets, total assets, turnover, and more importantly, the largest share in gross output."[52] Despite their overwhelming domination within the economy, the Rwandan party-statals have benefited from a variety of tax exemptions, reducing the impact of their revenue generation on the state's capacity. In Ethiopia, the relationship between the government and its party-statals differs in characteristics, but not in overall effect. The differences between the RPF and EPRDF's holding companies likely stem from the fact that:

> The EPRDF government has achieved a high degree of centralization of rent management and allocation, retaining control of a large proportion of available sources of rents and economic levers. These include the large state-owned enterprise sector, endowment-owned businesses, and substational regional development organizations; as well as tight regulation of financial institutions, including for micro-credit, and expansion of the tax base.... For instance, whilst party-associated companies in Rwanda seem to form a central key to long-horizon rent centralization, Ethiopia's Endowment owned businesses constitute only one strand of the government's strategy.[53]

Though Ethiopia maintains its reliance upon explicitly state-owned entities, just as in Rwanda, Ethiopian party-statals have acted as powerful catalysts for economic growth and have invested in a wide variety of sectors in the economy. EFFORT owns 16 enterprises, the "most important being Mesebo Cement, Guna Trading House, Almeda Textiles and Garmentine, TransEthiopia Transport & Logistics, and Mesfin Industrial Engineering, along with Sur Construction and Addis Pharmaceuticals."[54] Unlike in Rwanda, the EPRDF's party-statals contribute to the tax base and "are by far the largest regional tax-payer to the government, currently providing 60 percent of its regionally-generated revenues."[55] Given the ethno-federalist nature of the Ethiopian government, this creates "a situation that both promotes stability and mitigates political competition in the region," bolstering the importance of regions supportive of the EPRDF. This has the same effect as the absence of party-statal taxation in Rwanda, of putting such companies in politically important positions within the country's economy.[56] Further, the party has used these companies to reward loyal party members and induce broad-based membership; Azeb Mesfin, Meles Zenawi's widow, was appointed to the leadership of EFFORT and party members filled the endowment's ranks. One activist lamented that:

> Appointments within EFFORT are made based on politics. If you are a member of the ruling party, you can get a job no matter what your qualifications are. I can be a high school drop-out [sic], an elementary school drop-out [sic], but if I am connected to anybody who has a top leadership position within the ruling party, I will get into EFFORT with no questions asked…. It is politically charged, politically owned and… there is no transparency.[57]

Not only have party-led investments targeted the most promising sectors within each country, benefiting from the rapid gains to be made in underdeveloped economies, they have also benefited from the use of state power to fortify party investments. In both countries, legislation has been adopted to insulate these party-owned organizations from competition. In Rwanda, for example, publicly funded school programs have been instructed to purchase exclusively from the party-owned Inyange Industries.[58] In Ethiopia, the infamous Civil Society Ordinance of 2008/2009 recognized only four types of charities, charitable endowments (such as EFFORT), charitable institutions, charitable trusts, and charitable societies.[59] While private sector growth has been identified as a strength of these countries, it is clear that the parties have controlled (and have been strengthened by) the investment within their borders. Party-statals have contributed to the hegemonic domination of the political economies of Ethiopia and Rwanda by their ruling parties.

Developmental Authoritarianism, Repression, and Party Penetration

The legitimacy gleaned from the state's commitment to "accelerated development" and human development, channeled through government investment vehicles, should be considered a "carrot" in the state's coercive arsenal. More problematic and more visible are the "sticks" wielded by the countries' ruling parties, the EPRDF and the RPF. While Ethiopia and Rwanda both have adopted the veneer of democracy, including elections, such overtures should not be confused with viable opportunities for political contestation. These regimes have adopted strategies to contain domestic dissent through the limitation of the sorts of "coordination goods" necessary to organize successful opposition campaigns and to penetrate their respective societies.

These two countries seem to be star pupils in adapting to the authoritarian "learning curve," identifying which sorts of repressive policies are the most effective while attracting the least amount of international resistance or response. In Ethiopia and Rwanda, legislated limitations on free speech are accompanied by the suppression of the press and robust state initiatives to monitor society and to instill within the populace an official political narrative.

The limiting of "coordination goods" and the imposition of self-censorship bolsters the stability of these regimes in the short term by limiting opposition's ability to organize.

Much has been made of the "genocidal ideology" and "sectarianism" legislation in Rwanda that prohibits discussion of the 1994 genocide that falls outside of the RPF's official narrative and which has been used to jail political opponents and prevent Rwandans from using the terms "Hutu" and "Tutsi."[60] However, there is a broader framework of suppression at play in the country. Tellingly, when citizens were polled regarding the possibility of a third term for President Kagame, Rwandan lawmakers reported that "only 10 were against the idea," after consultations with "millions of Rwandans" about the constitutional changes necessary to legalize a third term.[61] In general, "the media is largely compliant, the opposition toothless, and the critical NGOs expelled," providing incentives to support the regime and encouraging self-censorship.[62]

Similar forces are at play in Ethiopia, both socially and legally; the country's Anti-Terrorism Bill, which was passed in 2009, has been used to persecute political opposition and to curtail journalists. The plight of the Zone 9 Bloggers, who were detained in 2014 under the law for their blog criticizing the government's treatment of certain regions, ethnic groups, and political opponents, illustrates how the EPRDF has leveraged the rhetoric of antiterrorism endeavors to stifle criticism. Though the Zone 9 Bloggers were released after more than a year in detention and an international effort to release them, the 34 other Ethiopians—including 15 journalists and political opponents—that were arrested under the law in the first three years of the law's implementation have enjoyed no such support.

While China may be the largest jailer of journalists globally in 2014, imprisoning 44 journalists, Ethiopia ranked fourth globally, with 17 imprisoned.[63] While there were no journalists jailed in Rwanda in 2014, since 1992 the Committee to Protect Journalists reports that 17 journalists have been killed in the country. Fifteen of those murders were committed with impunity.[64] Like all undemocratic regimes, these states have an interest in limiting the ability of citizens to organize freely and express dissent. More unique to these countries is their respective parties' ability to deeply penetrate their societies.

Implementation of these restrictions on free speech requires intensive state monitoring and an atmosphere of self-censorship. In Ethiopia, HRW has concluded that the government "has the technical capacity to access virtually every single phone call and SMS message in Ethiopia," including "mobile phones, landlines, and VSAT communications, and…all local phone calls made within the country and long-distance calls to and from local phones." Similarly, Ethio Telecom "controls access to the internet backbone that connects Ethiopia to the international internet," through service provision and regulation of internet cafés.[65] Activists, journalists, and politicians in Ethiopia are cognizant of being monitored; meetings are arranged through third-parties located outside of the country and many have a deep-seated suspicion of internet cafés and even their own mobile phones. Similarly, in Rwanda, there have been reports that the widespread monitoring of e-mail and internet chatrooms had led to detention and interrogation by Rwandan security forces.[66] Some websites, which published content critical of the ruling party, have been banned in the country.[67]

In conjunction with the limiting of freedom of speech and association both formally and through informal self-policing mechanisms, both Ethiopia and Rwanda have expanded their party presence. One means of expanding party presence has been through bolstering formal party membership. In Ethiopia, a party membership is referred to as a "green card," because it is the only way to get a job. Following the 2005 elections, the EPRDF undertook a mixture of cooperative and coercive efforts to expand party membership, resulting in an explosion in party ranks from 760,000 in 2005 to 4 million in 2008.[68] Similarly, in Rwanda, one civil servant interviewed stated plainly, "You cannot become an employee of [the] government if you are not a party member."[69] These efforts allude to an effort to bring all citizens into the party fold.

Other expansions have been couched in the language of improving bureaucratic accountability. In Rwanda, the RPF has made *imihigo* ceremonies a central part of its governance strategy. *Imihigo* is an attempt by the government to institute a measure of accountability; it "relates most closely to a performance contract. The concept developed as an idea of a public commitment from prominent military leaders…to achieve a specific object, such as the conquest of an enemy or region."[70] According to a review of the effects of these performance contracts, which are made in public ceremonies, by the European University Institute, "all respondents said that these issues have in fact been better addressed since the launch of the initiative, and that *imihigo* did indeed contribute directly to this progress."[71] However, such praise must be taken with a grain of salt, given the extent of self-censorship in Rwanda; other reviews have found that the system has "led to deeper permeation of society by the state; and represents openings for increased coercion."[72] This study also found that "more than half of the respondents confirmed some form of compulsion had been used to achieve the *imihigo* targets."[73]

In Rwanda in particular, the military (effectively the military wing of the RPF) is hegemonic and has facilitated the penetration of the RPF into Rwandan society. Jones et al. note that, "military involvement means the involvement of a top-down hierarchy answerable to the presidency and overseen by insiders;" thus, the military serves as the muscle behind the political decrees and plays a valuable role in surveilling its citizens.[74] The RPF has undertaken significant efforts to incorporate Rwandan citizens into the military, even if they are not soldiers; as a part of the government's ambitious Vision 2020 program, a minister in Kigali explained that the country hopes to have every citizen attend an *ingando* camp. *Ingando* camps, modeled after the RPF's military camps in Uganda pre-1994, are mandatory today for students attending university who are receiving government support.[75] At modern *ingando* camps, participants are expected to listen to lectures by RPF members regarding civic duty and Rwandan history and to participate in military training exercises. Opponents of the camps have likened it to indoctrination and have highlighted the dangers of promoting such a militaristic culture.[76]

A similar model of community monitoring and flawed accountability mechanisms can be found in Ethiopia. Though a relic of the communist Derg, the *kebele* system of local government organization, in which administrators are responsible for the oversight of a handful of households and report activities and developments up their "political chain of command," has been adopted and expanded by the EPRDF.[77] The development community, including the United States Agency for International Development (USAID), has found that the EPRDF's "five-to-one program," in which every set of five participants is monitored by one, in a cascading series of groupings, facilitates the implementation, monitoring, and evaluation of projects. Unarticulated by development partners, but deeply felt by Ethiopian citizens, is the ability of these "cascading networks" to transmit information about political activity and loyalty to the ruling party.[78]

Partially through the bolstering of the party ranks, both governments have ensured that their political hegemony will be unchallenged. The regularly-held elections in both Rwanda and Ethiopia are essentially political Kabuki theatre, a highly stylized and often surreal song and dance that is, ultimately, only a dramatic performance. *The Economist* noted wryly in 2013, "many things were in doubt when Rwanda held parliamentary elections…but not the outcome."[79] The RPF's absolute domination is reflected in electoral returns, in which the RPF received roughly 90 percent; such popular support is facilitated by the effective banning of opposition parties through legislative sleights of hand and outright repression. In 2010, two of the main opposition parties were prevented from registering and opposition politician Victoria Ingabire was arrested under the "genocidal ideology" laws, and even accused of funneling money to rebel groups in the Democratic Republic of the Congo.[80] Following the 2005 elections in Ethiopia, in which the opposition was able to win an unprecedented handful of seats in the national parliament, the EPRDF doubled down on its efforts to maintain control. District level elections, scheduled for 2005, were delayed until 2008. In the intervening years, the party changed the structure of the *kebele* councils, so parties had to field roughly 3.6 million candidates if they intended to run in all constituencies.[81] In the 2015 general elections, the EPRDF won all of the parliamentary seats. In response, the international community hardly raised an eyebrow.

This brief review of governance in Ethiopia and in Rwanda suggests that the EPRDF and RPF have crafted an alternative form of governance—inspired by the example of the CCP in China, but constrained by their own capacities and histories—that allows them to maintain their position as "donor darlings" of the West by promoting "private sector" growth through party-controlled vehicles, while maintaining coercive hegemony within the domestic political landscape. This model has been inspired, and in some small ways, enabled by the model of Chinese and East Asian development, though there are important implementational differences, as the next section will highlight.

Promotion or Emulation?

China has not actively promoted its purported model of development; however, its success has inspired emulation. Additionally, China's growing international presence has likely facilitated this ideational spread as well. Further, as China continues to rise within the international system, taking a more active role in international affairs, imitation of Chinese policies may become a means by which developing countries court China as a donor. Despite China's reticence to embrace the rhetoric of a "China Model," Joseph Nye notes that its pattern of governance "has become more popular than the previously dominant 'Washington Consensus' of market economics with democratic government."[82] Nye concludes that, "although China is far from America's equal in soft power, it would be foolish to ignore the gains it is making."[83] It would also be folly to ignore the various missteps of the West that have facilitated the popularity of the China model.

Presently, China is more comfortable discussing its role as a "partner" to developing countries, as opposed to proselytizing a particular set of policies. "In diplomatic terms, Beijing seems to have very little interest in exporting its political model," S.J. Cooper Knock notes, observing that, "Chinese foreign policy in Africa is rarely prescriptive, and is built on bilateral economic relationships with states." Ultimately, Knock observes that China's "concern with protecting sovereignty at home makes [it] reluctant to interfere with the internal politics of sovereign states abroad."[84] Though China does not present it as such, such policies do constitute an alternative model from Western engagement patterns.

China's emergence as the "leader of the [G]lobal South and champion of a progressive 'new international political and economic order featuring justice, rationality, equality and mutual benefit' and 'safeguarding legitimate rights and interests of developing countries,'" as stated in China's 2006 *African Policy,* has facilitated a number of partnerships with countries frustrated by the policies of the IFIs.[85] Undoubtedly, the Bretton Woods system's model of lending and its underrepresentation of developing countries' preferences needs revamping. As Soares de Oliviera et al. note, "Africa's dwindling interest in borrowing from the World Bank and European Investment Bank, which initially caused both institutions to decry Chinese lending practices as well as questioning their own relevance is seemingly another indicator of the West's putative marginalization on the continent."[86]

Bilateral relationships as well, have been an increasingly important aspect of Chinese foreign policy in the developing world. In 2014, Chinese aid was estimated to be nearly $5 billion, making China the world's 10th largest source of funding (or, in Chinese parlance "South-South cooperation provider") through interest-free loans, concessional loans, and grants.[87]

Chinese engagement is said to differ from traditional Western aid by focusing more heavily on infrastructure development and less on political conditionalities; de Oliviera et al. note that "the Chinese focus on turnkey infrastructure projects is far simpler and does not overstretch the weak capacity of many African governments faced with multiple meetings, quarterly reports, workshops, and so on."88 Interestingly, recent years have seen an uptick in the training programs that China has provided to bureaucrats from the developing world; an estimated 50,000 bureaucrats have been trained, with a particularly sharp uptick since 2010.89 The value of such training in norm diffusion remains to be seen, though it has the potential to be significant.

While there has been a great deal of unfounded fearmongering regarding the danger posed by so-called "no-strings-attached" lending by China, empirical reviews are finding that there are, in fact, differences between traditional sources of aid and investment from

China. Indeed, analysis of the characteristics of Chinese aid suggests that Chinese Overseas Direct Investment (ODI) has a slightly negative correlation with the World Bank's Rule of Law index; confirming anecdotes of Chinese willingness to invest in poorly governed and undemocratic states.[90] Other reports have found that Chinese aid is more likely than other assistance to be funneled to political leaders' birthplaces; Dreher et al. observe that, "Chinese official financing to a leader's birth region nearly triples after that individual comes to power."[91] China's emergence as a major source of aid and investment, thus, has ramifications for the architecture of international assistance and partnership as a whole. Deborah Brautigam, a noted scholar of Sino-African relations notes that, "the Chinese do not see themselves primarily as 'donors,' preferring the language of 'cooperation' and 'partnership,'" facilitating diplomatic bilateral relationships.[92] China's "going out" policy has galvanized companies to seek opportunities abroad and has fostered a self-image of China as an international investor, rather than a provider of assistance. However, the effects of these partnerships may be the erosion of conditionalities attached to assistance related to respect for human rights or the rule of law by the availability of alternative, and undiscerning flows.

Chinese rhetoric, multilateral efforts, bilateral efforts, and the mere existence of China as an economic success story contributes to an alternative development model that de-emphasizes political accountability in favor of delivering economic performance. The limiting of civil and political liberties by a dominant party, as well as the sort of "coordination goods" that could challenge the hegemony of the ruling party, are characteristics of this model; the model is also characterized by extensive intervention in the national political economy to benefit the ruling party.

Conclusion

Ethiopia and Rwanda are seen as emulating the "China model" insofar as they are authoritarian states that have private sectors and remain integrated into the global economy. However, a more precise definition of the China model and what has allowed it to preside over more than three decades of economic growth indicates that these African countries are very far from replicating Chinese success.

The China model is dependent first and foremost on the existence of a large, disciplined, and highly institutionalized Communist party. Over the years, this party has integrated itself into the Chinese government that it oversees, and constitutes a substantial portion of China's state capacity. Up through a provincial level, recruitment and promotion of cadres proceeds by strict rules and tends to be relatively meritocratic. The party has imposed term limits on itself and is not dependent on individual leaders for its continued functioning. The system is capable of exerting a huge amount of discipline on lower-level cadres. There have been several examples of this in recent years: the tax reform of the early 1990s, which stripped local governments of many of their resources; forcing the PLA to give up many of its economic privileges in the late 1990s; and most recently Xi Jingping's anticorruption campaign, which has seen the arrest of thousands of party officials, including a former minister of the interior, Zhou Youkang.

Ethiopia and Rwanda lack this bureaucratic tradition and rigor. Both are run by ruling parties organized on Leninist lines, but neither organization is institutionalized to nearly the degree of the CCP. Recruitment is less meritocratic and more patronage-based. Both parties are rooted in specific minority ethnic groups, and both are dependent on the qualities of individual leaders. While the EPRDF in Ethiopia survived the death of Zenawi, the latter was critical in establishing the developmental policies of the past two decades. Kagame has created a personal dictatorship; like other African autocrats, he has amended the constitution to permit himself to remain in office for a third term. While China has renewed its leadership three times since 1978 at regular 10-year intervals, there is no mechanism for leadership succession in Rwanda. It is very doubtful that the "Rwandan miracle" will survive Kagame's passing. In his absence, even the basic stability of the country is called into question, as he has made himself the axis of the country's political economy.

The way in which one-party rule is institutionalized will be critical for both African countries. China can draw on a tradition going back 2,500 years of public-spirited bureaucratic government. Ethiopia and Rwanda were lucky to have two charismatic leaders, in the forms of Meles Zenawi and Paul Kagame, who had a developmental vision and did not put personal gain first and foremost as objectives. However, neither country has China's deep cultural traditions, nor is there a deep cultural tradition of impersonal public service in either country.

In terms of U.S. policy, the challenge posed by the China model in Africa requires two different types of response. The first has to do with democracy promotion. Since at least the time of Woodrow Wilson, the United States has made the promotion of democracy outside the country an objective of U.S. foreign policy. Unfortunately, the 2003 Iraq War created an association of military invasion with "democracy promotion," but, in fact, the most important initiatives in this endeavor have been political and social in nature. The United States has tried to level the playing field in authoritarian countries by backing civil society organizations—labor unions, women's organizations, human rights monitors, anticorruption campaigners, and the like—through agencies like USAID and the National Endowment for Democracy. These kinds of activities, as well as rhetorical support for democracy and human rights, have played important roles in constraining arbitrary behavior on the part of authoritarian governments.

Global civil society is today under threat from governments around the world, including places like Ethiopia and Rwanda. Civil society promotion does not contradict the goals of development; indeed, it strengthens development in the long run by legitimizing governments and making them more accountable. There is no necessary tradeoff between developmental goals and democratic ones. The United States needs to push back against this trend, and stand as a principled supporter of the right of citizens to organize and to participate in the political life of their countries.

The norms of governance and international cooperation that the United States and its democratic allies have fostered are being undermined or manipulated by the rise of developmental authoritarianism; consider that before 1962, "international election observation" as we know it did not exist. Today, roughly 80 percent of elections have international observers, "but puzzlingly, many leaders invite foreign observers and orchestrate electoral fraud in front of them."[93] The United States and its allies must demonstrate a renewed commitment to the spread and institutionalization of democracy worldwide; complacency with the state of global democracy and American fatigue regarding international engagement—particularly after the failed attempts to install democracy in Iraq—threaten to allow the rollback of hard-fought gains made by democracy activists worldwide.

The question of whether the United States should condition aid on human rights and democracy performance is more complicated. Democracy promotion is an important goal of U.S. policy, but it is not the only one. Security, energy access, the need for diplomatic leverage, humanitarian relief, and other goals have often displaced democracy. While this often leads to charges of hypocrisy, it is unrealistic to think that democracy promotion will always be America's single priority. In the case of Ethiopia and Rwanda, the United States has been tempted to turn a blind eye to political abuses; in the first case because of Addis' support for American counterterrorism objectives, and in the latter because of its stellar postconflict economic performance and stability.

These may have been reasonable tradeoffs, but it is important to remember that support for friendly dictatorships often bears a hidden cost. Such countries are vulnerable to violent overthrow in coming years. The EPRDF and RPF's propensity to respond to criticism through violent repression only raises the likelihood of opposition being expressed cataclysmically. An emerging and rapidly growing field dedicated to the study of which factors facilitate political violence suggests that "violent repression" of political opposition, like that engaged in by the EPRDF and the RPF, "is likely to lead to escalation effects and defections."[94] Further, it is possible that the economic growth and human development these parties have cultivated will be insufficient to legitimize their rule in the long term. Examining political violence in Corsica, Xavier Crettiez finds that the grievances were a "rejection of a closed political system," that precluded peaceful reform.[95] Research also suggests that intercommunal divisions—characterized by "a feeling of fear vis-à-vis the 'other'"—facilitate political violence. This is a particular threat for countries like Ethiopia and Rwanda, which experience interethnic tensions, divisions, and inequality.[96] Difficult as it may be to identify causal linkages between violent political protest and civil discord, it seems clear that the respective ruling parties have created conditions in Ethiopia and Rwanda that are conducive to such disruptions over the long term. One World Bank report assessing Rwandan stability opines that, "regime capacity (the ability of the regime to control its population) is a different cause of internal order to regime legitimacy (the accepted right of the regime to govern a people)," and suggests that the Rwandan system of oppression was ultimately unstable; the report reminds readers that:

> While political liberalization then may seem perilous to the regime, in the longer-term the alternative may not be better.... In the absence of a change in political culture,

continued political exclusion may force the steam of ethnic or indeed other 'grievances' to simply continue to accumulate inside the pressure cooker.... It is this note of caution which must be sounded when assessing Rwanda's exit from violence, and it is a note reinforced by a reading of Rwanda's history. All three of Rwanda's previous regimes were regimes in which power was held by one ethnic group to the exclusion of the other and all three of these regimes came to an end through extra-constitutional and violent means.[97]

There is, however, a second lesson for U.S. policy to be drawn from the China model, which has to do with the type of development assistance being offered. China has, in effect, been exporting an approach to economic growth that has worked well for itself over the past three decades. This approach puts state spending on public goods, and particularly public infrastructure like roads, ports, electricity, water, and the like front and center. The United States and other Western countries did something similar at earlier stages of their economic development, though not on China's massive scale, when up to 50 percent of GDP was being reinvested.

The United States used to support large infrastructure projects in developing countries in the 1950s and 1960s; many of the dams and roads built with American assistance still exist in countries from Afghanistan to Lebanon. Infrastructure fell out of favor, however, in the 1970s and 1980s due to concerns over environmental consequences, impacts on indigenous communities, and governance problems associated with such large-scale investments. Western aid priorities shifted elsewhere, particularly to public health, where spending could be correlated with measurable results.

Over the past few decades, America and other Western development assistance has centered on issues like public health, women's empowerment, environmental sustainability, good governance, and the like. All of these are worthy goals, and their importance to development has been empirically documented. Nonetheless, however important they may be as components of overall development, no country ever got rich—that is to say, experienced a rapid increase in per capita GDP—through these kinds of measures alone. On the other hand, massive state-directed investments in public infrastructure have been key components of the "East Asian Miracle"—not just in China, but in Japan and the other East Asian Tigers.

Infrastructure today is one of the key binding constraints in African economic growth.[98] Inadequate and irregular supply of electricity, for example, has been a large obstacle to growth in countries from Nigeria to Kenya to Ghana. China has been able to spread its influence in Africa, because many countries in the region simply want what the Chinese can offer, more than the sorts of programs offered by the United States and other Western development agencies. In recognition of this, the United States launched a Power Africa initiative in 2013 to help electrify a number of countries in sub-Saharan Africa. It has found, however, that its own capacity to promote such projects is limited. The limitations have to do with inadequate sources of finance, missing capacity at agencies like USAID, and the simple lack of recent practice in managing major infrastructure projects.

The competition between Western and Chinese development models is, thus, not simply over abstractions like democracy, human rights, and free markets, but about the concrete types of assistance being offered to poor countries, with tangible ramifications for American influence abroad, and about the long-term stability of the international system. If the United States is to regain its influence in the region, it needs to restore its former capacity to provide public infrastructure, and listen more carefully to what developing countries themselves say they want.

These issues came to a head over the Asian Infrastructure Investment Bank (AIIB), a Chinese initiative begun when the U.S. Congress refused to act on IMF reform that would have given China a larger voting share in that organization. The United States tried to persuade its Western allies to boycott the AIIB, an effort that led to a diplomatic debacle when virtually all of them, with the exception of Japan, refused to go along. The United States argued that such an institution would weaken global standards with regard to safeguards in infrastructure projects. This was a very poorly thought-out position; first, the Chinese were doing these projects anyway without American support, and second, since the United States and its allies had a better chance of upgrading standards as founding members of the bank than as outside critics. A deeper, unaddressed question was whether the current global standards are, in fact, the right ones. Part of the reason that the United States finds it so difficult to promote infrastructure has to do with the fact that compliance with existing safeguards is extremely difficult, driving up costs and delaying project completion. Chinese policy in this regard is by no means the proper standard: Chinese companies have been willing to tolerate corruption, poor safety, and environmental harms in their investments. But facing up to the challenge of the China model will require more thought regarding the proper balance between safeguards and the developing world's need for public infrastructure and, beyond that, its need for economic growth.

Authoritarian systems in China, Ethiopia, and Rwanda have gained legitimacy through the provision of economic growth. If the norms of the liberal international system are to survive, the United States must identify and implement partnerships and policies that are both prodemocracy and pro-growth. A globe populated by governments inspired by the CCP may well bring economic progress without liberty and growth, without bureaucratic capacity, and courting instability in the long term. The security achieved today by enabling such governance models is sure to be undone in coming years as growth-related legitimacy wavers and brutal repression is met with ever more fervent protest.

Notes

[1] David Booth asserts that, in light of the growth records and human development improvements recorded in these two countries, the "most relevant distinction among African regimes is between the developmental-patrimonial types and the others." See David Booth, "Development as a collective action problem: Addressing the real challenges of African governance," *Africa Power and Politics*, (London: Overseas Development Institute, 2012).

[2] Kjeld Erik Brødsgaard, "Politics and Business Group Formation in China: The Party in Control?" *The China Quarterly* 211 (September 2012): 624-648.

[3] Andrew Szamosszegi and Cole Kyle, *An Analysis of State-owned Enterprises and State Capitalism in China* (Washington, DC: U.S.-China Economic Security Review Commission, 2001).

[4] Richard McGregor, "5 Myths About the Chinese Communist Party," *Foreign Policy*, January 3, 2011, available at <http://foreignpolicy.com/2011/01/03/5-myths-about-the-chinese-communist-party/>.

[5] Ibid.

[6] Fabrizio Gilardi, "Transnational Diffusions: Norms, Ideas, and Policies," in *Handbook of International Relations*, ed. Walter Carlsnaes, Thomas Risse, and Beth Simmons (Thousand Oaks, CA: SAGE publications, 2012), 436-467.

[7] Francis Fukuyama, "Dealing with China," Working Group in Foreign Policy and Grand Strategy, *Hoover Institution* (Stanford, CA: Hoover Institution Press, 2014), available at <http://www.hoover.org/sites/default/files/fukuyama_dealingwithchina.pdf>.

[8] Shaun Breslin, "The 'China Model' and the global crisis: from Friedrich List to a Chinese mode of governance?" *International Affairs* 87, no. 6 (2011): 1336.

[9] Larry Diamond, "The Democratic Recession," in *New Ideas on Development After the Financial Crisis*, ed. Nancy Birdsall and Francis Fukuyama (Baltimore, MD: Johns Hopkins University Press, 2011), 240-249.

[10] Ibid. Though there have been criticisms of the Freedom House rankings and methodology, it remains one of the more frequently cited rankings of political liberty. Additionally, given that many of the criticisms of the Freedom House rankings stem from their ideological affiliation with the United States government's conceptualization of democracy, these rankings are well-suited to our discussion of divergence from liberal, Western democracy. See Sara Bush, "The Politics of Rating Freedom: How Freedom House Became an Authority on Global Democracy" (presentation, "AGORA V," Centre for the Study of Globalisation and Regionalisation, Politics and International Studies at the University of Warwick, October 7, 2014).

[11] Our study contained only 48 countries, due to lack of data for a handful of nations.

[12] Elena Holodny, "The Thirteen Fastest-Growing Economies in the World," *Business Insider,* June 12, 2015, available at <http://www.businessinsider.com/world-bank-fast-growing-global-economies-2015-6>.

[13] Francis Fukuyama, "Democracy and the Quality of the State," *Journal of Democracy* 24, no. 4 (2013): 5-16, available at <http://cddrl.fsi.stanford.edu/sites/default/files/fukuyama_democracy_and_the_quality_of_state.pdf>.

[14] Axel Dreher, Andreas Fuchs, Roland Hodler, Bradley C. Parks, Paul A Raschky, and Michael J. Tierney, "Aid on Demand: African Leaders and the Geography of China's Foreign Assistance," *CESifo Group Munich*, March 30, 2015, available at <http://www.andreas-fuchs.net/uploads/1/9/8/9/19897453/chinese_aid.pdf>.

[15] Human Rights Watch, "They Know Everything We Do," *Human Rights Watch*, March 25, 2014, available at <https://www.hrw.org/report/2014/03/25/they-know-everything-we-do/telecom-and-internet-surveillance-ethiopia>.

[16] Deborah Brautigam, "Hearing before the United States Senate Committee on Foreign Relations Subcommittee on African Affiars," *Testimony on China's Growing Role in Africa* (Washington, DC: United States Senate Committee on Foreign Relations Subcommittee on African Affairs, 2011).

[17] Ibid.

[18] Freedom House, *Freedom in the World* (Washington, DC: Freedom House, 2015).

[19] Ibid.

[20] Xiao Ma, "Term Limit and Authoritarian Power Sharing: Theory and Evidence from China," *Journal of East Asian Studies* (2015).

[21] Kenneth G. Lieberthal, and David M. Lampton, *Bureaucracy, Politics, and Decision Making in Post-Mao China* (Berkeley, CA: University of California Press, 1992).

[22] Ibid.

[22] Andrew Szamosszegi and Cole Kyle, *An Analysis of State-owned Enterprises and State Capitalism in China* (Washington, DC: U.S.-China Economic Security Review Commission, 2001).

[24] Lieberthal and Lampton, *Bureaucracy, Politics, and Decision Making in Post-Mao China.*

[25] McGregor, "5 Myths About the Chinese Communist Party."

[26] Ibid.

[27] Maria Edin, "State Capacity and Local Agent Control in China: CCP Cadre Management from a Township," *The China Quarterly* 173 (2003): 35-52. Though the CCP's model of providing "state workplaces, health care, and other social services" has waned with its liberalization program, the political domination and hegemony it enjoys remains unchanged. The reduction in state provision of social services should not be considered a sign of a weakening party, but rather that the CCP has refocused its efforts elsewhere. Maria Edin argued such in *The China Quarterly,* writing that the 'reform era' actually increased the capacity of the CCP, writing "higher levels of the party-state have improved monitoring and strengthened political control through promoting successful township leaders to hold concurrent positions at higher levels and by rotating them between different administrative levels and geographical areas."

[28] Congressional-Executive Commission On China, *Information Control and Self-Censorship in the PRC and the Spread of SARS* (Washington, DC: Congressional-Executive Commission On China, 2003).

[29] Ibid.

[30] Ibid.

[31] Human Rights Watch, "How Censorship Works in China: A Brief Overview," *Human Rights Watch*, 2006, available at <https://www.hrw.org/reports/2006/china0806/3.htm>.

[32] McGregor, "5 Myths About the Chinese Communist Party."

[33] Ibid.

[34] David Shambaugh, "Civil-Military Relations in China," *Copenhagen Journal of Asian Studies* 16 (2002): 10-29. Though this relationship has evolved over the years, such that "senior PLA officers from the Central Military Commission down to Group Army commands are now promoted on meritocratic and professional criteria, while political consciousness and activism count for very little," the political allegiance of the PLA remains unquestioned.

[35] Dan Slater, "Strong-state Democratization in Malaysia and Singapore," *Journal of Democracy* 23, no. 2 (2012): 19-33.

[36] Ibid.

[37] Brødsgaard, "Politics and Business Group Formation in China: The Party in Control?"

[38] Ibid.

[39] Carl E. Walter, and Fraser J.T. Howie, *Red Capitalism: The Fragile Financial Foundations of China's Extraorginary Rise* (Singapore: John Wiley and Sons, 2011).

[40] Zhengxu Wang and Anastas Vangeli, "China's Leadership Succeson: New Faces and New Rules of the Game," *ISS Europe*, August 2, 2012, available at <http://www.iss.europa.eu/publications/detail/article/chinas-leadership-succession-new-faces-and-new-rules-of-the-game/>.

[41] Ma, "Term Limit and Authoritarian Power Sharing: Theory and Evidence from China."

[42] Ibid.

[43] Meles Zenawi, "States and Markets: Neoliberal Limitations and the Case for a Developmental State," in *Good Growth and Governance in Africa: Rethinking Development Strategies*, Akbar Noman, ed. (Oxford: Oxford University Press, 2012): 140-174.

[44] Emmanuel K. Dogbevi, "Replacing role of the state with externally funded NGOs left Africa in poverty – Kagame," *Ghana Business News*, August 22, 2015, available at <https://www.ghanabusinessnews.com/2015/08/22/replacing-role-of-the-state-with-externally-funded-ngos-left-africa-in-poverty-kagame/>.

[45] Ibid.

[46] Bruce Bueno De Mesquita and George W Downs, "Richer but Not Freer," *Foreign Affairs*, September 1, 2005, available at <https://www.foreignaffairs.com/articles/2005-09-01/development-and-democracy>.

[47] David Booth and Frederick Golooba-Mutebi, "Developmental Patrimonialism? The Case of Rwanda," *African Affairs* 111, no. 444 (2012): 379-403.

[48] Sarah Vaughan and Mesfin Gebremichael, "Rethinking business and politics in Ethiopia: The role of EFFORT, the Endowment Fund for the Rehabilitation of Tigray," *African Power and Politics*, August 2011, available at <http://www.institutions-africa.org/filestream/20110822-appp-rr02-rethinking-business-politics-in-ethiopia-by-sarah-vaughan-mesfin-gebremichael-august-2011>.

[49] Nilgun Gokgur, *Rwanda's Ruling Party-Owned Enterprises: Do they Promote or Impede Development?* (Belgium: Universiteit Antwerpen, Institute of Development Policy and Management, 2012).

[50] Ibid.

[51] Ibid.

[52] Ibid.

[53] Vaughan and Gebremichael, "Rethinking business and politics in Ethiopia."

[54] Ibid.

[55] Ibid.

[56] Ibid.

[57] Abebe Gellaw, "Tigrians outraged over EFFORT-led corruption," *EthioMedia*, July 29, 2009, available at <http://www.ethiomedia.com/adroit/2682.html>.

[58] Gokgur, *Rwanda's Ruling Party-Owned Enterprises: Do they Promote or Impede Development?*

[59] Vaughan and Gebremichael, "Rethinking business and politics in Ethiopia."

[60] Amnesty International, *Safer to Stay Silent: The Chilling Effect of Rwanda's Laws on Genocide Ideology and Sectarianism* (Washington, DC: Amnesty International, 2010).

[61] AFP, "Only 10 Rwandans against Paul Kagame's Third Term, says Lawmakers' Report," *The Monitor*, August 11, 2015, available at <http://www.nation.co.ke/news/africa/Paul-Kagame-third-term-bid-MPs-report/-/1066/2827968/-/e9v9m2z/-/index.html>.

[62] Will Jones, "Between Pyongyang and Singaore: the Rwandan State, Its Rulers, and the Military," in *Rwanda Fast Forward*, ed. Maddalena Campioni and Patrick Noack (New York, NY: Palgrave Macmillan, 2012), 230-250.

[63] Committee to Protect Journalists, "2014 Prison Census: 221 Journalists Jailed Worldwide," *CPJ*, December 1, 2014, available at <https://cpj.org/imprisoned/2014.php>.

[64] Ibid.

[65] Human Rights Watch, "They Know Everything We Do," *Human Rights Watch*, March 25, 2014, available at <https://www.hrw.org/report/2014/03/25/they-know-everything-we-do/telecom-and-internet-surveillance-ethiopia>. The EPRDF purchased telecom surveillance technology from the Chinese telecom giant ZTE; between 2006 and 2009, ZTE was the exclusive provider of telecom surveillance equipment, according to Human Rights Watch.

[66] Bureau of Democracy, Human Rights, and Labor, *Country Reports on Human Rights Practices for 2012: Rwanda,* (Washington, DC: Department of State, 2012).

[67] Ibid.

[68] Human Rights Watch, "One Hundred Ways of Putting Pressure," *Human Rights Watch*, 2010, available at <https://www.hrw.org/sites/default/files/reports/ethiopia0310webwcover.pdf>.

[69] Andrea Purdekova, "'Even if I am not here there are so many eyes:' Surveillance and State Reach in Rwanda," *The Journal of Modern African Studies* 49, no. 3 (2011): 475-497.

[70] Jesse McConnell, *Institution (un)Building: Decentralizing Government and the Case of Rwanda* (EUI Working Papers, 2010).

[71] Ibid.

[72] Purdekova, "'Even if I am not here there are so many eyes.'"

[73] Ibid.

[74] Will Jones, Ricardo Soares de Oliveira, and Harry Verhoeven, *Africa's Illiberal State-Builders* (Oxford: Oxford Department of International Development, Refugee Studies Centre, 2013).

[75] Purdekova, "'Even if I am not here there are so many eyes.'"

[76] Ibid.

[77] J. Abbink, "Discomfiture of Democracy? The 2005 Election Crisis in Ethiopia and Its Aftermath," *African Affairs* 105, no. 419 (2006): 173-199.

[78] Personal correspondence with author.

[79] "Rwandan Elections: Safe and Sorry," *The Economist*, September 21, 2013, available at <http://www.economist.com/news/middle-east-and-africa/21586597-president-tightens-his-grip-safe-and-sorry>.

[80] "Victoire Ingabire: Rwanda leader's jail term raised," *BBC News*, December 13, 2013, available at <http://www.bbc.com/news/world-africa-25371874>.

[81] Human Rights Watch, "One Hundred Ways of Putting Pressure," *Human Rights Watch*, 2010, available at <https://www.hrw.org/sites/default/files/reports/ethiopia0310webwcover.pdf>.

[82] Joseph Nye, "The Rise of China's Soft Power," *The Wall Street Journal*, December 29, 2005, available at <http://belfercenter.hks.harvard.edu/publication/1499/rise_of_chinas_soft_power.html>.

[83] Ibid.

[64] S.J. Cooper Knock, "Looking East: China and African Democracy," *Democracy in Africa*, August 7, 2012, available at <http://democracyinafrica.org/looking-east-china-and-african-democracy/>.

[85] Chris Alden, Dan Large, and Ricardo Soares de Oliveira, *China returns to Africa: Anatomy of an Expansive Engagement* (Madrid: Real Instituto Elcano, 2008).

[86] Ibid.

[87] Zhou Taidong, "China's Second White Paper on Foreign Aid Signals Key Shift in Aid Delivery Strategy," *Weekly Insight and Analysis in Asia*, July 24, 2014, available at <http://asiafoundation.org/in-asia/2014/07/23/chinas-second-white-paper-on-foreign-aid-signals-key-shift-in-aid-delivery-strategy/>.

[88] Deborah Brautigam, "Aid 'With Chinese Characteristics:' Chinese Foreign Aid and Development Finance Meet the OECD-DAC Aid Regime," *Journal of International Development* 23, no. 5 (2011): 752-764.

[89] Taidong, "China's Second White Paper on Foreign Aid."

[90] Wenjie Chen, David Dollar, and Heiwai Tang, *Why is China investing in Africa? Evidence from the Firm Level* (Washington, DC: Brookings Institution, 2015).

[91] Axel Dreher et al., *Aid on Demand.*

[92] Deborah Brautigam, "China's African Aid: Transatlantic Challenges," *German Marshall Fund of the United States,* 2007, available at <http://trends.gmfus.org/doc/Brautigam_WEB.pdf>.

[93] Susan D. Hyde, "Catch Us If You Can: election monitoring and International Norm Diffusion," *American Journal of Political Science* 55, no. 2 (2011): 356-369.

[94] Abel Escriba-Folch, "Repression, Political threats, and Survival under Autocracy," *International Political Science Review* 34, no. 5 (2013): 1-18.

[95] Xavier Crettiez, "Factors underlying political violence," *CRIMPREV Programme,* 2009, available at <http://lodel.irevues.inist.fr/crimprev/index.php?id=269>.

[96] Ibid.

[97] Omar Shahabudin McDoom, *Rwanda's Exit Pathway from Violence: A strategic Assessment* (Washington, DC: World Bank Background Case Study, 2011).

[98] Vivien Foster and Cecilia Briceño-Garmendia, *Africa's Infrastructure: A Time for Transformation* (Washington, DC: The World Bank, 2010).

5

Costs of Hedging Bad: The Global Threat Network and Impact on Financial Market Volatility

Jay Chittooran and Scott Helfstein

The illicit world of crime and terrorism seems far removed from everyday activity and seems especially divorced from legitimate commercial endeavors. Increasingly, tragic attacks or fictitious-sounding jailbreaks perpetrated by criminals and terrorists make headlines, but their day-to-day activities are often thought of as shrouded in darkness and best left to professionals in law enforcement, intelligence, and elite military units. Isolating illicit activities like crime and terrorism from everyday activity fosters the illusion that illicit activities have, at most, a limited impact on governance, commerce, and economics. The barriers between the licit and illicit, as well as the impact of illicit on licit markets, may be more permeable than often acknowledged.

A central aspect of the "convergence" literature is that the combination, or more accurately synergy, of terrorism and criminality amplifies threats beyond conventional law enforcement to legitimate national security concerns. Many of the chapters in this volume, particularly Matt Levitt's discussion of Hezbollah's global network of criminals and entrepreneurs, focus on that very issue. Rather than build upon the well-established contention that convergence poses a serious national security threat, this chapter uses an original dataset and analysis to argue that the synergistic challenges posed by the combination of crime and terrorism generates real challenges in the economic and governance spheres. The convergence of crime and terrorism fosters distortions in markets, creating real financial costs that damage countries' well-being and hinder their development.

This raises an important question. If this chapter focuses on the economic and financial implications of convergence, then why should the analysis appear in a volume on national security? Economics and markets may seem a step removed from national security concerns, but nothing could be further from the truth. Understanding how the connections between terrorist and criminal actors weigh on markets and economies is important for both policymakers and the military, as well as intelligence and civilian personnel, who address and counter the threats on the ground. Each is discussed briefly here in turn. Economic stability and prosperity is a critical foundation of national security; ignoring how convergence affects markets is, thus, to ignore one of the most insidious effects of this phenomenon on global stability.

The relationship between economic development and illicit activity is complicated. Terrorists and criminals rarely seek out economically broken or failed countries, but threats often manifest in countries facing economic challenges and struggling to achieve further development; some examples pertinent now include Iraq, Nigeria, and Pakistan. High unemployment and fewer legitimate employment opportunities, for example, may increase incentives to work in the illicit sectors, providing an ample pool of recruits for terrorists and criminals. Flow of investment capital to the public sector can also be a constraint by forcing an overreliance on government resources in the economy, increasing the chances of corruption and new political grievances among the disaffected. As policymakers consider different approaches to intervention, understanding the parts of illicit networks most likely to hurt investment should improve the efficacy of the financial statecraft toolkit.

Those operating against convergent threats around the world understand the importance of "working by, with, and through" host nations and local forces, which involves coalition building on the basis of mutual interests. Improving economic performance is almost always a priority for host nations, and this work potentially offers operators a guidepost for helping local security and economic officials think about the way certain threats may hinder investment and economic development. Enlisting support from local political, military, and law enforcement leadership can be challenging, but this shows that there may be concrete benefits to disrupting certain types of relationships. By finding common ground using the empirical backdrop laid here, operators and local forces may find prioritization, resourcing, and cooperation easier.

This chapter is based on a quantitative study of nearly 70 countries. Our findings contend that economic performance is meaningfully impacted by illicit activity and particular aspects of connectivity. The data driving this research, originally created to better understand crime-terror convergence from an empirical perspective, maps interpersonal connections in global illicit networks. The original analysis looked across 122 countries, but only 69 of those have equity markets (or "stock markets") sufficiently mature to include in this study.

Using equity volatility in the 69 countries studied, there is strong statistical evidence that a link between illicit network convergence and economic volatility exists. In fact, countries with a one-unit standard deviation increase in the convergence variable generates as much as 2.5 more volatility; put another way, this phenomenon increases average volatility by 17 percent. The analysis below provides an interesting perspective on global illicit activity and how it can affect the global financial system.[1]

Volatility, or the movement of prices in the equity market, is one common method of measuring risk for businesses and investors by calculating asset price fluctuations over time. Volatility is influenced by investors' expectations about future cash flows of companies. Riskier assets are usually more volatile, experiencing larger swings in price as investors struggle to assign value given the tradeoff between risk and return. Volatility can be calculated for individual assets like commodities or currencies, as well as entire markets. This chapter uses volatility measures of equity markets to see whether certain features of the illicit network are more or less associated with larger price swings, as volatility can have a significant impact on the propensity to attract investment or business partnerships.

While a small population of speculators often profit from price volatility, most international investors are concerned with the cost of hedging risk. Hedging is a means of limiting risk by buying certain types of financial instruments that protect against large price swings. As volatility increases, the cost of hedging downside risk increases. Given the empirical results here, the costs of hedging against illicit activity probably reaches into the billions of dollars. Market volatility increases as the ratio of relationships linking criminals and terrorists increases, referred to as convergence. As these two groups grow increasingly intertwined, governments and commercial enterprises face an increasingly complex and uncertain set of risks.

The second element of the illicit network that seems to increase equity volatility is the prevalence of individuals that link disparate parts of the network, a concept referred to as "betweenness" in graph theory. People with high betweenness are the glue that hold a network together, and without these boundary spanners, networks fall apart. Given the interconnected and global nature of the illicit network, these people are well-positioned to control the flow of scarce resources across borders and groups, while also moving between the licit and illicit economies.

Given the sheer magnitude of many asset markets and the globalization of the financial system, it might seem reasonable to assume that the world economy is immune to the activities of illicit actors involved in black markets. Hundreds of billions of dollars of products are exchanged in the global economy and on financial exchanges daily. The global illicit economy, however, is not insignificant. It is estimated to be between 8 and 30 percent of the world economy, amounting to a staggering $6 to $22 trillion.[2] While many people associate activities like narcotics smuggling and arms dealing with the illicit economy, organized crime, counterfeiting, theft, and financial crime are also significant components, as Karl Lallerstedt's contribution to this book illustrates. There seems ample reason, then, to revisit potential intersections of the licit and illicit economy in the midst of financial globalization.

There are many ways in which illicit activity might impact the basic economic forces of supply and demand. Countries with robust criminal networks are more likely to experience theft, smuggling, extortion, market manipulation, and other externalities frequently excluded from conventional economic modeling. All of these activities can impact the economy. While there are reasons to predict that the presence of a robust criminal network should impact an economy, there is much to learn about the particular mechanisms by which illicit networks impact licit financial activities. Unfortunately, there are only a few cross-sectional quantitative studies covering the subject.

Conventional wisdom suggests that policymakers should worry about less developed countries with weaker governance, poor rule of law, and economies built around natural resources or single commodities. This seems reasonable at first glance, but the relationship linking illicit and licit activities is nuanced. While governance certainly plays a role, modest increases in the connectivity of criminal and terrorist elements or the structural placement of individuals within the network can increase estimated volatility across global markets, increasing risks to investors, and likely impacting capital flows to economies in need of further development.

Our unique empirical study of the relationship between illicit networks and one small feature of the global economy provides statistical evidence that markets are not isolated from the evolving threats of the 21st century. The next section looks at the impact that criminal organizations can have on economic conditions. This is followed by a discussion of equity market mechanics and variables that can impact asset price volatility and the potential relationship between the illicit global network and markets. Attention will then turn to our dataset, analytical methods, and the empirical results of the study.

Lessons from the Global Illicit Network: Beware Real Dark Pools

In finance, a "dark pool" refers to large blocks of investment capital that can buy and sell assets outside of regular exchanges. Some argue that these pools increase risk by manipulating asset prices in an opaque fashion to benefit a handful of investors. Just as financial markets have resource pools that exist outside normal patterns of exchange, so do entire economies. There is a sector of the economy that operates in the dark, away from regulation, taxation, law enforcement, and official measurement.

Economists have long understood that illicit activity could adversely impact markets. Crime and violence were frequently treated as a local economic phenomenon. Al Capone's bootlegging enterprise cast a shadow on the Chicago economy in the 1920s, just as the Cosa Nostra did in New York 40 years later. The Revolutionary Armed Forces of Colombia (FARC) had a significant influence on the Colombian economy, just like the Taliban regulation of heroin production did in Afghanistan during the 1990s. The impact was not limited to those localities, as drugs produced found their way into American cities like Miami, casting a shadow over distant economies. Despite this global phenomenon and its tangible impacts, until recently no comprehensive picture of the illicit network existed. As a result, it was perfectly reasonable to focus on local effects.

A study by the Combating Terrorism Center (CTC) conducted in 2014 offered a good reason to set aside siloed, local studies to consider the global implications of an expansive criminal and terrorist network that capitalizes on the opportunities of increased globalization and regional connectivity. Rather than a series of unconnected parallel criminal and terrorist networks that coexist in different regions around the world, the CTC analysis showed that 98 percent of the 2,700 individuals in the study were subsumed in a single, expansive, cross-national network.[3] This was somewhat unexpected as the study started with a list of 40 leading criminals across narcotics, arms, and human trafficking. Instead of finding locally focused criminal networks, this study demonstrated that individuals maintain relationships between a variety of criminal, terrorist, and antistate enterprises across continents and oceans. Critical to this process is the ability of criminals in one illicit sector to maintain relationships with those involved in different criminal activities, as well as with individuals on global terrorist watch lists.

Focusing on a specific illicit actor or activity in a country or region can be helpful for local policy or law enforcement, but can be misleading in attempts to understand the broader socioeconomic ecosystem in which these individuals operate. Drug production in Colombia can impact crime and health care costs tied to addiction in Los Angeles. Arms made in Eastern Europe have found their way to Africa, Latin America, and Asia. In recent years, authorities have found money laundering and market manipulation crime syndicates operating at vast distances. Though this process seems relatively organized, there is no central command ordering interactions; there is no *Spectre*. The network is best described as a self-organizing complex system, or the outcome of self-interested opportunity-seeking social agents. The existence of an interconnected global network that leverages both licit and illicit marketplaces warrants an examination to better understand how the world's true pools of darkness impact legitimate economies and markets. This is meant to be a small step in legitimating that research.

How Do We Measure Economic Performance and What Moves Equity Markets?

In order to demonstrate the impact of illicit networks on the economy, we must first address how we measure the robustness, well-being, and stability of the economy. Economic performance can be measured by using macroeconomic indicators; the two most common are gross domestic product (GDP) and gross national income (GNI), using real, nominal, or per capita values. To clarify, GDP tracks all expenditures on goods and services produced domestically, while GNI is GDP plus income earned by foreign residents (and less income earned by nonresidents in the country). These types of metrics, while helpful, are subject to different interpretations and revisions because of the difficulty associated with aggregating data over an entire economy. More problematic is that these numbers are published infrequently. Indeed, in the United States, initial GDP numbers come out once every quarter, but the final numbers are lagged by up to three months. In other countries, particularly in the developing world, data is released far less regularly and may be unreliable. Because of this, it is much more difficult to track the effect of the crime-terror network with infrequent, lagged, and oftentimes, inaccurate data. Identifying the impact of illicit activity across statistics that aggregate infrequently across an entire economy is difficult at best. Most relevant here, such indicators shed little light on the volatility that activities like crime and terrorism can create.

Financial markets offer a different way of gauging the impact of illicit activity instead of the slow and opaque calculation that comes from relying on GDP indicators. Equity markets generally reflect investors' trust in business and the economic environment, specifically investors' willingness to risk capital in long-lived assets. Markets provide real-time price discovery, meaning there is regular feedback on the business environment and the probability of future cash flows. Another advantage is that equity values shift over time, and the price swings are one way to gauge investor uncertainty and risk. Larger price swings, or increased volatility, generally reflect greater uncertainty and risk to future cash flows.

Other asset classes, like sovereign bonds, could have been used to examine the macroeconomic conditions, but equities arguably provide the best metric for real-time sentiment regarding the business environment. Sovereign bonds reflect investors' belief in the government's ability to not only pay the bills, but also to remain in power. Corporate credits would provide a better sense of the business environment than sovereign bonds, but they value a fixed payment stream, as opposed to equity holders who face an uncertain path of cash flows. Bondholders also usually have a claim on assets that could be sold, limiting the downside whereas equity holders typically have less protection against total loss.

Looking at equity markets (like the New York Stock Exchange, the NASDAQ, the Shanghai Composite, the London Stock Exchange, or any others across the globe) can be a way to assess the economic performance of a country while also factoring in risk and investor sentiment.[4] Using equity markets also allows for microanalysis, offering real-time data on not just the broader economic performance, but also future expectations for the business environment.[5]

International equity market returns are highly variable from year to year, as investors often find one year's underperforming market attractive the following year. Investors try to identify markets most likely to yield returns, sometimes in risky markets that have sold off in previous years. In other words, investors may be compensated for risk through low entry prices. Even a risky market, given the right price, can be an attractive value-investing opportunity. International equity market returns, therefore, are implicitly risk-adjusted. If risky assets are priced attractively, the return could still be substantial. Equity returns can be highly variable and risk is only one of many factors that investors may consider.

By focusing on swings in asset prices over time, volatility offers a more straightforward method of thinking about risk. More mature, stable, and transparent markets experience lower levels of volatility as investors have better information and confidence in the market. By contrast, markets in developing countries with poor governance, rule of law, and economic foundations are likely to be quite volatile. These factors are not easily changed, and volatility levels generally change slowly over time unless impacted by a major exogenous shock.

What types of things can affect equity markets? An equity index can be affected by economic issues, financial conditions, geopolitical concerns, or exogenous factors, like the weather. At the base level, equity prices are determined by estimates of the growth of future cash flows and the cost of capital. Economic measurements like GDP help investors get a sense of the growth environment, and stock market returns usually correlate with future growth.[6] Investors use the data releases to improve their understanding of future conditions, and so incorporating an economic indicator on the health of the economy is an important control variable. When investors believe that the pathway of future economic performance remains strong, market values are likely to increase and volatility should decrease. Growth scares and poor economic activity should lead to increased volatility as investors struggle to price an uncertain set of cash flows.

Inflation also has a strong and significant relationship to equity market volatility. Inflation is the cost difference between buying a good today versus sometime in the future. At first this might seem to have little impact on financial markets, but it actually serves as a critical building block to asset valuation. The rate of inflation helps establish the cost of borrowing money, or the cost of capital. As inflation increases, investors demand higher compensation for allocating their funds today, since those same funds will have lower purchasing power in future environments experiencing high inflation. Investors struggle to accurately set rates of return, and thereby determine the true value of future cash flows. Research suggests that countries experiencing higher inflation do face higher levels of asset price volatility, and our research incorporates inflation as an additional control variable.[7]

Data and Methods

Our research relies on a number of data sources, but the most unique was a database developed by the Combating Terrorism Center at West Point based on open-source data compiled by Thomson Reuters' World-Check. Coders at World-Check relied extensively on court documentation, including indictments, from dozens of countries in over 60 languages as well as traditional open-source material. The initial database was compiled for commercial use as a due diligence tool. After the September 11 attacks, the United States adopted more stringent rules on money flows, raising the burden for financial services companies in particular. World-Check gathered information on individuals added to government watch lists along with their known associates, and the CTC used the data source to conduct an experiment on convergence in crime and terror across the global network of illicit actors and activities.

As noted above, the CTC study generated a list of the top 40 transnational criminals across narcotics, arms dealing, and human trafficking.[8] The project aimed at identifying the prevalence of linkages, or social distance, between the transnational criminals and terrorist actors based on known associates in the World-Check data. The researchers did not have to look very far. The initial 40 illicit actors linked directly to 754 known associates, and 86 were transnational terrorists on global watch lists. The frequency of terrorist elements in the network increased significantly when researchers moved out one degree.

As discussed, the most surprising conclusion in the CTC study was the interconnectedness of global illicit actors. This interconnectedness was not the work of any individual or group, but the outcome of a self-organizing complex system. The study then leveraged geographic data to identify potential drivers of crime-terror convergence. The cross-sectional analysis included a range of network variables across 120 countries. Rather than revisit the conclusions reached in that work, which focused on forces that may have influenced the formation of the network, the study conducted for this chapter leveraged the cross-sectional data to look at the way characteristics of the illicit network might impact the global financial system, using equity markets as a proxy. In other words, the CTC study used the network characteristics as a dependent variable, attempting to explain the patterns of illicit connectivity based on economic and political factors across countries.

Here, the network data serves as an independent variable to better understand whether illicit activities and networks impact licit economies and market functions.

The research conducted for this chapter leveraged three variables from the CTC data that summarize different aspects of the illicit network within each country. Networks can be characterized in a number of ways including number or density of connections, as well as structural features that help summarize an individual's role within the network. The three variables were chosen to reflect different ways that illicit behavior and the networks may weigh on governance and economic risks pertinent to financial markets. Each will be discussed here briefly. The main explanatory variables included factors for convergent relationships between criminals and terrorists, the average degree of illicit actors, and the average betweenness of those in the network. Each of these could impact the broader environment in which businesses and investors operate.

The convergence between criminal and terrorist elements was a critical aspect of the CTC study. Prior to building the network database, each individual in the network was assigned a role, or reason for inclusion. This was not a subjective decision by those that built the network graph, but a data field developed beforehand, which helped ensure that results were not driven by idiosyncratic or biased coding in the network study. Justification for the initial coding usually came from legal filings or watch list designations. As a result, some individuals were identified as terrorists while others were identified for their involvement in criminal activities such as narcotics or arms trafficking. In reality, the designations might not be mutually exclusive; consider individuals like Dawood Ibrahim. Though Ibrahim spans the worlds of terrorism and criminality, his primary interest is the D-Company criminal enterprise.

Convergence summarizes the density of ties crossing between individuals classified as criminals and terrorists in each country. The variable calculates the number of individuals that are criminals linked to terrorists, or terrorists linked with criminals, and divides that by the total number of illicit actors within the country. Higher levels of convergence reflect a greater propensity for terrorists and criminals to interact.

Degree centrality is perhaps the most basic metric within network science, and is generally thought of as a significant measure of importance in the network. The measure is simple and calculated by summing the total connections for each individual within the network. For example, if a network has nine people and one of the individuals has a social relationship with four other individuals, then the degree centrality equals four. The person who has two connections has a degree centrality of two, and is generally viewed as less connected than the individual with four. The more connections an individual has, or the higher the degree, the more influence that person could have in the network. Degree centrality does not incorporate unique structural aspects of the network or the placement of individuals within the infrastructure; it simply characterizes a network by the sum of connections of each individual. There are other measures of importance or influence within a network, as discussed below, but degree centrality is one commonly used metric, the easiest to calculate, and the most intuitive. In this study, degree centrality is converted into a country-level measure by taking the average for each individual operating within the country. This summary degree variable is one method of reflecting the level of connectivity that illicit actors have, irrespective of their respective illicit activities, within each country.

The final network measure included is betweenness, one of the many metrics for influence in graph theory. While degree centrality measures influence by the aggregate number of connections, betweenness incorporates network structure and positioning of an individual within that structure. The measure specifically captures the importance of an individual in linking disparate parts of the network. Returning to the nine-person network above, imagine there are two groups of four people, each of which know one another. Each of the eight individuals in the two groups has a degree centrality of four. The ninth person in the network knows one person from each of the two groups. In the example above, the individual with the two connections might seem less influential, however, betweenness incorporates their structural position. In this instance, anytime the two groups want to interact, they must go through the person with degree centrality of two. Despite having fewer connections than those within the four-person networks, that ninth individual plays an important role as an intermediary.

Technically, betweenness is calculated by looking at the shortest pathway between any two nodes in the network and calculating how many of the paths go through a single individual. Those with the most through traffic have the highest betweenness. Individuals with high betweenness connect parts of the network that would otherwise be unconnected. Like the degree centrality measure, in this study it is first calculated for each individual and then converted into a country-level variable by calculating the average betweenness for individuals in each country. Literature on network analysis often refers to those with high betweenness as boundary spanners. These people are the network glue, or bottlenecks, when connections grow sparse. By connecting groups that might be otherwise unconnected, they play an important role as brokers and intermediaries. In some ways, betweenness is the most interesting of the three network measures, reflecting the ability of illicit actors within each country to facilitate high value transactions in goods, information, skills, or people within or outside of the country. These are illicit brokers that generally have international reach.

Our study also included a political control variable to ensure that any relationships among network measures were not capturing other country-specific governance factors that could impact volatility. The first alternative hypothesis is that higher measures among the network variables really reflect the functioning or failure of the government. In other words, countries with robust illicit networks are really just those one would consider failed or failing states. The CTC report dealt with this at length and found the two were distinct. To ensure that the network characteristics are not just measuring governance, the study included the Fragile States Index. This metric uses 12 indicators of state fragility and assigns a number to each state based upon perceived risks; the higher the number, the more risk factors there are in the country.[9]

Since our research is primarily about licit finance, incorporating variables that drive markets is critical. Therefore, the study included economic growth and inflation rates.

Generally, countries with higher real economic growth should expect higher returns and lower volatility. Since returns are partially driven by expected growth, and markets in growing economies are more likely to move higher, there are fewer price swings. The final control used in this study is inflation. Work on the "diversionary war hypothesis," the idea that countries begin wars when internal turmoil rises, often uses inflation as a metric for internal turmoil.[10] From a market's perspective, higher inflation complicates efforts to price financial assets. Investors may grow concerned about political will or capability to control prices. The true value of an asset becomes more difficult to discern, and this can increase volatility.

We used two volatility measures over time to examine the potential relationship between illicit networks and licit markets. Equity market volatility generally refers to the standard deviation of closing prices for a given period of time. The analysis included volatility measures taken over 30-day and 260-day intervals in 2013 and 2014. Market data came from a commonly used financial database. The next section shows and discusses the results from the statistical test using the volatility measures as the dependent variables with the network characteristics and control variables as the independent variables.[11]

Analysis

The empirical analysis of equity volatility across the 69 financial markets reveals that certain aspects of illicit networks have a significant impact on licit economic and market activity. At the same time that the econometric results suggest that market participants and policymakers should consider the impact of illicit networks, only certain structural factors proved to correlate strongly with equity market volatility. The factors we identified as the most significant were the levels of interaction between criminal and terrorist actors and the level of betweenness of actors within the network; increases in both these factors were positively correlated with market instability. This illustrates the tangible economic ramifications of convergence.

Our research demonstrates that the threat posed by crime-terror convergence carries over into the licit economy. Convergence displayed the strongest positive correlation among the network factors incorporated in the analysis; thus, equity markets in countries where criminal elements and terrorists have higher levels of interaction are, on average, more volatile than those where criminals and terrorists were reasonably isolated from each other. The tendency, and ability, for illicit actors to cross the crime-terror boundary increases risks to businesses and investors, thereby weighing on the private sector. In short, criminal networks can have tangible effects on economic health.

One thought that immediately comes to mind is that convergence, or the tendency for terrorists and criminals to interact, is largely a feature of failed states. This is, after all, conventional wisdom. The CTC report on crime-terror connectivity empirically refuted this long-held notion. Poor and failing states did not necessarily have the highest rates of crime-terror connectivity as convergence is a unique variable distinct from measures of governance or lack thereof. Convergence is prominent in two conditions. First, poor and failed states, but only when the country is prone to initiating military conflict. Second, connectivity between terrorists and criminals is not isolated to poor and failed states, as rich countries can have high levels of convergence. Thus, assuming that convergence and failed states are synonymous is dangerous.

Our more recent research shows once again that convergence and failed state status are not equivalent concepts). While convergence is highly correlated with equity market volatility at statistically significant levels across time, the control variable for failed state status is relatively uncorrelated with equity volatility despite a small positive coefficient. The relationship between the failed states index control variable and volatility was not statistically significant. To ensure that the failed states index and convergence were not capturing the same phenomenon, correlation coefficients were run that showed a modest (below 0.20) relationship between the two variables.

Figure 5.1. Estimated Impact of One Standard Deviation Increase on Volatility

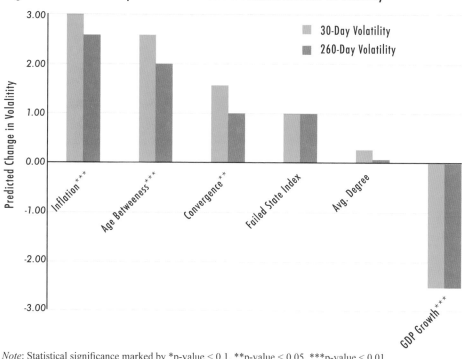

Note: Statistical significance marked by *p-value < 0.1, **p-value < 0.05, ***p-value < 0.01

Convergence in a country is clearly distinct from the failed state status in this sample as previous research suggested, and convergence is a much more powerful explanatory factor for equity market volatility. As the marginal analysis shows, a one standard deviation increase in the convergence variable generates an additional 1.1 to 1.9 more volatility, or put another way, increases average volatility by 7 to 10 percent. Using a theoretical options valuation model, that would increase the cost of hedging a 10 percent decline by 66 percent. That cost is built into the return calculations, meaning that expected returns have to increase enough to offset the higher hedging expense, which can have significant impact on allocation decisions, capital inflows, and private sector investment.

Convergence was not the only aspect of illicit connectivity that correlated with increased equity volatility. The average betweenness of the illicit actors within each country also seems to weigh on the markets. This was also consistent with the propositions offered earlier. Higher betweenness offers individuals unique standing and capabilities within a network. These are the individuals best capable of moving money and goods to parts of the network that might not otherwise have access. Just as in the case of licit economies, control of scarce resources is a source of power and wealth in the illicit world. By contrast, degree centrality did not correlate with equity volatility. Countries with illicit actors that have high betweenness scores are transit points, at least in the network sense, for the raw materials that enable illicit activity. The boundary spanners or brokers probably yield significant influence while also having access to networks and cells both inside and outside their country. The connected and transnational nature of the network often translates into the realm of illicit finance, with these individuals moving money across the illicit network as well as between the licit and illicit economies.

The relationship between betweenness and equity market volatility is even stronger than convergence, which is not surprising since the measure incorporates structural position in characterizing importance. The impact of betweenness actually outweighs that of connectivity. While a one standard deviation increase in convergence is associated with a 30-day volatility increase of 1.9, the predicted impact of a similar increase in betweenness increases volatility by 2.7. The glue that holds together the global illicit network, particularly disparate parts of the network, is the greatest threat to licit commerce as gauged by equity market volatility based on this analysis. At first glance, the increased volatility seems small. The mean 260-day volatility across the sample is 15.7. The marginal impact of a one standard deviation in convergence and betweenness is 1.1 and 1.8, respectively.

An increase in volatility from 15.7 to 18.6 might seem small, but the financial implications can be significant. These seemingly abstract measures can actually be translated into monetary costs. The costs of hedging such risks can increase from 50 percent to 230 percent. To justify the investment, asset allocators would have to overcome a drag of at least 0.5 percent on expected return, which could be sufficient to drive investment towards another market. For example, if an emerging market had a total market of $40 billion, the costs of hedging could increase from $400 million to $600 million on the low end of the estimate.

Policy Implications

In the last decade, the growing convergence between criminal and terrorist enterprises has gained attention in the foreign policy and military arenas, but as we showcase here, there is also a sufficient rationale to consider the economic impact as well.

Just like any of today's multinational companies operating with globalized supply chains, these illicit networks will continue to converge and work together, whether their missions overlap or not, in order to move resources, share knowledge, and raise funds to operate. The links between these groups will be stronger and the networks will naturally grow increasingly dense, as is the case with many naturally occurring complex systems. More convergence likely translates to greater economic effects locally as well as globally.

With higher volatility, there is an expectation that equity market returns will be affected as well. Businesses and investors frequently consider the costs of security and illicit activity, but rarely do they explicitly price this illicit network component into their calculations. This risk premium for businesses to operate in a country with stronger convergence remains higher than in countries with lower convergence rates, raising the costs of doing business in these countries. For countries looking to access foreign capital for commercial and infrastructure development, the costs to hedge risks from convergence can weigh heavily and limit capital inflow, which hurts development. Investor perception of convergent risk in developing markets may partially explain the predilection for foreign capital to flow to developed economies despite lower long-term economic growth rates.

Developed countries, however, also grapple with convergence threats. Less-developed countries do not necessarily have higher rates of convergence than developed countries. Development does not actually serve as a strong indicator of criminal or convergent activity (though this does not mean it is not an indicator). This is a global challenge, and undermines the conventional wisdom that poorly governed countries are hotbeds for illicit activity. The prominent nature of this misperception may be rooted in the vastly different capabilities that exist between developed and developing countries. Developed countries have highly institutionalized elements ready to deal with many of the problems, but there is no shortage of work for law enforcement and intelligence personnel. In some cases, developing nations have lower levels of convergence than developed countries, but they also have to do less with less.

Building on the CTC data that argued interaction in the global network was more common than previously expected, these relations seemingly affect a country's economic ecosystem. The rise of groups like the Islamic State of Iraq and the Levant (ISIL) that bridge gaps across the criminal-terrorism spectrum is a primary example of this. ISIL is selling oil on the black market and that can impact licit markets. This type of activity happened in the mid-2000s, with nearly 20 percent of Iraqi oil production (or up to 300,000 barrels per day) marked as unaccounted, which is industry parlance for likely stolen and smuggled.[12]

Following the September 11 terrorist attacks, then President George W. Bush created the Office of Terrorism and Financial Intelligence in the Department of Treasury and repurposed other departments and officials to work on criminal and terrorism financing. This included safeguarding the financial system against illegal use (i.e., money laundering) while also combating illicit actors, including rogue nations, terrorist cells, and drug traffickers. Even with these resources, the effect that illicit networks have on broader economic issues is not well-documented. By better understanding this relationship and seeing the extent to which illicit networks are a drag on a country's economy by producing increased volatility, policymakers can better assess the most economically damaging elements.

The crime-terror convergence is a growing, but still young, field of inquiry. Criminal enterprises from all facets of the illicit spectrum, from terrorism to financial criminals, often work together in some manner over time. The effects can be drastic and destabilizing. Criminal and terrorist groups aim for negative political control in that they benefit from denying or minimizing government and law enforcement operating capacity. Criminals want to pursue profit without fear of law enforcement, but criminal profits come at a cost in legitimate economies. When crime is rife, relying on contracts and other business conventions becomes tenuous. Criminal groups can also seek gain from levying illegitimate taxes on local businesses. Even in instances where criminals do not seek to directly profit, illicit activity redirects funds out of the licit economy. Moreover, terrorists often seek to overturn the political status quo and frequently rely in part on attacking the legitimacy of the ruling regime. Stifling economic growth and commercial activity is one method by which terrorists might look to cast doubt on the capability of current political leadership. The tactics employed, violence aimed at civilians, create a drag through damaged property, loss of life, and dampening of commercial sentiment. There are also direct costs associated with countering both criminal and terrorist elements, and those tend to increase as the problems grow more acute.

The impact of criminal and terrorist activity could be bleeding into and impacting the larger and more general (and licit) marketplaces. Criminal networks can interfere with trade routes or cause supply shortages. As Jessica Stern writes in this volume and others have noted elsewhere, ISIL lines its coffers with the sale of stolen oil (measured by the U.S. Treasury as over $1 million per day), impacting local and regional economies.[13] ISIL's success in raising illicit funds, working with criminal elements, controlling territory, and building an international strike capacity shows how dangerous and destabilizing the intersection of crime and terrorism can be.

The dynamics between illicit networks and national economic performance are not entirely new ground, but the data analyzed is somewhat novel. Extant scholarship suggests a number of things about criminal-terrorism convergence. Some have argued that the process of convergence has hastened in recent years, and the growing interconnection is a unique problem for and threat to U.S. national security.[14] Others have argued that convergence is overstated as a national security threat.[15] However, in a previous phase of this project, network analysis found that the illicit network is highly connected. Looking at more than 2,700 individuals operating in 3,600 locations and linked by 15,000 relationships that span 122 countries, the analysis showed that 98 percent of the individuals in the dataset were separated by a maximum of two degrees of association. In total, the CTC analysis found more than 1,000 country-to-country relationships.[16]

Questions about the types of interactions in the crime-terror network remain. These links can be to get money, but are primarily transactional in nature, based on partnerships of convenience and complementary business ties.[17] Some scholars have correctly identified that groups or individuals might work together for certain periods of time and then terminate their relations.[18] Some groups converge on an activity, when terrorists use criminal activities or criminals use terrorist tactics in pursuit of their respective political and economic ends, while others converge when a terror group works with a criminal enterprise. We have seen that with the Haqqani network's relationship with al-Qaeda and D-Company's relationship with Lashkar-e-Taiba. Related, groups may also transition along an apparent crime-terrorism continuum, transitioning from ideologically motivated groups that avoided involvement with criminal activities that now perpetuate crimes because of the attraction of the lucrative nature of criminal activities.[19]

Turning to the economic side of illicit networks, the debate is far less developed. Many questions arise when trying to assess the global economic footprint of illicit networks. Some have come up with estimates; however, not only do these estimates vary widely, but it is also difficult to determine their accuracy, with estimates ranging from 8 percent to as large as 30 percent of world GDP, or $6 trillion to $22 trillion. Even if these estimates are not fully accurate, the scope of these networks is jarring.

A handful of papers have studied the economic impacts of singular aspects of the illicit network. In the wake of the September 11 attacks, terrorism received most of the attention and research. It is clear that terrorism has a negative economic effect. Looking at the September 11 terrorist attacks, the direct costs were estimated at $27.2 billion, which represented about 0.25 percent of the U.S. GDP.[20] Looking more holistically, terrorist attacks have a stock price reaction of -0.83 percent, which corresponds to an average loss per firm per attack of $401 million in market capitalization, or the value of a company calculated by multiplying the current share price and the total number of outstanding stocks.[21] The correlations between regime type and development level with illicit networks are interesting and complex. The general consensus is that developing countries and nondemocratic regimes are more conducive to illicit networks that also damage the country's economy. However, transnational crime is strongest in the richest countries, but often obfuscated by the sheer size of the legitimate economy.[22]

Conclusion

Illicit networks leave a broad wave of destruction in their path. This clearly touches the community (and the broader region and globe), but these networks are also a headwind for national economies.

There are certainly limitations, as with any type of study delving into the clandestine. Now that we know these groups are well-connected and have a draining effect on the financial system, more effort needs to be paid to how these groups interact, and ultimately, how these groups first become connected. Lack of data (particularly in the 51 countries that lack an equity market that were included in the original study) limits this study, but additional attention needs to be focused on the aggregate effect over the long run. Our analysis lays out how the illicit and the licit economy bleed together. For illicit networks, destabilization is critical, and that is borne out in the numbers.

Appendix. Full Regression Results

	Ordinary Least Squares		Random Effects	
	30 Day Volatility	260 Day Volatility	30 Day Volatility	260 Day Volatility
Convergence	3.36 **	2.01 **	3.37 *	2.28 *
Std. Error	1.84	1.02	1.96	1.33
Average Between	1.65 ***	1.10 ***	1.66 **	1.20 **
Std. Error	0.75	0.41	0.80	0.54
Average Degree	0.05	-0.01	0.05	-0.03
Std. Error	0.25	0.14	0.26	0.18
Failed State Index	0.05	0.03	0.05	0.02
Std. Error	0.05	0.03	0.05	0.03
Inflation	0.49 ***	0.40 ***	0.48 ***	0.36 ***
Std. Error	0.17	0.09	0.18	0.11
GDP Growth	-0.88 ***	-0.89 ***	-0.85 *	-0.52 **
Std. Error	0.42	0.23	0.44	0.24
Constant	12.16 ***	12.86 ***	12.14 ***	12.65 ***
Std. Error	3.15	1.73	3.34	2.28

Note: Statistical significance marked by * p-value<0.10, ** p-value<0.05, *** p-value<0.01.

Notes

[1] For greater information on equity markets or volatility, please consult, among other resources, Frederic S. Mishkin and Stanley Eakins, *Financial Markets and Institutions* (New York, NY: Prentice Hall, 2014).

[2] Stephen Easton, "The Size of the Underground Economy: A Review of the Estimates," *Simon Fraser Working Paper* (2001), available at <http://www.sfu.ca/~easton/Econ448W/TheUndergroundEconomy.pdf>.

[3] Scott Helfstein and John Solomon, *Risky Business: The Global Threat Network and the Politics of Contraband* (West Point, NY: Combating Terrorism Center at West Point, May 2014).

[4] Brad Comincioli, "The Stock Market As A Leading Indicator: An Application Of Granger Causality," *The University Avenue Undergraduate Journal of Economics* 1, no. 1 (1996); Holger Sandte, "Stock Markets vs GDP Growth: A Complicated Mixture," *BNY Mellon Viewpoint,* 2012.

[5] Tracking volatility in equity markets can be relatively straightforward, but numerous studies reflect on methods of constructing and analyzing equity volatility. One common method of tracking volatility uses exchange indices like the Chicago Board Options Exchange Volatility Index (VIX). These index measures usually use derivative or hedging markets to assign a value to volatility, but most commonly the measure refers to the standard deviation of asset prices.

[6] Eugene F. Fama, "Stock Returns, Real Activity, Inflation, and Money," *The American Economic Review* 71, no. 4 (1981).

[7] Claude B. Erb, Campbell R. Harvey, and Tadas E. Viskanta, "Inflation and World Equity Selection," *Financial Analysts Journal* 51, no. 6 (1995); John H. Boyd, Ross Levin, and Bruce D. Smith, "The Impact of Inflation on Financial Sector Performance," *Journal of Monetary Economics* 47, no. 2 (2001).

[8] World-Check identified the primary activity for which illicit individuals were listed or indicted, and as result the Combating Terrorism Center team did not have to make coding judgments about the illicit activities with which individuals were involved. This helped avoid biased coding or results, creating a double blind experiment of sorts.

[9] The Fund for Peace, "Fragile States Index," 2015, available at <http://fsi.fundforpeace.org/>.

[10] Sara McLaughlin Mitchell and Brandon C. Prins, "Rivalry and Diversionary Uses of Force," *Journal of Conflict Resolution* 48, no. 6 (2004).

[11] Empirical assessment of the relationship between illicit networks and equity market volatility was

relatively straightforward. Volatility is a continuous measure with definable mean and standard deviation. Under these conditions, the most straightforward and simplest measure of estimation is the basic multivariate linear regression model or ordinary least squares. A number of other models and specifications were used to ensure the robust nature of the results. While the section below reports results from ordinary least squares, the Appendix with the regression results also includes a random effects model common in panel data.

[12] James Glanz, "Billions in Oil Missing in Iraq, U.S. Study Says," *New York Times,* May 12, 2007, available at <http://www.nytimes.com/2007/05/12/world/middleeast/12oil.html?_r=0>.

[13] David McCabe and Laura Barron-Lopez, "Treasury: ISIS makes $1M a day from oil sales," *The Hill,* October 23, 2014, available at <http://thehill.com/policy/defense/221644-treasury-isis-makes-1m-a-day-from-oil-sales>.

[14] Thomas M. Sanderson, "Transnational Terror and Organized Crime: Blurring the Lines," *SAIS Review* 24, no. 1 (2004); Bob Killebrew and Jennifer Bernal, "Crime Wars: Gangs, Cartels, and U.S. National Security," *Washington, DC: Center for a New American Security*, 2010.

[15] Chris Dishman, "Terrorism, Crime, and Transformation," *Studies in Conflict and Terrorism* 24, no. 1 (2001).

[16] Helfstein and Solomon, *Risky Business.*

[17] Vanda Felbab-Brown and James J.F. Forest, "Political Violence and the Illicit Economies of West Africa," *Terrorism and Political Violence* 24, no. 5 (2012); Michael Kenney, *From Pablo to Osama: Trafficking and Terrorist Networks, Government Bureaucracies and Competitive Adaptation* (University Park, PA: Pennsylvania State University Press, 2007); Peter Lowe, "Counterfeiting: Links to Organised Crime and Terrorist Funding," *Journal of Financial Crime* 13, no. 2 (2006); Louise Shelley and John Picarelli, "Methods Not Motives: Implications of the Convergence of International Organized Crime and Terrorism," *Police Practice and Research* 3, no. 4 (2002); John Picarelli, "Osama bin Corleone? Vito the Jackal? Framing Threat Convergence Through an Examination of Transnational Organized Crime and International Terrorism," *Terrorism and Political Violence* 24, no. 2 (2012).

[18] Shelley and Picarelli, "Methods Not Motives."

[19] Tamara Makarenko, "The Crime-Terror Continuum: Tracing the Interplay between Transnational Organized Crime and Terrorism," *Global Crime* 6, no. 1 (2004).

[20] Tilman Bruck and Bengt-Arne Wickstrom, *The Economic Consequences of Terror: A Brief Survey,* HiCN Working Paper No. 3 (Brighton: School of Social Sciences and Cultural Studies University of Sussex, 2004).

[21] Andrew Karolyi and Rodolfo Martell, "Terrorism and the Stock Market," *International Review of Applied Financial Issues and Economics* 2, no. 2 (2010).

[22] Patrick Radden Keefe, "The Geography of Badness: Mapping the Hubs of the Illicit Global Economy," in *Convergence: Illicit Networks and National Security in the Age of Globalization,* ed. Michael Miklaucic and Jacqueline Brewer (Washington, DC: National Defense

II. One Network

6

Terrorist and Criminal Dynamics: A Look Beyond the Horizon

Christopher Dishman

Criminal and terrorist labels provided a useful baseline for understanding nonstate actors for much of the 20th century. The Italian Mafia, left-wing European terrorist groups, and Colombia- and Mexico-based criminal enterprises, among others, usually fell cleanly into two categories: those motivated by greed and those motivated by grievances. To be sure, a spectrum of political and criminal activity existed within these two categories, particularly in the 1980s and 1990s, when some criminals participated in politics, served as local or regional patriarchs, or declared war against the state, using terrorism as a tactic, while others utilized criminal profits to replace revenues lost from state sponsorship or the unexpected absence of donations from abroad. In spite of this heterogeneity, analysts could usually discern a group's goals and motives and assess their impact on international security.

Analysts could do so because criminals and terrorists in the 20th century were typically organized hierarchically. Organizational constructs are an important and defining characteristic for understanding nonstate actors. A hierarchical organization provides more clarity to a group's intentions than a purely networked organization; such organizations retain a relatively clear mission and goal which autocratic leaders achieve by setting rules and boundaries for their business operations. A network, in contrast, can assume many forms and pursue multiple and seemingly competing or even conflicting goals. Polymotivated organizations present a challenge to analysts, who seek to understand the short-, medium-, and long-term threats posed by such organizations.

Traditional organized crime and terrorism paradigms, replete with hierarchical organizations, vertical illicit industries, and autocratic leaders, are giving way to a murkier and more complex environment. Successful counterterrorism and counter-transnational criminal organizations (TCO) efforts have dismantled hierarchies and compelled many TCOs and terrorist groups to disperse into networks. Lower and mid-level criminal and terrorist actors, once working for a single employer, now seek their own opportunities in regional or international networks. Ungoverned by hierarchical rules, today's networked actors are increasingly polymotivated and pursue a spectrum of criminal and terrorist activities.

Technological advances allow this potent combination of capabilities, skills, and products to thrive without the traditional "glue" of the underworld: face-to-face meetings, trusted introductions, autocratic leaders, and geographic centralization. The new environment is shaped by the erosion of state power, the staggering growth of shadow economies, lower entry barriers to criminal markets, and historic advances in communications technology and financial instruments. The "Deep Web" epitomizes this future: a real-time, virtual bazaar where illicit actors locate new partners, customers, and sellers, and conduct lightning-fast global transactions.

Modeling a criminal-terrorist dystopia produces alarming prospects without deviating significantly from global trend lines. In the future, terrorists and criminals will have easier access to funds due to lower entry barriers in global illicit markets and increasing engagement in the world's lawful economy. Terrorists and criminals will exchange goods and services at an unprecedented rate utilizing internet secrecy and covert messaging communications. Through internet communications, terrorists will be able to acquire weapons, explosive precursors, fraudulent documents, and other necessary instruments more quickly than in the past. Criminals, acting as buyers and sellers in this virtual environment, will see staggering revenues as the internet grows the black market in the coming decades. Geography will matter less in the coming years as criminals and terrorists utilize advances in communications technology to build transnational relationships. Alliances of criminals and terrorists, dispersed across the globe, may target weak or collapsing states so they can profit from the state's decline and the emerging war economy. Networks will also obscure the objectives of nefarious actors who become increasingly polymotivated, and in turn, more difficult to label a "terrorist" or "criminal."

This chapter will assess the relationships between terrorists and criminals, and present a dystopian view of future collaboration. It will assess the impact of the shift from hierarchical to networked organizations and address the implications of fragmenting illicit industries. Advances in communication technology, namely the Deep Web, will enable "hyper-collaboration" between anonymous nefarious actors, and challenge law enforcement's ability to identify people and activities. The chapter concludes that these trends will pose major challenges for law enforcement. Polymotivated networks will pair more "buyers and sellers" within illegal and informal markets who identify new global opportunities in weak states, conflict zones, and ungoverned areas. Fractured illicit industries will offer new opportunities to criminal entrepreneurs previously shut out from vertically organized markets. Polymotivated networks will utilize new communications technology to connect like-minded individuals that seek to profit economically or politically from subverting global rules and norms.

Hierarchy as Dominant Organizational Construct

Hierarchies remained the dominant organizational construct for criminal and terrorist enterprises for most of the 20th century. A hierarchy:

> uses authority (legitimate power) to create and coordinate vertical and horizontal division of labor. Under hierarchy, knowledge is treated as a scarce resource and is therefore concentrated, along with the corresponding decision rights, in specialized functional units and at higher levels of the organization.[1]

Hierarchies allowed kingpins to retain profits at the "CEO" level and ensure that the leader maintained tight control of the organization. These kingpins organized their businesses into various branches including financial, operations, legal, security, and counterintelligence, among other areas of responsibility. Some terrorist groups also utilized hierarchical elements to manage military operations, finance, media, and propaganda activities. Al-Qaeda core, for example, consisted of four committees that ran various aspects of the organization. Osama bin Laden implemented standard Islamic business practices when al-Qaeda relocated to Sudan and Afghanistan.

Insurgent groups also utilize hierarchies because they require a "command and control" structure to regulate territory and govern. Guerrilla leaders issue rules and directives that they expect to be followed, and a hierarchy enables them to enforce these directives. Insurgents also undertake governmental functions to include education, health care, and law, which are initiatives best implemented through a hierarchy. Guerrillas who control commodity production, like diamonds and oil, rely on a hierarchy to regulate production and transportation of the product. An internal security apparatus ensures that rank-and-file members and paid laborers do not steal from their organization. The Islamic State in Iraq and the Levant (ISIL), for example, retains a cabinet which manages the group's finances and oil production, in addition to other duties such as media, security, prisoners, and recruitment.[2]

Most importantly for understanding terrorist and criminal collaboration, hierarchies provide a mechanism for leaders to control their members and eliminate those who contravene or undermine the group's mission. Kingpins and terrorist leaders decide the organization's rules and operational boundaries. If a maverick TCO associate conducts independent criminal business, for example, the organization's internal security apparatus will find and punish the offender.[3] A hierarchy, of course, can contain multiple, quasi-independent groups that pursue interests that undermine the organization. Typically, however, the autocratic leader would steer the organization back to its original mission by eliminating the source of the problem.

Until recently, most TCO leaders have avoided conducting business with terrorists.[4] Criminals, seeking a stable, status quo environment, avoided partnerships with organizations seeking to overthrow or undermine the state. Most TCOs, embedded in weak states, invest and profit from a state's corrupt machinery and they do not want to disrupt that profitable dynamic. A TCO kingpin's lifestyle, replete with ranches, family, cars, zoo animals, and other materialistic goods, reflects his interest in preserving the status quo. A terrorist or insurgent group, in contrast, is normally disinterested in materialistic comfort and is instead driven by the desire for political change.

Throughout the 20th century, some hierarchically structured criminal and terrorist groups collaborated on a one-time or ad hoc basis. Criminals utilized terrorist groups for explosives and weapons training, assassination, and other services, while terrorists utilized the fraud and smuggling services of criminal groups. Some longer-term business partnerships existed for services and equipment, such as between left-wing terrorist groups in Europe and the Italian Mafia during the 1980s, but these long-term associations were unique in the global landscape. Strong leaders, through a centralized organization, could steer their organizations away from relationships they believed were detrimental to their goals.

Terrorists Turn to Organized Crime for Funding

When funding for terrorist organizations and insurgents dried up at the end of the Cold War, many violent extremist groups turned to organized and petty crime to raise revenue. State-sponsored funding, combined with a global crackdown on terrorist financing, forced terrorists and insurgents to seek alternate sources of revenue to continue operations. Insurgents, for example, with thousands of soldiers, required money to buy weapons, food, and pay salaries, while smaller terrorist organizations required funding to purchase fraudulent documents, pay smugglers, buy weapons, and conduct other organizational activities. Contemporary terrorists and insurgents exploit the local economy to raise funds for their activities. These groups will initially seek quick, ad hoc methods of acquiring funds but they will eventually locate more stable sources of income.[5] Urban-based terrorists rely on kidnapping, bank robberies, or schemes that abuse tax or other legal loopholes. Insurgents located in conflict zones exploit a war economy by selling weapons, arms, and security services. Rural terrorists or insurgents will extort commodity producers or overtake production of the commodity themselves. These commodities can include drug production and distribution; gold, gem, and diamond mining; oil production; and illegal logging, among others.

Drug trafficking is a particularly lucrative income source because it is often the most profitable industry in rural or areas lacking a robust state presence. The Taliban, for example, have generated millions of dollars through their participation in the white heroin trade. Taliban factions tax farmers, drug convoys, and opium lab operators and they receive steady payments from drug trafficking organizations that operate in their area of control.[6] Taliban leaders also maintain large opium stockpiles that serve as the equivalent of a bank account that can be drawn upon when needed.[7] In areas of Afghanistan where poppy production is absent, the Taliban extort gem and mining operators, conduct illegal logging operations, and shake down mobile phone businesses.[8]

For other terrorist groups, trade in legitimate commodities has funded their illicit activities. Terrorist groups in Africa and South America rely heavily on commodity production for generating revenue. Natural resources like gold, timber, hard rock minerals, diamonds (especially diamonds that can be retrieved through alluvial mining), gems, and illicit crops, provide a stable income stream for insurgents. Commodity production, with large extraction machinery and in-ground resources, is an attractive insurgent target because commodity managers cannot easily move their operations. Insurgents extort commodity producers, tax the product's pathways to export, or overtake production of the commodity themselves.[9]

In a war zone or a collapsed state, criminal and terrorist groups generate revenue by conducting business in the shadow economy which emerges during the conflict. Often a shadow economy is the only productive market in a war-torn country; thus, criminals and terrorists, as well as some licit businessmen, will turn to the shadow economy to raise funds. Many of these war-zone criminal groups, as noted by Louise Shelley, invest their profits overseas instead of into the state's economy.[10] Since violence against the state will not harm their investments, they are less hesitant about attacking the state and damaging its legal economy than traditional organized crime groups. Unlike traditional organized crime groups, these criminals may even ally with a local terrorist group to destabilize the state and perpetuate the war economy from which they profit.[11] These relationships threaten to create cycles of perpetual violence in modern conflict zones.

Modern-day terrorists diversify their funding sources to avoid dependence on a single source of income. They understand that relying on a single funding stream like state sponsorship or charities can leave their organization financially stranded if that pipeline is closed by authorities. Hezbollah, for example, reportedly receives funding from a broad range of activities, including conflict diamonds, donations from Lebanese Shia businessmen in Africa, cocaine smuggling in West Africa, credit card schemes, and cigarette smuggling in America.[12] Osama bin Laden received revenue from conflict diamonds, drug trafficking, investments in Sudan's state infrastructure, and state charities and benefactors.[13] A diversified investment portfolio provides a relatively stable funding source, since alternate revenue streams can compensate for the unpredicted loss of support from a single source.

Endangered wildlife is an example of portfolio diversification for terrorist groups, with the global illegal wildlife trade estimated to generate between $7 and $23 billion annually.[14] Some insurgent groups organize hunting parties to kill elephants for their ivory tusks or they task independent poaching teams to acquire wildlife.[15] Wildlife parts are easy to conceal and transport and the guerrillas do not require sophisticated machinery to extract the resource.[16] These military-style hunting parties are often equipped with night vision goggles, heat-seeking telescopes, GPS satellite receivers, helicopters, and RPGs.[17] Joseph Kony, the messianic leader of the Lord's Resistance Army (LRA), tasked his rebels with acquiring ivory to sell for weapons.[18] Al-Shabaab leaders have also ordered soldiers into Kenya to fetch the horns.[19]

Organized criminal groups have also diversified their income streams to ensure constant revenue. Mexico-based drug trafficking organizations, for example, have evolved into "poly-criminal" groups that generate revenue from human smuggling and trafficking, kidnapping, extortion, and the illegal extraction and sale of oil.[20] Some Mexico-based drug cartels raise funds from illegal red snapper fishing in the Gulf of Mexico and through the sale of exotic fish bladders to Asia, each of which can fetch between $7,000 and $14,000 on the black market.[21]

Hierarchical Terrorist Groups Transform Motivations

Many terrorist groups that raised money through organized crime kept the financial activity "in-house" rather than outsourcing the operation to an external criminal party.[22] When state-sponsored funding dwindled after the Cold War, terrorists and insurgents looked for new sources of funding and many found crime to be a lucrative source of income. These groups already maintained a sophisticated, hierarchical structure which they now utilized to profit from illicit activity. By controlling the criminal activity within their organization, these groups retained relatively tight security over their transactions and maximized profits. Over time, these violent groups invested significant resources and expertise into raising money, and unprecedented profits resulted from their efforts.

The influx of new cash into these organizations produced unintended consequences for terrorist leaders. Some members, relishing the new cash flow, became more interested in profits than politics. This dynamic is especially prominent within groups that recruit new members with bonuses and pay salaries to their soldiers. Money, rather than ideology, becomes the glue that holds these "transformed" organizations together. These groups may maintain a political or ideological veneer, but their underlying goal is to raise money for personal gain, rather than to advance the political mission.[23] As Paul Collier concludes, the presence of a commodity often leads to civil war as various groups vie for the resource. A violent extremist group may retain a superficial grievance like Kony's LRA, but ultimately the outlaw group is driven by greed and survival more than politics. According to Collier's research, ideologically based terrorist groups will transform from "grievance" to "greed" rebellions once they become dependent on a commodity for survival.[24]

The Revolutionary Armed Forces of Colombia (FARC) is perhaps the best example of such a group. The loss of its key ideological leaders, combined with increasing drug revenues, prompted many bloc leaders to choose profit over politics.[25] The FARC originally maintained a steady stream of income by extorting local businesses, kidnapping foreign workers, and taxing ranchers, but soon they seized upon Colombia's most profitable rural industry—cocaine. They shifted their criminal operations "upstream" to obtain maximum profits from the industry; at one point even meeting with Mexico-based drug traffickers to sell Cocaine HydroChloride.[26] The influx of cash created a handful of criminal bloc leaders, interested in lining their coffers with illicit funds, and transformed much of the organization into a criminal enterprise.

The activities of a transformed terrorist group can undermine its effectiveness as a terrorist organization. Like a TCO, these transformed entities are less inclined to pursue the violent overthrow of the state because they are increasingly interested in preserving the status quo. Criminal pursuits can also hurt the group's ability to attract popular support, as they are viewed more as a criminal band than a violent political organization seeking reform.[27]

Many factors can induce a terrorist or insurgent entity to transform into a criminal organization. If key political leaders are removed, their successors may choose to focus the organization's effort on profit over ideology. Abu Sayyaf in the Philippines, for example, transitioned to a criminal group with the death of its ideological leader, Abdurak Janjalani. Bandit leaders filled his void and steered Abu Sayyaf toward kidnapping-for-ransom operations and away from political objectives.[28] One could also argue that some

Taliban leaders are more interested in money than politics. In February 2009, the Pakistani government allowed the Taliban Movement of Pakistan (TTP) to occupy portions of the North-West Frontier area, where militants deforested rare hardwoods at an unprecedented rate, opened marble factories, restarted emerald mines, and ransacked the territory.[29] During the cease-fire, TTP's actions showed that their primary goal was to generate revenue rather than to establish an Islamic state.

Some terrorist groups also transform into criminal bands because they have been defeated by the state or because the conflict ended through negotiation or stalemate.[30] A civil war often leaves a ruined economy in its wake, which offers few economic prospects for former guerrilla fighters. Demobilized insurgents often resort to criminality to generate income, but they may do so under the veneer of a political banner.[31] Paramilitary groups in Colombia, for example, turned to drug trafficking when their anticommunist mission subsided; thus, we can expect members of the FARC to shift seamlessly to criminality after the group's demise. Central American anticommunist groups underwent a similar change and devolved into bandits and gangs seeking profits and power through crime.

A Hybrid Criminal-Terrorist Organization

Some terrorist groups did not transform completely into criminal groups, but instead changed into polymotivated entities pursuing both political and financial ends.[32] These hybrid organizations pursue seemingly divergent goals under the same insurgent banner.[33] A hybrid organization can be a precursor to a transformed terrorist group, as an increasing number of members and leaders seek profit over politics. A hybrid's leaders acknowledge the new reality and allow members to undertake criminal activities for personal gain, with the understanding that a portion of the profits will be funneled to the organization's leaders. The leader of a hybrid organization might also have a criminal background, but decide to become involved in politics, such as Dawood Ibrahim, leader of the so-called "D-Company." Hybrid organizations like D-Company will regularly shift resources from criminality to politics, and vice versa, depending on the environment and leadership priorities at any given time.

Hybrid organizations thrive in conflict zones. They raise money for their operations through the shadow economy and local criminals join the group to seek profits and security from the military wing of the organization. War legitimizes a hybrid's criminal actions since it requires income to continue the war.[34] It may seize commodity assets, sell weapons, or engage in other gray economy activities. Simultaneously, the hybrid will try to establish an autonomous zone, displace the state, or keep the country in perpetual conflict. Since economic control is intimately related to political power, a terrorist group that manages commodity production is one step closer to achieving its political aims.[35]

The Emergence of Networks

During the latter part of the 20th century, terrorist groups that mutated into hybrid organizations or transformed into criminal groups were sometimes propelled to such a change because of the arrest or death of a formidable ideological leader. Successful law enforcement and military operations have captured or killed many high-profile terrorist and criminal leaders in the last three decades. These autocratic managers maintained an iron grip on their organizations through a command and control structure where deputies controlled security, finances, and operations, among other functions. Pablo Escobar, the Orejuela Brothers, Benjamin Arellano Felix, and Amado Carrillo Fuentes, exemplified this type of "narco-leader." Terrorist leaders with similar cache included Osama bin Laden, Abdurajak Janjalani (the founder of Abu Sayyaf), Manuel Marulanda (one of the original leaders of the FARC), among others.

Successful government efforts to kill or arrest these individuals led to unintended consequences. In some cases, a less effective manager replaced the original, capable leader, and gradually lost control of the organization to internal fighting and external pressures. In such a scenario, the fractured hierarchy allowed lower and mid-level leaders to pursue alternate objectives under the umbrella of the organization.[36] As hierarchies disintegrated, networks and independent operators filled the void. Some groups retained the trappings of a hierarchy, but the leadership lost control of its members' activities. The organization's name might be used for branding purposes, but no structured organization existed after the hierarchy collapsed. In such an environment, the traditional organized crime concept of "closing ranks," where mafia dons temporarily refuse new members while they inventory the group's membership, became a distant and impossible directive. Networked criminals, both within and outside of the old structure, became virtual entrepreneurs, self-motivated and unbounded by hierarchical rules.

One result of decentralization was that terrorist leaders required their followers to locate their own sources of funding. Al-Qaeda, for example, could not serve as a "central bank" for Salafist terrorists around the world. Locally generated revenue could be used for area operations or conveyed to the organization's leadership. The Taliban, for example, began raising money through criminality, instead of receiving funds from overseas donations.[37] In a similar vein, devolved criminal organizations mutated into loose associations of self-sufficient criminal entrepreneurs, some of whom utilized a TCO brand name for marketing advantage, but otherwise retained little association with the disassembled group.[38] Los Zetas, for example, devolved after a 2012 split between Zeta leadership, and the fracturing spawned multiple Zeta franchises that claimed the Zeta brand name while committing a range of nontraditional Mexico-TCO activities like arson, extortion, random kidnapping, and petty theft.[39]

Vertically Integrated Industries (VII) Fragment

Equally important as the entrance of new actors into the black market is the fragmentation of vertically integrated industries (VIIs). Western Hemisphere drug markets, in particular, have fractured into segments and spawned independent "micro-economies." VIIs are markets that are regulated by a single TCO that controls the production, transport, and sale of a global product. The TCO utilizes its own employees to execute the movement of the drug through the supply chain. When a VII fractures, new gangs, organizations, and niche operators emerge to manage newly independent segments of the industry. These opportunists will maximize profits by purchasing and reselling the product to customers in the next segment of the market. They assume more risk by retaining ownership in case the product is seized or stolen, but they also generate higher profits than if they worked on salary for a TCO.

A segmented industry is an economic form of revenue sharing since more criminals split the profits from the industry's overall revenue. Criminal entrepreneurs thrive in this new economic environment. They no longer need to ascend a mafia's hierarchical ladder to be successful. They can develop narrow specialties and connections that allow them to gain influence over one aspect of the overall market. As a result, more criminals and terrorists— no longer shut out of industries controlled by one organization—will have access to new funding sources to support violent operations or accrue personal income.

The Western Hemisphere cocaine market, for example, fractured after the removal of Pablo Escobar and the Orejuela Brothers. The cartels' hierarchical structure and ruthless leadership helped preserve a vertical "farm-to-market" industry.[40] The cartels' dismantlement, combined with other factors, began to segment this VII. Mexico-based drug trafficking organizations exploited the fracture and paid for cocaine outright from Colombian suppliers. As a result, the cocaine industry segmented into pieces; more actors, organizations, and networks were required to maintain the cocaine pipeline to the United States.

The Western Hemisphere white heroin industry underwent a similar change. In Colombia, a small number of groups specialized in refining morphine base into heroin and exporting the drug to distributors in the United States. One report estimated that 28 of these organizations controlled the export of heroin to the United States.[41] Now, Mexican TCOs also produce white heroin from poppies grown in Mexico. This new dynamic has fractured a once vertical industry and allowed additional actors to profit from the growing white heroin trade to the United States. In today's fractured heroin market, more farmers are required to produce poppies, and independent gangs oversee morphine base production in Mexico.[42]

In each case, the breakdown of the VII allowed new actors to profit from an industry that had been previously closed off to them. The barriers to entering the heroin or cocaine markets have been lowered to allow more entrepreneurs, gangs, and networks to benefit from industry-wide revenue. Mexico-based TCOs, independent operators, gangs, and other groups acquired industry profits (and risk) from Colombian syndicates that had previously retained most of the industry's revenue for themselves. Thus, the size of the industry "pie" is the still same, but it is now cut into more slices with the income generated more broadly distributed.

Research on networked organizations suggests that they are poised to succeed in a global, segmented economy. Network nodes maintain a specific function and purpose which allows them to flourish in the "micro-economies" created by a segmented market.

Transporters, money launderers, and weapons dealers, among others, bring functional expertise to a network designed for a larger, economic purpose. Initially, one would think that a segmented market would be less efficient than a VII since responsibility in a network is diffused across multiple entities. In segmented markets however, networks ensure efficiency because the organizational construct fosters rapid coordination and communication between nodes. Specialists maintain transaction-based relationships which pass shared knowledge quickly through the network.[43] Information that is synthesized in networks often produces new knowledge which allows the nodes to succeed in difficult circumstances.[44]

In the illicit marketplace, the risk of law enforcement disruption or theft by another organization forces networks to be adaptive in order to survive. They also organize themselves into the forms necessary to maximize efficiency for a required function or specialty.[45] Phil Williams describes the advantages of criminal networks:

> Networks can vary in size, shape, membership, cohesion, and purpose. They can be large or small, local or global, cohesive or diffuse, centrally directed or highly decentralized, purposeful or directionless. A network can be narrowly focused on one goal or broadly oriented toward many goals, and its membership can be exclusive or encompassing. Networks are at once pervasive and intangible, everywhere and nowhere. More prosaically, they facilitate flows of information, knowledge, and communication as well as more tangible commodities. They operate in licit as well as illicit sectors of the economy and society. This enormous variability makes the network concept an elusive one; at a practical level, it also makes networks difficult to combat.[46]

The operations of Ghanaian smuggler, Mohammed Kamel Ibrahim, provide some insight into network efficiency. Ibrahim spearheaded a complex human smuggling pipeline that brought East Africans to the United States through Mexico. By any account, this was a complex global operation, spanning multiple continents, and involving numerous actors. Ibrahim's smuggling routes crossed through Ethiopia, Eritrea, Sudan, Kenya, South Africa, United Arab Emirates, Cuba, Brazil, Bolivia, Panama, Nicaragua, Honduras, Guatemala, and Mexico. In his court indictment, Ibrahim also stated that he could smuggle people through Europe to America (instead of Mexico) if that route was desired by the customer.[47] Ibrahim utilized a vast network that included recruiters in Africa, smugglers in South and Central America, a corrupt embassy employee in Belize, as well as transporters, guides, and money collectors in Mexico. He did not control or meet most of the individuals in his network; the organization itself executed the functions necessary to smuggle East Africans to Mexico (and the United States), and Ibrahim monitored or managed its activities as necessary.

Networks like Ibrahim's facilitate the movement of goods or services from the underdeveloped to developed countries.[48] Commodity chain nodes, for example, retain specialized roles that enable diamonds, gems, and other materials to be smuggled out of Africa to worldwide markets. Douglas Farah describes three roles that enable commodities to move from their source to market: fixers, super fixers, and shadow facilitators. Local fixers involve elites who understand and maintain the necessary relationships in their environment, but possess little awareness outside their area of operations. Super fixers assist the local elite move the product to a broader global market, while shadow facilitators enable these nodes to succeed by smuggling contraband, laundering money, and obtaining fraudulent documents, among other activities.[49]

Networks which incorporate specialized roles like the ones described by Farah, are poised to exploit the breathtaking growth of the black market. The profits from some illicit industries already rival that of major global businesses. In 2006, Mexico's marijuana industry generated $2.9 billion more in revenue than Google that year, and Mexico's methamphetamine industry's revenue roughly equaled Google's 2006 revenue. Mexico's drug trafficking industry also possessed net margins that towered over the country's licit industry. The marijuana industry in Mexico retained a net margin of 42 percent, which exceeded the home improvement industry at 7 percent and even the application software at 21 percent.[50] Other illicit global industries like the stolen art and antiquities trade, wildlife poaching and smuggling, and human trafficking, also generate staggering revenues.

A Dystopian Future of Criminal-Terrorist Convergence

The growth of the illicit economy, coupled with the rise of segmented markets and networks, has significant implications for terrorist and criminal collaboration. Globalization will continue to provide the tools necessary for criminals to succeed in the underworld, while also challenging law enforcement's ability to disrupt illicit markets. In a dystopian future, revenue from the black market and the informal sector will begin to outpace the licit economy and will become the economic bedrock for most of the world's weak states.[51] Opportunists and entrepreneurs, with few legal pathways to economic success, will thrive in this environment by developing new and innovative methods to complete timely and secure global transactions.[52] Sometimes referred to as "deviant globalization," this economic and political force will create new opportunities for criminals, terrorists, and the range of actors in between.[53]

In a dystopian future, these opportunities will be more evenly shared than before. Segmented markets will create a global criminal middle class that disperses criminal profits evenly throughout the industry. Lower barriers to entry will allow more terrorists and criminals to generate income from this "democratized" shadow economy and will encourage entrepreneurs to choose opportunities in the informal or black market over the licit sector. Thus, while the black market grows its earning potential, those profits will be shared with more actors than ever before.

In this profitable environment, terrorists and criminals, operating independently without autocratic rules, will be increasingly polymotivated like Ibrahim, to seek profit through criminality and also engage in violent extremist politics. They will maintain diverse relationships that change based on their operational needs.[54] Organizational affiliation and "hierarchy-to-hierarchy" alliances will devolve into individual, ad hoc activities.[55] In this regard, the dystopian future is largely devoid of criminal and terrorist labels. Networks will form around specialized services with little regard for organizational affiliation. Terrorists, criminals, government officials, and licit businessmen, among others, will rotate through networks as needed to execute transactions and functions. Globally networked terrorists, for example, with access to dozens of licit and illicit contacts, could easily shift their interests to pursue multiple and seemingly divergent goals. "Nodal motivation" will be thus obscured and networks will be increasingly difficult to label as "terrorist" or "criminal." Intelligence analysts may be surprised to find that a "terrorist network" contains varied membership including corrupt officials from multiple states, criminals, and nongovernmental organization personnel, among others.

Continued advances in communication technology will enable this dynamic to flourish. Collaboration will be increasingly conducted in the virtual world, which allows criminals and terrorists to execute anonymous and secure transactions. Today's encryption is already difficult for authorities to penetrate. In remarks given in 2014, Federal Bureau of Investigation (FBI) Director James Comey noted that:

> Unfortunately, the law hasn't kept pace with technology, and this disconnect has created a significant public safety problem. We call it "Going Dark," and what it means is this: Those charged with protecting our people aren't always able to access the evidence we need to prosecute crime and prevent terrorism even with lawful authority. We have the legal authority to intercept and access communications and information pursuant to court order, but we often lack the technical ability to do so.[56]

Covert messaging applications like Kik, Wickr, WhatsApp, and dozens of other similar programs challenge law enforcement's ability to gather intelligence necessary to disrupt illicit plots. In one example, authorities could not decipher 109 messages sent by one of the ISIL-inspired terrorists who attempted an attack in Garland, Texas.[57] Other "off-network" programs utilize Wi-Fi networks to communicate and include in-game chats through common gaming platforms or through tablets or other devices. In June 2015, Belgian authorities detained two groups of violent Chechen extremists who had been communicating with each other through WhatsApp, for example.[58]

Advances in communication technology have other implications for terrorist and criminal collaboration. The internet allows criminals and terrorists to locate buyers and sellers in rural areas, conflict zones, slums, and other areas which contain markets that are traditionally difficult to penetrate. Criminals and terrorists use software like The Onion Router (TOR), which enables anonymous communication while they troll the "Deep Web"—the unindexed internet—to advertise illicit services and products and identify potential buyers. Customers search the same forums to locate goods and services that can advance their terrorist or criminal ventures. The Deep Web, and its more secretive counterpart, the "Dark Web," allow sellers to advertise their products and services anonymously and for their customers to purchase securely those services or goods. In 2013, for example, the FBI dismantled "Silk Road," an online marketplace for marijuana, methamphetamine, ecstasy, and fake driver's licenses, among other items. In another example, an illegal weapons dealer in Germany converted nonlethal weapons into firearms and sold them online. Other weapons dealers purchased parts online to repair old assault rifles and sell them illegally to terrorists and other buyers.[59] In forums like "Silk Road," criminals and terrorists do not meet in person and are probably unaware of their counterpart's affiliation. The dystopian result of such interactions is the equivalent of multiple one-time arrangements that occur at an unprecedented breadth and pace. "Hyper-collaboration" shrinks the connectivity time between suppliers and customers from months to minutes.

Contemporary terrorist groups already recognize and exploit the power of the internet for their operations. Many violent extremist groups have established cyber departments responsible for propaganda, hacking, and cyber-attacks, among other activities. This capability allows terrorists in infrastructure-starved environments the opportunity to engage in the global economy. They can seek new sources of revenues, identify customers for an illicit product, or locate required services. Terrorists can also recruit criminals to join their organization through social media forums. These "criminals-turned-terrorists" bring their underworld connections and expertise to the group.

Criminal and terrorist collaboration serves as a force multiple for both groups. For a terrorist, interacting with criminals in a networked world is no longer a "zero-sum game." Terrorists will increasingly operate transnationally and will not be exclusively focused on their local areas of operations. They may sell weapons or explosive expertise to a distant criminal syndicate with little fear of repercussion. The supplier and consumer are connected virtually, not geographically, and the terrorist is shielded from prosecution through a computer, as well as layers of national and international laws. Similarly, criminal groups will increasingly collaborate with terrorists outside their area of operations by developing new contacts through the internet.

The efficiency of global networks will place more lethality in the hands of terrorists in the future. Viktor Bout, the Russian weapons magnate, epitomized a global logistical genius when he moved items ranging from weapons to chickens to conflict zones around the world.[60] Bout, however, was an essential node in his operations (a spoke-and-wheel network), and his removal dismantled his organization's operations. Most networks, in contrast, will ensure redundancy and efficiency in the face of a disturbance. If a network fails to deliver on a terrorist's order, the organization will be replaced by another set of nodes that is capable of executing the task.

The dystopian criminal-terrorist future, however, will not completely reside in cyberspace. Criminals will continue to establish themselves in areas where they can avoid or co-opt law enforcement. Conflict, postconflict, and underdeveloped countries provide hospitable areas where criminals can establish zones of autonomy to conduct their operations.[61] Criminal relocation is the human form of "habitat selection," during which plants and animals locate a preferred habitat and develop specialized functions to maximize performance.[62] Ungoverned spaces and conflict zones will serve as incubators for deepening terrorist-criminal collaboration. Slums, free trade zones, and border regions will foster terrorist and organized crime links, while megacities will continue to nurture collaboration between groups who seek political change, security, and profit.[63] Some of these locations will serve as geographic hubs for segmented illicit markets and will allow for quick in-person meetings to exchange goods and services.

Criminals and terrorists may find common cause in destabilizing a weak state in order to jumpstart or promote a war economy. Upstart criminal organizations may be recruited by terrorists to assist in attacking a state with the promise that the criminal group will profit from the emerging shadow economy. States that contain raw materials and resources could be the targets of a hybrid criminal-terrorist group that seeks to divide up the spoils once the conflict begins. These groups may even overtake the state and maintain a front of statehood while continuing their criminal or terrorist activities.[64]

One can also envision criminal and terrorist groups finding common cause in attacking a first-world country's financial system. A criminal group could hack into a major financial system to steal or extort money from a company, while the terrorist group conducts its own physical or cyber-attack against the country's economy. The attack would result in profit for the criminal group and a destabilized economy and headlines for the terrorist. A hybrid organization would be able to plan such an attack internally, which would facilitate operational planning and enhance security.

Conclusion

Many of the dystopian themes presented in this chapter could be reasonably considered likely scenarios for the future. Very few of the themes in this essay are "science fiction." Many of the dynamics discussed already exist in Latin America and Africa. A less dystopian view would hope for better outcomes of current global trend lines: the strengthening of the state; the growth of the licit economy in developing countries; the continued reluctance of criminals and terrorists to collaborate; and the presence of clear, identifiable organizational structures for nonstate actors.

All of these scenarios, however, seem unlikely, especially in a dystopian scenario. Polymotivated individuals and networks will complicate law enforcement's ability to identify and discern a network's wide spectrum of activity. These networks will thrive in the criminal-terrorist chasm found in most global law enforcement agencies that maintain separate structures and laws for criminals and terrorists. New, independent criminal entrepreneurs, profiting in a fragmented illicit market, will grow the informal and illegal economies and become critical actors in ungoverned spaces and conflict zones. Unburdened by a hierarchy and autocratic boss, these empowered autonomous agents will utilize the internet and virtual financial instruments to bridge new connections with terrorists, criminals, and businessmen who operate on the fringes of the licit economy. In a dystopian future, polymotivated networks will harness the skills, capabilities, and resources of a vast array of individuals who are determined to undermine and co-opt global rules and norms for their own benefit. As a result, states will face unprecedented challenges to maintaining and developing instruments of power in the face of a nimble and resourceful competitor.

Notes

[1] Paul Adler, "Market, Hierarchy, and Trust: The Knowledge Economy and the Future of Capitalism," *Organizational Science* 12 (2001): 216.

[2] Gregor Aisch, Joe Burgess, C.J. Chivers, Alicia Parlapiano, Sergio Peçanha, Archie Tse, Derek Watkins, and Karen Yourish, "How ISIS Works," *The New York Times*, September 16, 2014, available at <http://www.nytimes.com/interactive/2014/09/16/world/middleeast/how-isis-works.html?_r=0>; Ben Hubbard and Eric Schmitt, "Military Skill and Terrorist Technique Fuel Success of ISIS," *The New York Times*, August 27, 2014, available at <http://www.nytimes.com/2014/08/28/world/middleeast/army-know-how-seen-as-factor-in-isis-successes.html>.

[4] Chris Dishman, "The Leaderless Nexus: When Crime and Terror Converge," *Studies in Conflict and Terrorism* 28, (2005): 237–252.

[5] Kimberly Thachuk, "Transnational Threats: Falling Through the Cracks?" *Low Intensity Conflict and Law Enforcement* 10, no. 1 (2001): 51.

[6] R.T. Naylor, *Wages of Crime: Black Markets, Illegal Finance, and the Underworld Economy* (Cornell, NY: Cornell University Press, 2005), 63.

[7] Gretchen Peters, *Seeds of Terror: How Heroin Is Bankrolling the Taliban and al Qaeda* (New York, NY: Thomas Dunne Books, 2009), 13.

[8] Ibid., 114.

[9] Ibid., 13, 132-34, 138-39.

[10] Michael Ross, "The Natural Resource Curse: How Wealth Can Make You Poor," in *Natural Resources and Violent Conflict* (Washington, DC: World Bank, 2003), 17-43.

[11] Louise Shelley, *Dirty Entanglements: Corruption, Crime, and Terrorism* (Cambridge: Cambridge University Press, 2014): 104, 111.

[12] Ibid., 104, 111.

[14] Matthew Levitt, *The Global Footprint of Lebanon's Party of God* (Washington, DC: Georgetown University Press, April 2015), 248, 257, 261-62, 317, 320.

[15] Douglas Farah and Richard Shultz, "Al Qaeda's Growing Sanctuary," *Washington Post*, July 14, 2004, available at <http://www.washingtonpost.com/wp-dyn/articles/A48221-2004Jul13.html>; Douglas Farah, "Digging Up Congo's Dirty Gems," *Washington Post*, December 30, 2001, available at <https://www.washingtonpost.com/archive/politics/2001/12/30/digging-up-congos-dirty-gems/1cfa01f6-e23f-46d0-94f4-64d95c971517/>.

[16] United Nations Environment Programme, "The Environmental Crime Crisis," 2014, available at <http://pfbc-cbfp.org/news_en/items/Environnmental-Crime-en.html>.

[17] Ibid., 273.

[18] Leo Douglas and Kelvin Alie, "High-Value Natural Resources: Linking Wildlife Conservation to International Conflict, Insecurity, and Development concerns," *Biological Conservation* 171 (2014): 273.

[19] Ibid., 273.

[20] Jeffrey Gettleman, "Elephants Dying in Epic Frenzy as Ivory Fuels Wars and Profits," *The New York Times*, September 3, 2012, available at <http://www.nytimes.com/2012/09/04/world/africa/africas-elephants-are-being-slaughtered-in-poaching-frenzy.html>; Kasper Agger and Jonathan Hutson, "Kony's Ivory: How Elephant Poaching in Congo Helps Support the Lord's Resistance Army," *Enough Project*, June 2013, available at <http://www.enoughproject.org/files/KonysIvory.pdf>.

Ibid.

[21] CBS News and Associated Press, "Feds Bust Mexico-U.S. Oil Smuggling Scheme," *CBS News,* August 10, 2009, available at <http://www.cbsnews.com/news/feds-bust-mexico-us-oil-smuggling-scheme/>.

[22] Shannon Tompkins, "Gulf poachers Threaten to Deplete Fisheries," *Houston Chronicle*, June 8, 2013, available at <http://www.houstonchronicle.com/sports/outdoors/article/Gulf-poachers-threaten-to-deplete-fisheries-4589290.php>; "Fish Bladder, Coveted In Asia, Is The New 'It' Among Mexican Cartels," *Fox News Latino*, August 7, 2014, available at <http://latino.foxnews.com/latino/news/2014/08/07/mexican-traffickers-move-into-asia-pricey-fish-bladder-market/>. Industry is valued conservatively at $11.6 million, according to *FishWatch: U.S. Seafood Facts*, accessed at <http://www.fishwatch.gov/seafood_profiles/species/snapper/species_pages/red_snapper.htm>.

[24] Chris Dishman, "Terrorism, Crime, and Transformation," *Studies in Conflict & Terrorism* 24, no. 1 (2001), available at <http://www.tandfonline.com/doi/pdf/10.1080/10576100118878>.

[25] Ibid.

[26] Ian Bannon and Paul Collier, "Natural Resources and Conflict, What Can We Do," *Natural Resources and Violent Conflict* (Washington, DC: World Bank, 2003), 1-17; Paul Collier, *Greed and Grievance,* Policy Research Paper 2355 (Washington, DC: World Bank, 2000). Makarenko outlines how ideological war can be hijacked by criminal interests; see Tamara Makarenko, "Crime-Terror Continuum: Tracing the Interplay between

Transnational Organised Crime and Terrorism," *Global Crime* 6, no. 1 (2004): 129-145, available at <http://www.iracm.com/wp-content/uploads/2013/01/makarenko-global-crime-5399.pdf>.

[27] Dishman, "Terrorism, Crime, and Transformation."

[28] "Colombians Held In Drug Raid," *Orlando Sentinel*, August 14, 2002, available at <http://articles.orlandosentinel.com/2002-08-14/news/0208140080_1_mexico-city-colombia-felix-drug>.

[29] Audrey Cronin, *How Terrorism Ends: Understanding the Decline and Demise of Terrorist Campaigns* (Princeton, NJ: Princeton University Press, 2011), 146.

[30] McKenzie O'Brien, "Fluctuations Between Crime and Terror: the Case of Abu Sayyaf's Kidnapping Activities," *Terrorism and Political Violence* 24, no. 2 (2012): 328-29.

[31] Peters, *Seeds of Terror.*
Audrey Cronin, *How Terrorism Ends: Understanding the Decline and Demise of Terrorist Campaigns* (Princeton: Princeton University Press, August 2011): 146.

[32] Naylor, *Wages of Crime,* 82. Naylor points out that when guerrilla groups a criminality converge it is usually during the failing part of an insurgency.

[33] Chris Dishman, "The Leaderless Nexus: When Crime and Terror Converge," *Studies in Conflict and Terrorism* 28, no. 3 (2005): 247-248.

[34] Some analysts refer to this dynamic as "convergence;" see Makarenko, "Crime-Terror Continuum."

[35] Ibid., 141.

[36] Ibid.

[37] Dishman, "The Leaderless Nexus."

[38] Quote from Lorenzo Vidino in Peters, *Seeds of Terror,* 23.

[39] Dishman, "The Leaderless Nexus," 237-252.

[40] Will Grant, "Mexico's Zetas drug gang split raises bloodshed fears," *BBC News Mexico City*, September 11, 2012, available at <http://www.bbc.com/news/world-latin-america-19543286>.

[41] The general business model was for the cartel to purchase cocaine base from a rural independent operator at markets where it is consolidated. They would then refine the base into cocaine HcL with their own organization or using contractors, and hired their own transporters to smuggle the drug to the United States to their own distributors, who sold wholesale amounts of the drug and sent the profits back to the cartel.

[42] "The Way Heroin Networks Move Heroin," *El Tiempo*, May 21, 2012.

[43] Kyra Gurney, "Mexico Poppy Production Feeds Growing [U.S.] Heroin Demand," *InSight Crime: Organized Crime in the Americas*, February 11, 2015, available at <http://www.insightcrime.org/news-analysis/mexico-poppy-production-feeds-growing-us-heroin-demand>; Mark Stevenson, "Mexican opium farmers expand plots to supply [U.S.] heroin boom," *Associated Press*, February 2, 2015, available at <https://www.yahoo.com/news/mexican-opium-farmers-expand-plots-supply-us-heroin-145711238.html?ref=gs>.

[44] Linda deLeon, "Embracing Anarchy: Network Organizations and Interorganizational Networks," *Administrative Theory & Praxis* 16, no. 2 (1994): 234-253.

[45] Joel Podolny and Karen Page, "Network Forms of Organization," *Annual Review of Sociology* 24 (1998): 57-76.

[46] Shelley, *Dirty Entanglements,* 99.

[47] Phil Williams, *"The Nature of Drug Trafficking Networks,"* Current History 97, no. 618 (1998): 154-59.

[48] "Foreign National Pleads Guilty to Conspiracy and Alien Smuggling Charges," *USDOJ Press Release*, September 22, 2008, available at <http://www.justice.gov/opa/pr/foreign-national-pleads-guilty-human-smuggling-charges>.

[49] Shelley, *Dirty Entanglements,* 134.

[50] Douglas Farah, "Fixers, Super Fixers, and Shadow Facilitators: How Networks Connect," in *Convergence: Illicit Networks and National Security in the Age of Globalization* (Washington, DC: National Defense University Press, 2013), 75-97.

[51] Calculations conducted by the author in 2007, utilizing open-source data and basic accounting methods. Calculations included the total revenue and net margins for annual wholesale sales in the United States, and the model outlined specific business models for each of the four major drugs sold by Mexican drug trafficking organizations.

[52] Phil Williams, *From the New Middle Ages to A New Dark Age: The Decline of the State and U.S. Strategy* (Carlisle, PA: Strategic Studies Institute, U.S. Army War College, 2008), 31.

[53] This point is raised in Nils Gilman, Jesse Goldhammer, and Steven Weber, "Deviant Globalization," in *Convergence: Illicit Networks and National Security in the Age of Globalization* (Washington, DC: National Defense University Press, 2013), 3-15.

[54] Ibid., 5. See also Gilman, "The Twin Insurgency: Plutocrats and Criminals Challenge the Westphalian State," in this book.

[55] This point is raised in John T. Picarelli, "Osama bin Corleone? Vito the Jackal? Framing Threat Convergence Through an Examination of Transnational Organized Crime and International Terrorism," *Terrorism and Political Violence* 24, no. 2 (2012): 180-198.

[56] Ibid., 184-185.

[57] FBI Director James B. Comey, remarks at the Brookings Institution, October 2014, available at <https://www.fbi.gov/news/speeches/going-dark-are-technology-privacy-and-public-safety-on-a-collision-course>.

[58] Pierre Thomas, "Feds Challenged by Encrypted Devices of San Bernardino Attackers," *ABC News*, December 9, 2015, available at <http://abcnews.go.com/US/feds-challenged-encrypted-devices-san-bernardino-attackers/story?id=35680875>.

[59] Nick Veasey, "Belgium arrests in anti-terror raids targeting Chechens," *BBC Europe,* June 8, 2015, available at <http://www.bbc.com/news/world-europe-33046258>.

[60]"How Europe's Terrorists Get Their Guns," *Time Magazine*, December 7, 2015, available at <http://time.com/how-europes-terrorists-get-their-guns/>.

[61] Douglas Farah and Stephen Braun, "The Merchant of Death," *Foreign Policy*, no. 157 (2006): 52-61.

[62] Duncan Deville, "The Illicit Supply Chain," in *Convergence: Illicit Networks and National Security in the Age of Globalization* (Washington, DC: National Defense University Press, 2013), 63-75.

[63] David D. Ackerly, "Community Assembly, Niche Conservatism, and Adaptive Evolution in Changing Environments," *International Journal of Plant Sciences* 164, no. S3 (2003): 165-184.

[64] Ibid., 19-23; Picarelli, "Osama bin Corleone? Vito the Jackal?"

[65] Williams, *From the New Middle Ages to A New Dark Age*, 31.

7

Hezbollah's Criminal Networks: Useful Idiots, Henchmen, and Organized Criminal Facilitators

Matthew Levitt

Hezbollah's Nodal Model of Criminal Enterprise

Since its inception, Hezbollah has leveraged a worldwide networks of supporters—from formal operatives to informal sympathizers—to provide financial, logistical, and sometimes even operational support. Through these networks, the group is able to raise funds, procure weapons and dual-use items, obtain false documents, and more. Some of these are formal Hezbollah networks, run and supervised by Hezbollah operatives on the ground and back in Lebanon. But most are intentionally structured to be more informal than formal. That is, the links back to formal Hezbollah structures are opaque, providing distance and a measure of deniability for Hezbollah leaders. Generally, the Hezbollah criminal enterprise tends to be organized around loosely connected nodes, and does not depend on hierarchical links up the Hezbollah chain of command. The types of operatives and supporters involved in criminal activities benefiting Hezbollah run the gamut from trained and committed Hezbollah henchmen to what one Federal Bureau of Investigation (FBI) official experienced in Hezbollah investigations describes as "useful idiots who want to be associated with a glorious cause but don't need or want to know more."[1] Hezbollah also plugs into preexisting criminal networks that provide specialized services to any client who pays.

But make no mistake: Hezbollah operates as a transnational and transregional criminal organization. To cite just one recent finding, note the March 2013 conviction in the Cypriot criminal court of Hussam Yaacoub, a Hezbollah operative involved in planning attacks against Israeli tourists visiting the island nation. The head of the three-judge panel concluded, "It has been proven that Hezbollah is an organization that operates under complete secrecy. There is no doubt that this group has multiple members and proceeds with various activities, including military training of its members. Therefore, the court rules that Hezbollah acts as a criminal organization."[2]

Hezbollah has, since its inception, benefited from Iranian largesse, including the assurance of hundreds of millions of dollars in cash and generous material support in the form of weapons and other goods. But over the years, due to the impact of sanctions, a drop in the price of oil, and other considerations, Iran has cut back its direct support in favor of proxy support—sometimes very suddenly.[3] As a result, Hezbollah has, in recent years, significantly expanded both its formal and informal criminal enterprises, as a means of diversifying its financial portfolio and insulating its budget from the impact of Iranian belt-tightening. And while the group does raise funds through charities fronting for the group, the big money comes from a range of criminal activities, from arms smuggling and counterfeiting operations to credit card fraud and narcotics trafficking.

To facilitate its criminal activities, Hezbollah and Hezbollah-affiliated supporters cooperate with a broad range of facilitators who have less definite ties to the group itself, if any. In other words, Hezbollah often leverages the services of cut-outs, people only loosely connected to the organization, and even completely unaffiliated criminal middlemen and facilitators. The Hezbollah fingerprint is often more opaque than was the case with Yaacoub in Cyprus. Indeed, Hezbollah purposefully structures the command and control of its covert operations and illicit criminal activities to be as opaque as possible.

Hezbollah is deeply involved in criminal enterprises, including running illicit networks of its own and plugging into those of other criminal entities. There remains, however, deep disagreement within U.S. government circles about the extent to which some of these are truly Hezbollah enterprises or merely the criminal activities of individual Hezbollah members or supporters. These members or supporters share the proceeds of their crimes with Hezbollah, but they do not always act under the direction of Hezbollah leadership. Indeed, while both situations exist, the question may be based on a distinction that makes little difference. As far back as the Lebanon hostage crisis in the 1980s, Hezbollah's involvement in crimes and operations has often been driven less by a hierarchical organization chart than by particular relationships. And yet, these still benefit the group overall. A similar dynamic is at play today, governing a loosely structured, untransparent, but organized Hezbollah effort to leverage criminal enterprises for financial, logistical, and even operational benefit. At a time when Hezbollah is deeply involved in the Syrian civil war, has been caught planning terrorist attacks in three continents, and is active on the ground in battlefields as far afield as Iraq and Yemen, identifying and targeting Hezbollah's fundraising and logistics operations through criminal enterprise is a priority for intelligence, law enforcement, and policymakers alike.

Hezbollah's Useful Idiots: Small-Scale Crime

In the diaspora in particular, Hezbollah has successfully motivated and bullied fellow Lebanese and Shia (whether from Lebanon or not) to donate funds to the group and sometimes to engage in criminal activities specifically intended to support it. Often working with only a few people with actual ties to the organization, committed Hezbollah operatives and supporters abroad have created conditions within which they can facilitate and leverage the criminal activities of a larger community to fund Hezbollah.

Consider, for example, the tri-border area (TBA) in South America where the borders of Brazil, Argentina, and Paraguay meet at the famous Iguazu Falls. Described as "the United Nations of Crime," the region is a counterfeiting hot spot where "just about everything that is not biodegradable is fake."[4] There, Hezbollah found a vastly under-regulated free trade zone ripe for exploitation. According to a study conducted for the

U.S. Special Operations Command, "Hezbollah clerics reportedly began planting agents and recruiting sympathizers among Arab and Muslim immigrants in the TBA at the height of the Lebanese Civil War in the mid-1980s."[5] The result was the establishment of both formal Hezbollah cells and the comparably amorphous networks of individuals of Lebanese descent, particularly Shia Muslims, who provided some measure of financial support to Hezbollah.

As the Muslim community in the TBA grew, so did its need for educational, cultural, and religious institutions catering to the local Arab and Muslim communities. One such institution, the Profeta Mahoma mosque in Ciudad del Este, was reportedly built by a prominent member of the local Arab community, Mohammad Yousef Abdallah, who had been living in the city since July 1980. According to Argentine intelligence officials, Abdallah was one of the first Hezbollah members to settle in the TBA, though they did not know it at the time.[6] Only four years later, in April 1984, would the Argentine Federal Police see the first indications of a Hezbollah network in the TBA.[7] By mid-2000, experts would put the estimated number of Hezbollah operatives living and working in the TBA at several hundred.[8] The number of supporters or sympathizers who are neither trained operatives nor official group members, but still provide services or support (out of a more general affinity or to improve the lot of their families back home in Lebanon), is believed to be much larger.

The State Department's "Country Reports on Terrorism 2009" summarized global concerns about terrorist supporters and a flourishing black market economy being co-located in the TBA. Terrorist supporters "take advantage of loosely regulated territory" in the TBA "to participate in a wide range of illicit activities," including "arms and drugs smuggling, document fraud, money laundering, trafficking in persons, and the manufacture and movement of contraband goods through the TBA."[9] Some of that was carried out by active Hezbollah supporters, but far more of that was the product of criminals of Lebanese descent who were happy to donate a portion of their profits to Hezbollah, as the price for plugging into Hezbollah networks in the TBA and those back home in Lebanon.

For example, the "Barakat network" in the TBA was led by Hezbollah Secretary-General Hassan Nasrallah's personal representative to the region, Assad Ahmad Barakat. Some criminal supporters of Hezbollah—like Barakat—maintained close and meaningful ties to the organization. But Barakat's network also created space for others to support Hezbollah more passively. Consider, for example, the Galeria Page shopping center in Ciudad del Este, Paraguay, which served as both a major source of fundraising and as the local headquarters for Hezbollah in the TBA. According to the U.S. Department of Treasury, Galeria Page "is managed and owned by TBA Hezbollah members, including members of the Barakat network."[10] The U.S. Treasury left no doubt as to Assad Barakat's place in Hezbollah, describing him as "a key terrorist financier in South America who has used every financial crime in the book, including his businesses, to generate funding for Hezbollah.... From counterfeiting to extortion, this Hezbollah sympathizer committed financial crimes and utilized front companies to underwrite terror."[11] But the shopping center—and others like it—also served as a place where less committed supporters and "useful idiots" seeking association with a glorious cause could also work in the area's gray economy and donate some of their profits to the cause.

As in South America, the large Lebanese diaspora in Africa has long made the continent a rich source of financial and logistical support for Hezbollah, allowing it to effectively hide its operatives and criminal activities in plain sight. Crime and corruption are endemic in Africa, making the continent an attractive place for individuals and organizations engaged in illicit activities. In one representative assessment, a senior law enforcement advisor for Africa at the United Nations Office on Drugs and Crime (UNODC) explains that Africa attracts international organized crime because it consists of weak states, often characterized by corruption, dominated by weak and uncoordinated law enforcement agencies, and accustomed to the involvement of high-level officials in criminal activity.[12] By 1988, support for Hezbollah had already become so entrenched within the Lebanese immigrant community in Africa that the Central Intelligence Agency's (CIA) Directorate of Intelligence wrote an 18-page analytical report on "the economic and political roles of Lebanese communities in the region [sub-Saharan Africa]."[13]

Tactics used by Hezbollah (and Amal before it) to raise funds from Lebanese communities in Africa have included "appealing to their religious convictions, appealing to their Lebanese identity or using threats and even outright violence," according to one nongovernmental organization's report.[14] For those expatriates who have resisted solicitations to support various groups back home, including Hezbollah, the response has often been attacks on commercial properties by organized groups of Lebanese thugs.[15] The more continuous expense for legitimate Shia businessmen involves "taxes" levied by Hezbollah, with payment enforced through mafia-style racketeering. In 2004, U.S. diplomats in West Africa publicly charged that Hezbollah was "systematically siphoning profits" from the region's lucrative diamond trade, in part by threatening Lebanese merchants. The deputy chief of mission at the U.S. Embassy in Sierra Leone was emphatic: "One thing that's incontrovertible is the financing of Hezbollah." Citing interviews by embassy staff with the targets of Hezbollah racketeering schemes, the deputy mission chief said, "It's not even an open secret; there is no secret. There's a lot of social pressure and extortionate pressure brought to bear: 'You had better support our cause, or we'll visit your people back home.'"[16]

Hezbollah sympathizers have proved able to function just as easily within fully developed economies with functioning regulatory and law enforcement oversight. Hezbollah has long seen North America as a cash cow, where its supporters and operatives have run charities and engaged in a vast array of criminal activities to raise money and procure material for the organization. Consider the case of a group of North Carolina-based Hezbollah operatives in 2002, who were convicted of channeling some of their profits from an interstate cigarette smuggling scheme—an estimated $1.5 million—back to a Hezbollah commander in Lebanon. Investigators calculated that the Charlotte Hezbollah cell purchased more than $8.5 million worth of cigarettes, making an estimated $2 million in profits.[17] Members of the cell also collected donations to be funneled to Hezbollah through the group's charities, and in some cases, held onto receipts they received for their donations.[18] The group had well-established connections back to Hezbollah, to be sure, but most of the people indicted in the Charlotte case for their roles in the group's various criminal schemes—especially visa fraud and other basic crimes—were useful idiots.

In the authorities' successful prosecution of the Charlotte Hezbollah network, one witness stood out. Said Harb was, as one agent describes him, "a one-man crime wave."[19] Wherever Harb could make a buck, he would. He was involved in an internet pornography business that, he conceded, was "religiously…wrong," adding, "I'm not a religious person."[20] Some of the drivers who transported cigarettes to Michigan for Harb complained that he ripped them off. As one driver put it, "I mean, what was I going to do, go to the local police?"[21] When a female driver threatened to quit, Harb retorted that he would go to the restaurant where she worked, "kick everybody's ass," and blow it up.[22] Neither particularly religious nor a blood relative of other members of the Charlotte network (as several others were), Harb was at the center of nearly all of the network's criminal enterprises, from credit card and bank fraud to cigarette smuggling to dual-use procurement efforts.

Given his profile, the amount of jail time he faced, and the evidence U.S. and Canadian authorities had compiled, investigators were keen to "flip" Harb—that is, offer him a reduced sentence in return for testifying against the rest of the network. As it turned out, Harb was the critical link between the Charlotte cigarette smuggling ring and a parallel Canadian dual-use procurement ring. One of the major players in the Canadian Hezbollah network was a childhood friend of Harb's, and Harb provided him with false IDs that the Hezbollah operatives in Canada used to replicate Harb's credit card "bust out" scams. An expert in using multiple identities, Harb had four different ringtones on his cell phone, each for a different identity.[23]

In Charlotte, Harb proved to be more of a criminal logistician and opportunist than a committed Hezbollah henchman. But a few key people, like ringleader Mohamad Hammoud, were well-connected Hezbollah operatives. In the end, the financial investigation into what was dubbed Operation *Smokescreen* included some 500 bank and credit card accounts, and the overall investigation encompassed so many agencies that it led to the formation of the North Carolina Joint Terrorist Task Force (JTTF).[24]

In Los Angeles, another Hezbollah-linked case offers a representative example of the kind of criminal enterprise loosely tied to Hezbollah that the FBI typically finds across the country. These often involve investigations into petty criminal activities, including food stamp fraud, watered-down baby formula, misuse of grocery coupons, and sale of unlicensed T-shirts, and also involve criminal fundraising plots with ties to Hezbollah.

In November 2007, after a two-year investigation centered on the Los Angeles garment district, a multiagency task force arrested a dozen suspects on narcotics trafficking, sale of counterfeit goods, and money laundering charges. The investigation, dubbed Operation *Bell Bottoms*, culminated in the arrests of Ali Khalil Elreda and his associates. Elreda was detained at the Los Angeles International Airport (LAX) while attempting to smuggle $123,000 in money orders and cashier's checks to Lebanon, all of which were stuffed in a child's toy.[25] Counterfeiting crimes by Hezbollah supporters predate the Elreda group's arrest, as affirmed in a 2005 congressional testimony by John Stedman of the Los Angeles Sheriff's Department, who recounted two notable instances in particular. In one instance, during a search of a suspect's home in which thousands of dollars in counterfeit clothing were seized, Stedman saw small Hezbollah flags displayed next to a photograph of Nasrallah. When Stedman identified Nasrallah in the photo, the suspect's wife said, "We love him because he protects us from the Jews." Also in the home were dozens of audiotapes of Nasrallah's speeches and a locket containing Nasrallah's picture. In 2004, while serving a search warrant at a Los Angeles County clothing store, detectives recovered thousands of dollars in counterfeit clothing and two unregistered firearms. The suspect—typical of the "useful idiot" model—was found to have a tattoo of the Hezbollah flag when he was booked into custody.[26]

The financial rewards from petty crime like counterfeiting can be immense, but the practice makes it difficult for a criminal network to use traditional banking practices. In a Los Angeles case, investigators discovered more than $800,000 in cash throughout the suspect's home, hidden under the bed in trash bags and stashed in trash cans and the attic, with over $10,000 in a child's piggy bank. In another case, U.S. Customs officers at LAX stopped a Lebanon-bound woman with $230,000 in cash strapped to her body. The woman told the customs officers that she was heading to Lebanon for vacation. According to Stedman, authorities learned that the woman owned a chain of cigarette shops. In their raids, the authorities seized more than a thousand cartons of counterfeit cigarettes and an additional $70,000 in cash, along with documentation of wire transfers to banks throughout the world.[27]

Hezbollah's Henchmen: Running Criminal Networks

Not all Hezbollah supporters are small-time crooks donating a portion of their profits to the cause. Others are either themselves actual Hezbollah operatives or close associates who can, by virtue of their ties to Hezbollah operatives, plug into Hezbollah networks, as in the case of the Galeria Page shopping center in the TBA.

Assad Barakat's network was largely run out of the Galeria Page shopping center, where he also maintained stores of his own. Authorities determined that Galeria Page served "as a cover for Hezbollah fundraising activities and as a way to transfer information to and from Hezbollah operatives." Barakat used one of his companies, Barakat Import Export Ltd., to raise money for Hezbollah "by mortgaging the company in order to borrow money from a bank in a fraud scheme."[28] The extent of Barakat's criminal activity in support of Hezbollah was impressive. From distributing and selling counterfeit U.S. dollars to shaking down local shopkeepers for donations to Hezbollah, Barakat was accused by the U.S. Treasury Department of engaging in "every financial crime in the book" to generate funds for the group.[29]

"Barakat is more than a financier," explains a Paraguayan investigator.[30] According to Argentine authorities, Barakat was a card-carrying member of Hezbollah's Islamic Jihad Organization (IJO) terrorist wing.[31] He and his network collected sensitive information about the activities of other Arabs in the TBA, including those who traveled to the United States or Israel. This was of particular interest to the group and duly collected and passed along to Hezbollah's Foreign Relations Department (FRD) in Lebanon.[32] Even in his fundraising role, Barakat's heavy-handed tactics underscored the fact that he was no mere fundraiser. His threats to shopkeepers are a case in point: instead of threatening local store owners themselves, Barakat threatened that their family members in Lebanon would be put on a "Hezbollah blacklist" if they did not pay their quota through him.[33]

As a result of evidence collected during raids of Assad Barakat's store in 2001, Brazilian police eventually arrested Barakat in June 2002 on tax evasion charges, just as Barakat was making plans to flee to Angola. The arrest prompted a leader of the Husseinia Iman al-Khomeini mosque in Foz, where Barakat served as deputy financial director, to ban all non-Hezbollah members from attending services.[34]

In the wake of the 1994 bombing of the Asociación Mutual Israelita Argentina (AMIA), a Jewish community center in Buenos Aires, early reports from Argentinean intelligence focused on Mohammad Youssef Abdallah and Farouk Abdul Omairi, among others, as two of the "main activists" suspected of being members of an Islamist terrorist organization that provided support for the bombing.[35] According to an FBI informant, Mohammad Abdallah was regarded as "the principal leader of Hezbollah in the region," and his brother, Adnan Yousef Abdallah, functioned as his deputy. As the "second most important Hezbollah figure in the tri-border area," the FBI reported, Adnan "oversees the clandestine extremist activities of the Abdallah clan."[36] Only much later, in December 2006, would the U.S. government publicly designate Mohammad Abdallah as a Hezbollah terrorist.[37]

Prosecutors were especially interested in Omairi's contacts with the recently minted Iranian diplomat Mohsen Rabbani.[38] In time, these connections would be leveraged by Rabbani to help execute the AMIA bombing, as demonstrated through telephone calls traced to and from the travel agency co-owned by Omairi and Abdallah, Piloto Turismo. Intelligence officers would later determine that Piloto Turismo was not just a business that provided a convenient cover for illicit conduct, but rather that it was opened with start-up funds supplied by Hezbollah and that it was explicitly meant to serve as a Hezbollah front company.[39]

After the AMIA attack, Omairi would run a new travel agency for a time before authorities arrested him and his son on charges of using the travel agency and a money exchange business to falsify documents and launder drug money.[40] The arrests were the result of a 2006 Brazilian Federal Police counternarcotics investigation dubbed Operation *Camel*, which determined that Omairi provided travel support to drug mules transporting cocaine and that he had obtained Brazilian citizenship illegally.[41] The arrests came within months of the U.S. Treasury Department's designation of Omairi and Abdallah as Hezbollah operatives. Omairi, the U.S. Treasury stated, served as a regional coordinator for Hezbollah and procured false Brazilian and Paraguayan documentation with which he helped people illegally obtain Brazilian citizenship. He was also involved in trafficking narcotics between South America, Europe, and the Middle East.[42]

Abdallah's role included personally carrying monies raised in the region to Hezbollah in Lebanon, where he met with senior Hezbollah officials and members of Hezbollah's security division, the U.S. Treasury revealed. Sometimes, money flowed in the reverse direction, and Hezbollah would send Abdallah back with funds intended to support the Hezbollah network in the TBA.[43]

As with the useful idiots, similarly organized Hezbollah networks operate elsewhere around the world, including in Africa. While the precise numbers are unknown, Hezbollah raises significant funds within the Lebanese diaspora communities in Africa. One person who reportedly facilitated such donations was Sheikh Abd al-Menhem Qubaysi, a Lebanese national living in the Ivory Coast. In May 2009, the U.S. Treasury Department designated Qubaysi as a terrorist financier who played a public and prominent role in Hezbollah activities in the Ivory Coast. According to intelligence released by the U.S. Treasury, Qubaysi served as Hassan Nasrallah's personal representative in West Africa. "Qubaysi communicates with Hezbollah leaders and has hosted senior Hezbollah officials traveling to Cote d'Ivoire and other parts of Africa to raise money for Hizballah," the U.S. Treasury added. Beyond fundraising, the U.S. Treasury found that Qubaysi helped establish an official Hezbollah foundation in the Ivory Coast "which has been used to recruit new members for Hizballah's military ranks in Lebanon."[44]

According to U.S. intelligence, "Hezbollah maintains several front companies in sub-Saharan Africa."[45] In May 2009, the U.S. Treasury Department designated Kassim Tajideen as "an important financial contributor to Hezbollah who operates a network of businesses in Lebanon and Africa."[46] According to the U.S. Treasury's fact sheet, Tajideen contributed tens of millions of dollars to Hezbollah and funneled money to the group through his brother, a Hezbollah commander in Lebanon. Tajideen, a dual citizen of Lebanon and Sierra Leone, was joined by his brothers in running "cover companies" for Hezbollah in Africa, the Treasury revealed.[47]

In fact, Tajideen had already been under investigation six years earlier. In May 2003, after a four-month international investigation by Belgium's Economic Crimes Unit, Belgian Judicial Police raided the Antwerp offices of Soafrimex, a Lebanese export company owned by Kassim Tajideen, arrested several of its officials, and froze its bank accounts on charges of "large-scale tax fraud, money laundering, and trade in diamonds of doubtful origin, to the value of tens of millions of euros." Tajideen and his wife were also arrested in connection with the charges. A few months later, Belgian authorities informed officials from the Congolese Embassy that an investigation conducted on the ground in the Democratic Republic of the Congo (DRC) demonstrated that "the company systematically undervalued its imports, shipping and insurance costs and that it filed false customs declarations."[48]

In December 2010, the Treasury Department targeted two more Tajideen brothers— Ali and Husayn—as Hezbollah financiers, designating them and several of their companies, including Arosfram.[49] Described as "two of Hezbollah's top financiers in Africa," the two brothers ran a multinational network that generated millions of dollars for Hezbollah, according to the U.S. Treasury. The businesses targeted by the Treasury Department were located as far afield as The Gambia, Sierra Leone, the DRC, Angola, the British Virgin Islands, and Lebanon. The Treasury added that Ali alone provided huge cash payments to Hezbollah, in amounts as large as $1 million. And while he was apparently a major donor to Hezbollah, Ali Tajideen was no mere fundraiser, the Treasury stressed; he was also "a former Hezbollah commander" in Lebanon.[50]

Direct Hezbollah connections would be found in North America as well. According to a former senior Canadian intelligence official, by 1997, a covert network of 50 to 100 Hezbollah operatives—above and beyond the significantly larger pool of Hezbollah sympathizers and supporters—was "directly involved in Hezbollah activities in Canada." Just a week before his comments, the Canadian Security Intelligence Service informed a Canadian court that Hezbollah had established an "infrastructure" in Canada involving individuals who "receive and comply with direction from the Hezbollah leadership hierarchy in Lebanon."[51]

For example, consider the Hezbollah procurement network in Canada that was tied to the cigarette smuggling Hezbollah fundraising networks based in Charlotte and Detroit. The Hezbollah procurement agents in Canada operated under the command of Haj Hasan Hilu Laqis, Hezbollah's chief military procurement officer. Funded in part with money that Laqis sent from Lebanon, in addition to their own criminal activities in Canada (primarily credit card and banking scams), the cell obtained night vision goggles, global positioning systems, stun guns, naval equipment, nitrogen cutters, and laser rangefinders to be smuggled into Lebanon.[52]

One of the largest series of Hezbollah investigations in the United States—Operation *Bathwater*—involved no small number of bit criminal players who fell short of actual Hezbollah ties, but the FBI focused its effort on higher-level figures with ties to the group who were running the network. *Bathwater* grew out of a 1999 U.S. Secret Service case investigating what proved to be one of the largest credit card fraud schemes in the country at the time.[53] The investigation would turn up bank fraud, credit card fraud, mortgage fraud, cigarette smuggling, and much more, and would lead investigators to several suspects with impressive Hezbollah resumes.

Consider Elias Akhdar, who was involved in cigarette smuggling and other crimes in the United States. Akhdar first received military training from Islamic Amal and later engaged in Hezbollah military campaigns within Lebanon. Once he built up his criminal enterprise, prosecutors added, he contributed a portion of his illicit profits to Hezbollah.[54] One of Akhdar's partners, Salim Awde, threatened a potential government witness, saying, "If you are working with the FBI, I will blow you away."[55] Getting to the heart of the matter—the group's ties to Hezbollah—the government requested that Awde be detained without bond pending trial, stating that Hezbollah "would be motivated to assist Awde in fleeing the United States."[56] Prosecutors also opposed bail for another co-conspirator, Hassan Makki. One of the reasons Makki had joined the criminal conspiracy in the first place, prosecutors maintained, was to raise money for Hezbollah.[57] Once, he was stopped at the U.S.-Canada border with half a million dollars in checks and cash, some of which were meant for Hezbollah.[58] He solicited money for Hezbollah from other members of the smuggling ring and admitted to holding "membership/official status with Hezbollah." He would "telephone Sheiks in Lebanon and in Iran to clear criminal acts that he was committing." Materials seized in a raid of Makki's home included a photomontage of Hezbollah leaders and spiritual figures, militants in battle fatigues, funeral processions, celebrations, tanks, rockets, and firearms.[59]

Later, another one of Makki's criminal partners, Imad Hammoud, was publicly indicted along with 18 other individuals in a parallel criminal conspiracy case. Hammoud's criminal enterprise crisscrossed the United States from Michigan, California, Florida, Georgia, Illinois, Kentucky, Missouri, New York, and North Carolina to West Virginia (and, prosecutors stressed, points in between) and spanned the globe from the United States and Canada to Lebanon, Brazil, Paraguay, and China. Members obtained counterfeit cigarette stamps from Paraguay and Brazil, prompting Detroit JTTF agents to travel to the TBA to investigate that end of the conspiracy's activities. The U.S. agents' efforts enabled Brazilian authorities to apprehend a major counterfeit ring that manufactured not only false cigarette stamps, but also passports, national identity cards, and more.[60] Other elements in the conspiracy involved the import of counterfeit Viagra pills from China and Zig-Zag-branded cigarette paper from Indonesia.[61]

Individuals who were critical of Hezbollah were confronted and kicked out of Hammoud's crew. Hammoud and his partner, Hassan Makki, both charged a "resistance tax," an additional fee over the black market cost per carton of contraband cigarettes, which customers were told would go to Hezbollah. They also maintained collection boxes for Hezbollah donations and set aside additional chunks of money for transfer to the group in Lebanon.[62]

Another reason why prosecutors saw Hammoud as a priority target was his enterprise's close ties to senior Hezbollah leaders. One sign of this intimacy was the involvement of co-conspirator Hassan Ali Moussawi, the Beirut-based brother of Islamic Amal founder and then Hezbollah leader Hussein Moussawi.[63] Now a Hezbollah Member of Parliament in Lebanon, Hussein's brother was caught raising money for Hezbollah by selling counterfeit Viagra in the United States. According to prosecutors, some of the funds raised for Hezbollah were sent to Lebanon through Hassan Moussawi, "who had a close personal relationship with upper echelon Hezbollah officials." While some of the conspirators were driven by personal gain and supported Hezbollah as a bonus, "Al-Mosawi's participation with the conspiracy was expressly for the purpose of benefitting Hezbollah; and virtually all of the conspirators' collaboration with al-Mosawi was to garner favor from Hezbollah."[64] In September 2012, the Moussawis would be tied to a similar scheme producing another counterfeit drug back in Lebanon.[65]

Formal Hezbollah operatives continue to be actively involved in Hezbollah's criminal activities around the world, sometimes even in direct support of Hezbollah's military or terrorist activities. Several examples are discussed below.

The United States designated Hezbollah operative Ali Mohamad Saleh as both a narcotics kingpin (in 2011) and a global terrorist (in 2012). Saleh operated as a money launderer for an established criminal organization, but at the same time also served as "a key Hezbollah facilitator who has directed and coordinated Hezbollah activity in Colombia." A former Hezbollah fighter, Saleh led a Hezbollah support cell in Maicao, Colombia, that raised funds for the group and also served as a contact for Hezbollah's FRD. He maintained communication with suspected Hezbollah operatives around the world, including in Venezuela, Germany, Lebanon, and Saudi Arabia.[66]

In February 2015, a Hezbollah support network and its businesses in Nigeria were sanctioned by the Treasury Department. Members of the Nigeria cell included Mustapha Fawaz, a known member of the IJO; his brother Fouzi Fawaz, a Hezbollah FRD official; and Abdallah Tahini, a permanent representative of Hezbollah's FRD in Abuja, Nigeria.[67]

Four months later, the Treasury sanctioned Adham Tabaja, a Hezbollah member and majority owner of the Lebanon-based real estate and construction firm, Al-Inmaa Group for Tourism Works, which maintains direct ties to senior Hezbollah organizational elements, including the terrorists group's operational component, the Islamic Jihad.[68] With the assistance of Kassem Hejeij, a Lebanese businessman tied to Hezbollah, and Husayn Ali Faour, a known member of the IJO, their overseas terrorism unit, they exploited the firm's Iraqi subsidiaries to fund Hezbollah.

In November 2015, the Department of Treasury blacklisted three different Hezbollah procurement agents and their companies for obtaining various electronic equipment, unmanned aerial vehicles (UAVs), and accessories from the United States, Europe, Asia, and the Middle East for Hezbollah, and material for improvised explosive devices (IEDs) meant for the Houthis in Yemen. Included in this set of designations were Adel Mohamad Cherri, Fadi Hussein Serhan, and Ali Zeaiter. Zeaiter was previously designated in 2014 for using the China-based subsidiary of Stars Group Holding, Stars International Ltd., to acquire components for Hezbollah's UAVs and other projects.[69]

The U.S. Department of Treasury's relentless and successful campaign of targeting Hezbollah's global criminal networks and procurement operations seems to be striking a nerve with Hezbollah. Speaking in July 2015, Hezbollah Secretary-General Nasrallah felt the need to push back against recent U.S. designations of Lebanese businessmen and companies exposed for their criminal business ties and for operating for or on behalf of Hezbollah. Keeping the illicit nature of these activities secret is important for Hezbollah, as the group struggles to maintain public support against the backdrop of its support for the Assad regime next door in Syria. "We don't have projects for profit," Nasrallah insisted. "We don't deal or conduct business with companies. We don't have employees in Lebanese companies." Nasrallah complained that "America targets businessmen, companies, corporations, and tradesmen in Lebanon and tries to negatively affect them and destroy them," but insisted such actions "only hurt the Lebanese people and not Hezbollah. They work to destroy this country and the economy." Nasrallah openly conceded Hezbollah gets significant support from Tehran, and took a swipe at the United States at the same time: "We get our money from Iran and we publicly and proudly say that. We don't have WikiLeaks. We don't need to hide."[70]

But in fact, Hezbollah goes to great lengths to hide its criminal activities, both to preserve operational security and to shield itself from the political backlash it could incur if its less altruistic activities were exposed.

Hezbollah's Organized Criminal Facilitators

Beyond the group's useful idiots and more formal henchmen, Hezbollah also leverages the specialized skills of well-placed facilitators for a wide array of purposes. At the low end of the spectrum, idiots and henchmen leverage their relationship for small, personal gain. But on the high end, "super-facilitators" open doors that enable Hezbollah to engage in multimillion-dollar criminal enterprises.

Transnational criminal organizations (TCOs) pose a variety of threats, all of which are made possible by facilitators who sit at the crossroads of the licit and illicit worlds. These are the bankers, attorneys, brokers, and more who provide services to legitimate and criminal customers alike, hiding the criminal enterprise—money laundering, smuggling—behind the veneer of their otherwise legitimate business activities. At one end of the spectrum, criminals employ otherwise legitimate professionals as cover for their illicit activities, while at the other extreme they rely on "specialists" whose unique skill sets are subsumed into criminal networks.[71] In the case of Hezbollah, the group itself, as well as some of its most prominent supporters who are also involved in criminal activities, leverages connections with facilitators of all shapes and sizes to execute operations.

Consider an example along the Lebanese-Israeli border. In August 2012, 8 residents of Nazareth and Ghajar (a small town straddling the "Blue Line" demarcating Israel's northern border with Lebanon) were charged in an Israeli district court in connection with a June 2012 attempt to smuggle 20 kilograms of explosives into Israel. Israeli authorities foiled the plot, which included enough C4 to wage "a wave of serious attacks in Israel."[72] Israeli authorities said the explosives crossed the border with the help of a Ghajar resident, a criminal facilitator who was not only known to authorities primarily as a narcotics smuggler, but also for his ties to Hezbollah. The explosives were then passed along in a series of exchanges between smugglers who Shin Bet believed were smuggling drugs, not explosives.[73] The case is not exceptional, as Hezbollah has long been known to trade drugs to Israeli drug smugglers (both Jewish and Arab) in return for intelligence on Israeli army positions and more.[74]

But Hezbollah will work with criminals in other contexts and in other regions as well. In November 2014, Brazilian police reports revealed that Hezbollah helped a Brazilian prison gang, the First Capital Command (PCC), obtain weapons in exchange for the protection of prisoners of Lebanese origin detained in Brazil.[75] The same reports indicated that Lebanese traffickers tied to Hezbollah reportedly helped sell C4 explosives that the PCC allegedly stole in Paraguay.[76]

In June 2005, Ecuadorian police broke up a cocaine-smuggling operation run by members of the Lebanese diaspora community who were involved in Hezbollah fundraising activities.[77] From a Middle Eastern restaurant in northern Quito, Rady Zaiter ran a drug ring that brought in $1 million per shipment of cocaine smuggled to Europe and Asia. An internal police report confirmed that the operation was connected with Hezbollah, which reportedly received upwards of 70 percent of the ring's profits. Zaiter was arrested in Colombia, and six others were arrested in Ecuador for involvement in his network.[78] The operation reportedly also recruited airport officials to get their merchandise past airport security checkpoints.[79]

In April 2009, 17 people were arrested in the Dutch Caribbean island of Curacao for their involvement in a "Hezbollah-linked drug ring."[80] The suspects included locals, as well as individuals from Lebanon, Cuba, Venezuela, and Colombia. According to the Dutch prosecution service, the network shipped vast amounts of cocaine from the Caribbean to the Netherlands, Belgium, Spain, and Jordan. More were shipped from Venezuela to West Africa, and from there onward to the Netherlands, Lebanon, and Spain. Profits were invested in real estate in Colombia, Venezuela, and Lebanon. Prosecutors asserted, "The organization has international contacts with other criminal networks that financially supported Hezbollah in the Middle East. Large sums of money flooded into Lebanon, from where orders were placed for weapons that were to have been delivered from South America."[81]

Hezbollah has also sought to leverage support from Lebanese diaspora communities in Europe. According to a 1992 report to the U.S. Senate Committee on Foreign Relations, in the 1980s and early 1990s, Hezbollah leveraged the business acumen and connections of sympathetic Lebanese businessmen in Europe to build a "web of import-export companies in Western Europe as part of its dormant network," with the purpose of inserting "large quantities of explosive and related equipment into target countries."[82] In October 2009, German authorities arrested two Hezbollah members with ties to Nasrallah and other senior Hezbollah officials. According to the report, the two had trained in Hezbollah camps, but were not arrested for terrorist or militant activities, rather for cocaine trafficking on behalf of the group. The arrests followed a pitched investigation that began in May 2008, when customs agents at Frankfurt Airport seized €8.7 million in cash and another half million euros during a subsequent investigation at the suspect's apartment. To their surprise, investigators found traces of cocaine on the bills, along with the fingerprint of an infamous Dutch drug kingpin.[83]

In the Philippines, a Hezbollah network planning attacks in Asia and trying to infiltrate operatives into Israel plugged into a criminal document forgery network to procure false passports for its operatives. In 1999, two senior Hezbollah operatives, Pandu Yudhawinata and Abu al-Ful, traveled to Zamboanga City, Mindanao, to procure false passports, only to find that their usual contact was out of the country.[84] In his absence they turned to Talib Tanjil, another of Pandu's sources for false documents, who agreed to provide the necessary passports.[85] Like Pandu's usual contact, Tanjil, too, was more of a collaborator than a member of the Hezbollah network, though he was sympathetic to its mission.[86]

In one case typical of Hezbollah's modus operandi, Hezbollah supporters in the United States paid a ground services coordinator at Chicago's O'Hare International Airport to smuggle bulk cash and other packages (e.g., a cellular jammer, night vision goggles, and rifle scopes) onto airplanes, circumventing security inspections. Although he was fully aware where the packages were going, Riad Skaff was not a Hezbollah operative himself, just a well-placed facilitator. Prosecutors noted Skaff's conduct was "in essence that of a mercenary facilitating the smuggling of large amounts of cash and dangerous defense items for a fee."

So long as the facilitators can provide the service in question, Hezbollah operatives do not tend to have any qualms about working with them. Consider the U.S. case dubbed Operation *Phone Flash*, in which a Hezbollah procurement agent based out of Europe with extensive import-export businesses around the world attempted to procure small arms and antiaircraft missiles from individuals he believed were affiliated with the Philadelphia mafia (they were actually FBI informants and undercover agents).[87]

But the area where Hezbollah has relied most heavily on the services of criminal facilitators—some of whom are also affiliated with the terrorist group but are primarily illicit businessmen, some of whom are linked to the group only by virtue of the services they provide Hezbollah as a client—is drug trafficking. The UNODC estimates the international narcotics market to be a $435 billion a year industry. Hezbollah has had a finger in the business since at least the mid-1990s, when U.S. officials testified that "Hezbollah activities, which include narcotics and smuggling as well as terrorism, are supported in the tri-border area."[88] Fifteen years later, in March 2010, Anthony Placido, chief intelligence officer at the U.S. Drug Enforcement Administration (DEA), echoed the sentiment when he told the U.S. Congress: "Drug trafficking organizations based in the tri-border area have ties to radical Islamic organizations such as Hezbollah."[89] By 2011, the U.S. Department of State's Bureau of International Narcotics and Law Enforcement reported, "Lebanese citizens with links to [criminal and drug trafficking] organizations are a major presence among international drug trafficking and money laundering organizations in South America, and are tied into the highest levels of Colombian traffickers moving cocaine throughout the world."[90]

There is, however, no one model for Lebanese criminals involved in drugs and tied to Hezbollah. In the words of one investigator, "Some belong to families tied to Hezbollah, some just pay money to Hezbollah because it represents 'the cause' [of resistance against Israel and the West]. Some of what we see is Hezbollah actively involved in drugs [as a group], some is [sic] just Lebanese Shia involved in drugs who happen to be sympathetic to Hezbollah" and support the group but are neither members of nor directed by Hezbollah.[91] Wherever they fall on the Hezbollah spectrum, all of these characters interface extensively with criminal facilitation networks to carry out their business, the majority of whom are not themselves affiliated with the group beyond their working relationship.

One of the most significant takedowns to date was dubbed Operation *Titan*, a two-year investigation of a cocaine smuggling and money laundering operation in Colombia run by a Hezbollah figure named Chekry Harb, who used the alias, "Taliban." The case began when agents overheard Harb talking to cartel members on wiretaps targeting a Medellín, Colombia, cartel called La Oficina de Envigado. By June 2007, an undercover DEA agent met Harb in Bogotá and learned details about Harb's smuggling routes, including one that involved shipping cocaine to Jordan's Aqaba port and then smuggling it overland to Syria. At one point, Harb bragged to the undercover agent that he could get 950 kilos of drugs into Lebanon within hours, prompting the agent to casually suggest he must have Hezbollah connections in order to operate so freely there. According to the agent, Harb just smiled and nodded.[92]

In time, the undercover agent got close enough to the cartel to serve as one of its money launderers. The agent laundered some $20 million, enabling the DEA to follow the money and map out much of the cartel's operations. But before Harb could identify his Hezbollah contacts to the DEA undercover agent, the operation broke down, reportedly due to CIA interference.[93]

Over the course of the operation, Colombian and U.S. agents arrested more than 130 suspects, seized $23 million, deployed 370 wiretaps, and monitored 700,000 conversations. Harb's network reportedly paid Hezbollah 12 percent of its narco-income. "The profits from the sales of drugs went to finance Hezbollah," Gladys Sanchez, lead investigator for the special prosecutor's office in Bogotá, commented after the October 2008 arrests. Among those reportedly cooperating in this narcotics and money laundering enterprise were members of the Northern Valley Cartel, right-wing paramilitary groups, and the Revolutionary Armed Forces of Colombia (FARC).[94] Despite the collapse of the intelligence operation, Harb was ultimately convicted on drug trafficking and money laundering charges and placed, along with several of his associates, on the U.S. Treasury Department's list of Specially Designated Narcotics Traffickers.[95]

"Mark my words," warned Michael Braun, the former head of the DEA's Special Operations Division, in congressional testimony in the summer of 2009, "As we speak here today, operatives from al-Qaeda, Hezbollah, and, Hamas—perhaps others—are rubbing shoulders with Latin American and Mexican drug cartels, including the FARC, in West African countries and other places on the [African] Continent."[96] Together, these illicit actors have become what Braun describes as "hybrid terrorist organizations," groups that he described, from firsthand experience, as "meaner and uglier than anything law enforcement or militaries have ever faced."[97] And they have developed mutually beneficial relationships of convenience. In Africa, Braun testified, "they are frequenting the same seedy bars and sleazy brothels, and they are lodging in the same seamy hotels. And they are 'talking business.' They are sharing lessons learned, sharing critically important contacts and operational means and methods."[98]

A dramatic exposé of the full scope of both the African drug problem and Hezbollah's role in it appeared in January 2011, when the Treasury Department blacklisted Lebanese narcotics trafficker Ayman Joumaa along with an additional 9 individuals and 19 businesses involved in his drug trafficking and money laundering enterprise. With criminal associates and front companies in Colombia, Panama, Lebanon, Benin, and the DRC, Joumaa's organization was truly transnational. But while his drug ties to South America and his terrorist ties to Lebanon were unsurprising, the African narco-terrorism link—and its direct connection to the United States—was a wake-up call. An extensive DEA investigation revealed that Joumaa laundered as much as $200 million a month from the sale of cocaine in Europe and the Middle East through operations located in Lebanon, West Africa, Panama, and Colombia, through money exchange houses, bulk cash smuggling, and other schemes.[99]

Another prominent Hezbollah operative involved in laundering drug money through Africa is Oussama Salhab, according to U.S. officials. Court documents citing evidence collected during DEA investigations identify Salhab as "a Hizballah operative who, among other things, controls a network of money couriers who have transported millions of dollars in cash from West Africa to Lebanon." A close associate of his, Maroun Saade—who belongs to the Free Patriotic Movement, a Lebanese Christian group allied with Hezbollah—ran another major drug trafficking organization tied to Hezbollah in Africa. Along with several other defendants, Saade was indicted in February 2011, on narco-terrorism and other charges related to his alleged agreement to sell cocaine to people he believed were affiliated with the Taliban and to transport and distribute Taliban-owned heroin in West Africa.[100] In several instances, Saade reportedly bribed officials to close investigations into narcotics trafficking or to obtain the release of money couriers working for Salhab who were arrested by Togolese authorities for smuggling bulk currency across the border. In one case in July 2010, Salhab himself was arrested along with one of his couriers, and Saade paid the bribes for their release. In another case, Saade bribed officials to close a narcotics investigation into the activities of Imad Zbib, described in U.S. court documents as "a prominent Hezbollah representative in Togo." According to U.S. officials, Zbib is a close associate of Salhab and has transported loads of two to three metric tons of cocaine from South America to Togo, concealing the hauls in used cars purchased through lots he owns and transporting the drugs to Europe for sale.[101]

People like Ayman Joumaa and Oussama Salhab appear to have been in the right place at the right time. As demand for narcotics increased in Europe and the Middle East, South American cartels started looking for new routes to these growing markets. One route went to Europe through West Africa, and another through Syria and Lebanon. The Middle East route benefited from Hezbollah's ties to Iran and Venezuela. According to Lebanon's drug enforcement chief, Colonel Adel Mashmoushi, one way drugs were sent to Lebanon was on board the weekly Iran Air flight from Venezuela to Damascus and then overland by trucks to Lebanon. Confirming this route, U.S. officials stressed that "such an operation would be impossible without Hezbollah's involvement."[102]

Thanks to its support networks on both sides of the Atlantic, Hezbollah has a natural advantage vis-à-vis Latin American drug kingpins looking to transport their product to or through Africa. The only Spanish-speaking country on the African continent is Equatorial Guinea, and even the Portuguese dialect spoken in Angola differs from that spoken in Brazil. Long active as cargo aggregators, Lebanese merchants—including those with ties to Hezbollah—are well-placed to use their existing logistical machinery to facilitate the movement of other products. From a trafficking perspective, it matters little whether the products being moved are frozen chickens, cigarettes, gasoline, or drugs.[103]

Similarly, Hezbollah plugs in to purely criminal networks to launder the proceeds of its narcotics smuggling, moving beyond the close-knit sectarian and familial bonds that tend to define the group's own networks. Consider some of the money launderers engaged by Joumaa, himself a Lebanese Sunni Muslim, to process the proceeds of the drug sales he coordinated with Hezbollah. Luis Santiago Calle Quiros (Spanish-Peruvian dual national) and his Spanish wife, Maria Paloma Rodriguez Badillo, collected cash proceeds from narcotics sales (including Joumaa's) at their Madrid office.[104]

These two, along with several others in Madrid, operated as a cell of a larger money laundering network run by Colombian money launderer Isaac Perez Guberek Ravinovicz. Designated a narcotics kingpin for his narcotics money laundering, the Guberek network laundered hundreds of millions of dollars in drug money, including from the Joumaa drug trafficking organization. Announcing Guberek's kingpin designation and federal indictment, the U.S. government revealed that the actions were part of its "campaign to target narcotics networks, including [those] run by Joumaa whose global narcotics enterprises have stretched from South America to Africa and have benefited terrorist groups such as Hezbollah." For Joumaa and his Hezbollah partners, these were strictly business relationships into which they would enter regardless of ideology. The Guberek network offered excellent money laundering services, so neither Joumaa nor his Hezbollah partners were concerned that Guberek was Jewish, or that his son, Henry Guberek Grimberg, was a dual Colombian-Israeli national living in Israel. (Nor does it appear to have bothered the Gubereks that their client worked with Hezbollah, a terrorist group pledged to Israel's destruction.)[105]

Conclusion

Today, Hezbollah criminal enterprises—some formal, some informal—span the globe from drug-running and organized crime in the Balkans, to procuring false passports in Southwest Asia, to trafficking in stolen goods and trade-based money laundering in South America, and operating corporate front organizations and extorting financial support in Africa. U.S. officials believe "a substantial portion" of revenue raised by Middle Eastern terrorist groups in general comes from the $20 to $30 million brought in annually by the illicit scam industry in the United States alone.[106] Little, if any, of this is believed to be directed in any top-down form of command and control by Hezbollah leaders. But often, if one follows the loose threads of a Hezbollah-linked criminal investigation, connections emerge with at least one person with formal, often high-level Hezbollah connections, including people with Hezbollah military training or people serving somewhere in the diaspora as the local representative of senior Hezbollah officials like Secretary-General Nasrallah. Occasionally, a case will reveal the involvement of a senior Hezbollah leadership figure as well. In 2009, the U.S. Immigration and Customs Enforcement agency revealed a series of Hezbollah criminal schemes in the United States which ranged from stolen laptops, passports, and gaming consoles to selling stolen and counterfeit currency, procuring weapons, and a wide range of other types of material support, pointing to the wide range and scale of Hezbollah's criminal activities.[107] In those cases, senior Hezbollah officials from both the organization's public and covert branches played hands-on roles in the planning and execution of many of the criminal schemes.[108] According to U.S. investigators, following the loose threads in several recent cases led straight back to a Hezbollah representative in Iran, Abdallah Safieddine. That should not surprise, given that a United Nations criminal investigator found that Safieddine—a cousin and close associate of Nasrallah himself—is "considered one of the Hezbollah's top moneymen." His brother, Sayyed Hashem Safieddine, is a senior Hezbollah official sometimes mentioned as a possible successor to Nasrallah.[109]

Recently, Hezbollah's criminal activity has turned away from the petty crime of sympathizers. Today, Hezbollah seeks to leverage the services of organized criminal networks to facilitate its own entry into large-scale organized criminal enterprise. From narcotics to international used car sales, and from money laundering to procurement front companies, U.S. law enforcement now sees multiple cases where the Hezbollah-affiliated criminal operators are themselves acting as "super-facilitators" for a tremendously wide range of crimes and clients.

Consider again the case of Ayman Joumaa and the Lebanese Canadian Bank (LCB). In January 2011, Joumaa was designated a drug kingpin for "laundering as much as $200 million per month" through LCB and other channels tied to Hezbollah.[110] The following month, the U.S. Treasury designated LCB a primary money laundering or terror financing concern under Section 311 of the USA PATRIOT Act, effectively shutting down the bank. But several months later, in December 2011, the Southern District of New York unsealed a $483 million civil forfeiture action against LCB, which read like a cross between a criminal indictment and a suspense thriller. Since then, investigators around the world have followed the threads of the LCB case from the United States to Europe, Australia, Africa, and South America, arresting not only garden-variety drug traffickers and money launderers but key "super-facilitators" with ties to senior Hezbollah figures as well.[111] For example, one of the senior Hezbollah officials implicated in the LCB case: Abdallah Safieddine.[112]

Investigators have pursued so many Hezbollah-related cases that the group can no longer pretend to ignore them, and the press has taken notice. "The U.S. government is intensifying efforts to stop the flow of money to the terror group Hezbollah as officials work more closely with their European counterparts to stymie the organization's international network," the *Wall Street Journal* reported in December 2015. Citing the case of an American suspected money launderer in Colombia with ties to Hezbollah, and cases of suspects arrested in Lithuania, France, and the United States, the *Journal* cited U.S. officials describing how "Hezbollah uses a variety of means to make, move, and hide cash," aided by "super-facilitators—moneymen who help various criminal enterprises disguise the sources of their funding."[113]

The challenge Hezbollah poses has become global in nature. In this brief review we have discussed criminal operations in five continents; North and South America, Africa, Europe, and Asia. They include terrorist operations in Cyprus, the Philippines, and along the Lebanese-Israeli border; money laundering in the TBA, North Carolina, Belgium, Sierra Leone, and plenty more; trafficking of drugs and other contraband in Canada, Colombia, Germany, and Africa, to name a few; and waging war in Syria on behalf of the brutal Assad regime. The lesson to be learned here is that the Hezbollah threat has evolved and is no longer limited in scope to politics in the Middle East or their mission to wipe Israel from existence. It is a threat that must be recognized everywhere it is manifest, and challenged on a global scale. Hezbollah can no longer be seen as an Iranian proxy and terrorist organization alone; it is now a powerful military force, a globally lethal terrorist organization, and a complex criminal and money laundering network. To effectively thwart Hezbollah's anti-American and anti-Western operations will require a much more accurate understanding of how Hezbollah has evolved over the three decades since its establishment.

As the White House strategy to combat transnational organized crime warns, illicit networks of all kinds—drug traffickers, weapons dealers, terrorist groups, and more—share key "convergence points"—in particular people, places, or organizations that launder their money.[114] In the case of Hezbollah, super-facilitators—some with ties to the group, some just criminal middlemen—are attractive convergence points for their ability to quickly and efficiently move and launder massive illicit money flows. But they are also Hezbollah's Achilles' heel, exposing a group presenting itself as a noble "resistance" for the criminal enterprise it has become. This presents Hezbollah both with a grassroots legitimacy problem at home and exposes the group to law enforcement scrutiny around the world.

Notes

[1] Former FBI official, interview by author, Washington DC, February 23, 2011.

[2] Menelaos Hadjicostis, "Cyprus Court Convicts Hezbollah Member," *Associated Press*, March 21, 2013, available at <http://www.timesofisrael.com/cyprus-convicts-hezbollah-member-in-terror-plot/>.

[3] Matthew Levitt, "Hezbollah: Party of Fraud," *Foreign Affairs*, July 27, 2011, available at <https://www.foreignaffairs.com/articles/2011-07-27/hezbollah-party-fraud>.

[4] Sebastian Rotella, "Jungle Hub for World's Outlaws," *Los Angeles Times*, August 24, 1998, available at <http://articles.latimes.com/1998/aug/24/news/mn-16046>; U.S. Department of State, "Country Reports on Terrorism 2009," *Office of the Coordinator for Counterterrorism*, August 5, 2010, available at <http://www.state.gov/documents/organization/141114.pdf>.

[5] Universal Strategy Group Inc., *Directed study of Lebanese Hezbollah* (Nashville, TN: Universal Strategy Group, October 2010), 54.

[6] Marcelo Martinez Burgos and Alberto Nisman, "Argentina Investigations Unit of the Office of the Attorney General, Office of Criminal Investigations: AMIA Case," October 25, 2006, 311, available at <http://albertonisman.org/wp-content/uploads/2015/03/2006-Nisman-indict-AMIA-full-ENG_.pdf>.

[7] Buenos Aires, Argentina Judicial Branch, AMIA Indictment, Office of the National Federal Court No. 17, Criminal and Correctional Matters No. 9, Case No. 1156, March 5, 2003, 153.

[8] Universal Strategy Group Inc., *Directed study of Lebanese Hezbollah*.

[9] U.S. Department of State, "Country Reports on Terrorism 2009."

[10] U.S. Department of Treasury Press Center, "Treasury Targets Hizballah Fundraising Network in the Triple Frontier of Argentina, Brazil, and Paraguay," press release, December 6, 2006, available at <https://www.treasury.gov/press-center/press-releases/Pages/hp190.aspx>.

[11] U.S. Department of Treasury Press Center, "Treasury Designates Islamic Extremist, Two Companies Supporting Hizballah in Tri-Border Area," press release, June 10, 2004, available at <https://www.treasury.gov/press-center/press-releases/Pages/js1720.aspx>.

[12] Flemming Quist, "Is Africa Under Attack?" *United Nations Office on Drugs and Crime*, available at <http://www.un.org/en/events/tenstories/08/westafrica.shtml>.

[13] Directorate of Intelligence, "The Lebanese in sub-Saharan Africa," *Central Intelligence Agency*, January 1988, approved for release in June 1999; for more information, see Matthew Levitt, *Hezbollah: The Global Footprint of Lebanon's Party of God* (Washington, DC: Georgetown University Press, 2013), 246-276.

[14] "For a Few Dollar$ More: How al Qaeda Moved into the Diamond Trade," *Global Witness*, press release, April 17, 2003, available at <https://www.globalwitness.org/en/archive/few-dollar-more-how-al-qaeda-moved-diamond-trade/>.

[15] Lansana Gberie, "War and Peace in Sierra Leone: Diamonds, Corruption and the Lebanese Connection," Occasional Paper #6, *Partnership Africa Canada*, November 2002, available at <http://www.pacweb.org/Documents/diamonds_KP/6_War-Peace_sierraleone_Eng-Nov2002.pdf>.

[16] The Associated Press, "Hezbollah Extorting Funds from West Africa's Diamond Trade," *Haaretz* (Tel Aviv), June 30, 2004, available at <http://www.haaretz.com/hezbollah-extorting-funds-from-west-africa-s-diamond-trade-1.127125>.

[17] United States of America v. Mohamad Youssef Hammoud and Chawki Youssef Hammoud, "Appendix: Summary of Cigarette Purchases, after Allocation of Unidentified Wholesale Account Purchases," June 10, 2002; Kenneth Bell, "Hizballah Fundraising in the American Heartland," Policy #700, *The Washington Institute*, January 15, 2003, available at <http://www.washingtoninstitute.org/policy-analysis/view/hizballah-fundraising-in-the-american-heartland>.

[18] United States of America v. Mohamad Youssef Hammoud and Chawki Youssef Hammoud, United States District Court for the Western District of North Carolina, Charlotte Division, Docket No. 300-CR-147, June 2002.

[19] David Kaplan, "Homegrown Terrorists: How a Hezbollah Cell Made Millions in Sleepy Charlotte, N.C.," *U.S. News and World Report*, March 10, 2003, 30.

[20] United States of America v. Mohamad Youssef Hammoud and Chawki Youssef Hammoud, June 10, 2002, 1457.

[21] United States of America v. Mohamad Youssef Hammoud and Chawki Youssef Hammoud, May 29, 2002, 599.

[22] Ibid., 552-53.

[23] David Kaplan, "Homegrown Terrorists," 30.

[24] Robert Fromme and Rick Scwein, "Operation Smokescreen: A Successful Interagency Collaboration," *FBI Law Enforcement Bulletin* 76, no. 12 (December 2007): 20-25.

[25] U.S. Department of Justice, United States Attorney's Office, Central District of California, "Operation Bell Bottoms Targets Counterfeiting, Drug Operation in Los Angeles–Area Clothing Stores," press release, November 6, 2007.

[26] Statement of John C. Stedman, "Counterfeit Goods: Easy Cash for Criminals and Terrorists," hearing before the Committee on Homeland Security and Governmental Affairs, U.S. Senate, May 25, 2005.

[27] Ibid.

[28] U.S. Department of Treasury Press Center, "Treasury Designates Islamic Extremist."

[29] Ibid.

[30] Marc Perelman, "U.S. Hand Seen in Paraguay's Pursuit of Terrorism Suspect," *Forward*, January 17, 2003, available at <http://forward.com/news/9127/us-hand-seen-in-paraguay-s-pursuit-of-terrorism/>.

[31] Buenos Aires, Argentina Judicial Branch, AMIA Indictment, Office of the National Federal Court No. 17, Criminal and Correctional Matters No. 9, Case No. 1156, March 5, 2003, 156.

[32] U.S. Department of Treasury Press Center, "Treasury Designates Islamic Extremist."

[33] Ibid.

[34] Ibid.

[35] Buenos Aires, Argentina Judicial Branch, AMIA Indictment, Office of the National Federal Court No. 17, Criminal and Correctional Matters No. 9, Case No. 1156, March 5, 2003, 154.

[36] United States of America v. Bassam Gharib Makki, United States District Court, Southern District of Florida, Miami Division, CR 98-334-01, February 12, 1999, Affidavit of FBI Agent James Bernazzani Jr., 5.

[37] U.S. Department of Treasury Press Center, "Treasury Targets Hizballah Fundraising Network."

[38] Buenos Aires, Argentina Judicial Branch, AMIA Indictment, Office of the National Federal Court No. 17, Criminal and Correctional Matters No. 9, Case No. 1156, March 5, 2003, 190; Burgos and Nisman, "Argentina Investigations Unit," 239.

[39] Ibid., 314–15; Buenos Aires, Argentina Judicial Branch, AMIA Indictment, Office of the National Federal Court No. 17, Criminal and Correctional Matters No. 9, Case No. 1156, March 5, 2003, 154–55, 232.

[40] Larry Rohter, "South America Region under Watch for Signs of Terrorists," *New York Times*, December 15, 2002, available at <http://www.nytimes.com/2002/12/15/world/south-america-region-under-watch-for-signs-of-terrorists.html?pagewanted=all>; "Hizbullah Gets Hit with the Conviction of a Collector," *ABC Color* (Paraguay), September 30, 2007; "Drop a Course Link in the Drug Trade to the East," *ABC Color* (Paraguay), May 6, 2006.

[41] "Hizballah Captive in Brazil Participated in 1994 Argentine Attack," *Estadao* (Brazil), October 24, 2008.

[42] U.S. Department of Treasury Press Center, "Treasury Targets Hizballah Fundraising Network."

[43] Ibid.

[44] U.S. Department of Treasury Press Center, "Treasury Targets Hizballah Network in Africa," press release, May 27, 2009, available at <https://www.treasury.gov/press-center/press-releases/Pages/tg149.aspx>.

[45] U.S. intelligence official, interview by author, Washington, DC, July 2003.

[46] U.S. Department of Treasury Press Center, "Treasury Targets Hizballah Network in Africa."

[47] Ibid.

[48] "Coin in the Congo: The Moral Bankruptcy of the World Bank's Industrial Logging Model," *Greenpeace*, April 11, 2007, 34, available at <http://www.greenpeace.org.uk/files/pdfs/forests/carving_congo_3.pdf>.

[49] U.S. Department of Treasury Press Center, "Treasury Targets Hizballah Financial Network," press release, December 9, 2010, available at <https://www.treasury.gov/press-center/press-releases/Pages/tg997.aspx>.

[50] Ibid.

[51] Bill Gladstone, "Ex-official: Hezbollah Network of Operatives Active in Canada," *Jewish Telegraphic Agency*, April 9, 1997, available at <http://www.jta.org/1997/04/10/archive/ex-official-hezbollah-network-of-operatives-active-in-canada-2>.

[52] Matthew Levitt, "Hizballah's Canadian Procurement Network," in *Terror in the Peaceable Kingdom: Understanding and Addressing Violent Extremism in Canada*, ed. Daveed Gartenstein-Ross and Senator Linda Frum (Washington, DC: FDD Press, 2012).

[53] State of Ohio v. Ali Abdul Nasrallah, Court of Appeals of Ohio, Sixth District, Lucas County, No. L-99-1194, September 1, 2000; Michael D. Sallah, "Terror Links Sought in Credit Card Scam," *The Blade* (Toledo, OH), December 15, 2002, available at <http://www.toledoblade.com/World/2002/12/15/Terror-links-sought-in-credit-card-scam.html>.

[54] United States of America v. Elias Mohamad Akhdar, Government's Written Proffer; United States of America v. Elias Mohamad Akhdar et al., Indictment.

[55] PoliceOne Critical Alert, "Terrorism-Related Arrests in Va., Mich; One Officer Shot," *Policeone.com*, February 4, 2003, available at <http://www.policeone.com/terrorism/articles/59229-Terrorism-Related-Arrests-in-Va-Mich-One-Officer-Shot/>.

[56] Jim Irwin, "Cigarette Smugglers Fed Money to Hezbollah," *Associated Press*, February 5, 2003, available at <http://abcnews.go.com/US/story?id=96402>.

[57] United States of America v. Hassan Moussa Makki, Government's Written Proffer in Support of Its Request for Detention Pending Trial, Criminal No. 03-80079, United States District Court, Eastern District of Michigan, Southern Division, April 4, 2003; United States of America v. Elias Mohamad Akhdar et al., Indictment.

[58] David Shepardson, "Dearborn Resident Pleads Guilty to Terror Charges," *Detroit News*, September 19, 2003, available at <http://www.detnews.com/2003/metro/0309/19/a02-275736.htm>.

[59] United States of America v. Hassan Moussa Makki, Government's Written Proffer in Support of Its Request for Detention Pending Trial, Criminal No. 03-80079, United States District Court, Eastern District of Michigan, Southern Division, April 4, 2003; United States of America v. Elias Mohamad Akhdar et al., Indictment.

[60] Former FBI official, phone interview by author, September 25, 2009.

[61] United States of America v. Imad Mohamad-Musbah Hammoud et al., Indictment; former FBI official, phone interview by author, September 25, 2009; see also United States of America v. Tarek Makki, Plea Agreement, Criminal No. 03-80617, Eastern District of Michigan, Southern Division, October 5, 2005.

[62] United States of America v. Imad Mohamad-Musbah Hammoud et al., Indictment.

[63] Former FBI official, email interview by author, April 29, 2010.

[64] United States of America v. Imad Mohamad-Musbah Hammoud et al., Indictment.

[65] Roi Kais, "Hezbollah Funding Terror with Fake Medicine," *YnetNews*, September 10, 2012, available at <http://www.ynetnews.com/articles/0,7340,L-4279293,00.html>.

[66] U.S. Department of Treasury Press Center, "Treasury Targets Major Money Laundering Network Linked to Drug Trafficker Ayman Joumaa and a Key Hezbollah Supporter in South America," press release, June 27, 2012, available at <https://www.treasury.gov/press-center/press-releases/Pages/tg1624.aspx>.

[67] U.S. Department of Treasury Press Center, "Treasury Targets Hizballah Financial Network."

[68] U.S. Department of Treasury Press Center, "Treasury Sanctions Procurement Agents of Hizballah Front Company Based In Lebanon with Subsidiaries In The UAE And China," press release, July 10, 2014, available at <https://www.treasury.gov/press-center/press-releases/Pages/jl2562.aspx>.

[69] U.S. Department of Treasury Press Center, "Treasury Sanctions Hizballah Procurement Agents and Their Companies," press release, November 5, 2015, available at <https://www.treasury.gov/press-center/press-releases/Pages/jl0255.aspx>.

[70] "Sayyed Nasrallah: Before and After Nuclear Deal, U.S. Remains the Great Satan," *Al Manar*, July 25,

2015, available at <http://english.alarabiya.net/en/News/middle-east/2015/07/25/Hezbollah-U-S-remains-great-Satan-after-nuke-deal.html>.

[71] U.S. Strategy to Combat Transnational Organized Crime, Addressing Converging Threats to National Security, *White House*, July 2011, available at <https://www.whitehouse.gov/sites/default/files/Strategy_to_Combat_Transnational_Organized_Crime_July_2011.pdf>.

[72] Quoting ISA official.

[73] Yaakov Katz, Yaakon Lappin, and Ben Hartman, "Shin Bet nabs explosive-smuggling Israeli Arabs," *Jerusalem Post*, August, 9, 2012, available at <http://www.jpost.com/Defense/Shin-Bet-nabs-explosive-smuggling-Israeli-Arabs>.

[74] Levitt, *Hezbollah*, Chapter 8.

[75] "Hezbollah has ties to Brazil's Largest Criminal Gang; Group also Found Active in Peru," *Fox News Latino*, November 11, 2014, available at <http://latino.foxnews.com/latino/news/2014/11/11/hezbollah-has-ties-to-brazil-largest-criminal-gang-group-also-found-active-in/>.

[76] Ibid.

[77] "Ecuador Arrests: Suspects arrested in connection with drug smuggling to raise funds for Hezbollah," *Associated Press*, June 21, 2005, available at <http://www.aparchive.com/metadata/Ecuador-Arrests/9ef863bc50d869875ef0e894d8def7a4?query=nigeria&current=6&orderBy=Relevance&hits=9&referrer=search&search=%2Fsearch%3Fquery%3Dnigeria%26allFilters%3DArrests%3ASubject%2CCrime%3ASubject%2CSmuggling%3ASubject&allFilters=Arrests%3ASubject%2CCrime%3ASubject%2CSmuggling%3ASubject&productType=IncludedProducts&page=1&b=def7a4>.

[78] "Hezbollah drugs ring' broken up," *BBC News*, June 22, 2005, available at <http://news.bbc.co.uk/2/hi/americas/4117960.stm>.

[79] "Ecuador Arrests: Suspects arrested in connection with drug smuggling to raise funds for Hezbollah."

[80] Associated Press, "17 arrested on Curacao for involvement in Hezbollah-linked drug ring," *The Guardian*, April 29, 2009, available at <http://www.theguardian.com/world/2009/apr/29/curacao-caribbean-drug-ring-hezbollah>.

[81] "Police crackdown on Curacao drugs ring with ties to Hezbollah," *Agence-France Presse*, April 29, 2009, available at <https://now.mmedia.me/lb/en/archive/police_crackdown_on_curacao_drugs_ring_with_ties_to_hezbollah>.

[82] U.S. Senator John Kerry and U.S. Senator Hank Brown, "The BCCI Affair," a report to the *Committee on Foreign Relations United States Senate*, 102nd Congress, 2nd session, December 1992, 445.

[83] "Hezbollah will profit from the Cocaine Trade in Europe," *Der Spiegel*, January 9, 2010, available at <http://www.haaretz.com/news/report-hezbollah-funded-by-drug-trade-in-europe-1.261091>.

[84] Philippine intelligence report, "TIR on Pandu Yudhawinata," 9, cited in Levitt, *Hezbollah,* 132.

[85] Philippine intelligence report, "TIR on Pandu Yudhawinata," 9-10, cited in Levitt, *Hezbollah,* 132.

[86] Levitt, *Hezbollah*, 132.

[87] Levitt, *Hezbollah,* 336-337.

[88] Statement of Philip Wilcox, Jr., "International Terrorism in Latin America," testimony before the Committee on International Relations, U.S. House of Representatives, September 28, 1995.

[89] Statement of Anthony Placido, "Transnational Drug Enterprises (Part II): Threats to Global Stability and U.S. Policy Responses," before the Subcommittee on National Security and Foreign Affairs Committee on Oversight and Government Reform, U.S. House of Representatives, March 3, 2010.

[90] Bureau of International Narcotics and Law Enforcement, "2011 International Narcotics Control Strategy Report," *U.S. Department of State*, available at <http://www.state.gov/j/inl/rls/nrcrpt/2011/>.

[91] DEA Agents, interview by author, Washington, D.C., June 18, 2010.

[92] Jo Becker, "Beirut Bank Seen as a Hub of Hezbollah's Financing," *New York Times*, December 13, 2011, available at <http://www.nytimes.com/2011/12/14/world/middleeast/beirut-bank-seen-as-a-hub-of-hezbollahs-financing.html>.

[93] Ibid.

[94] Chris Kraul and Sebastian Rotella, "Drug Probe Finds Hezbollah Link," *Los Angeles Times*, October 22, 2008, available at <http://articles.latimes.com/2008/oct/22/world/fg-cocainering22>.

[95] U.S. Department of Treasury Press Center, "Treasury Designates Medellin Drug Lord Tied to Oficina de Envigado Organized Crime Group," press release, July 9, 2009, available at <https://www.treasury.gov/press-center/press-releases/Pages/tg201.aspx>.

[96] Prepared statement of Michael A. Braun, "Confronting Drug Trafficking in West Africa," hearing before the Subcommittee on African Affairs of the Committee on Foreign Relations of the U.S. Senate, June 23, 2009.

[97] Michael Braun, "Drug Trafficking and Middle Eastern Terrorism Groups: A Growing Nexus?" Policy #1392, *The Washington Institute*, July 25, 2008.

[98] Prepared statement of Michael A. Braun, "Confronting Drug Trafficking in West Africa."

[99] United States of America v. Lebanese Canadian Bank SAL et al., Verified Complaint, 11 CIV 9186,

United States District Court, Southern District of New York, December 15, 2011.

[100] Ibid.; United States v. Maroun Saade et al., Indictment, 11 Cr. 111 (NRB), United States District Court, Southern District of New York, February 8, 2011.

[101] United States of America v. Lebanese Canadian Bank SAL et al., Verified Complaint, 11 CIV 9186, United States District Court, Southern District of New York, December 15, 2011.

[102] Becker, "Beirut Bank Seen as a Hub of Hezbollah's Financing."

[103] West Africa expert, interview by author, Washington, D.C., November 28, 2011.

[104] U.S. Department of Treasury Press Center, "Treasury Targets Spanish Cell of Guberek Money Laundering Network," press release, October 29, 2013, available at <https://www.treasury.gov/press-center/press-releases/Pages/jl2193.aspx>.

[105] U.S. Department of Treasury Press Center, "Treasury Targets Major Money Laundering Network Operating Out of Colombia," press release, July 9, 2013, available at <https://www.treasury.gov/press-center/press-releases/Pages/jl2002.aspx>.

[106] John Mintz and Douglas Farah, "Small Scams Probed for Terror Ties: Muslim-Arab Stores Monitored as Part of post-Sept. 11 Inquiry," *Washington Post*, August 12, 2002, available at <http://www.washingtonpost.com/wp-dyn/articles/A6565-2002Aug11.html>.

[107] U.S. Immigration and Customs Enforcement, "Indictment charges 4 with conspiracy to support Hezbollah," press release, November 24, 2009, available at <https://www.fbi.gov/philadelphia/press-releases/2009/ph112409.htm>.

[108] Levitt, *Hezbollah,* 337-342.

[109] Michael Isikoff, "Bank Accused of Laundering Drug Money for Hezbollah," *NBC News*, February 10, 2011, available at <http://www.nbcnews.com/id/41512092/ns/world_news-mideast_n_africa/t/bank-accused-laundering-drug-money-hezbollah/#.VnxjC_krLIU>.

[110] U.S. Department of Treasury Press Center, "Treasury Targets Major Lebanese-Based Drug Trafficking and Money Laundering Network," press release, January 26, 2011, available at <https://www.treasury.gov/press-center/press-releases/Pages/tg1035.aspx>.

[111] Statement of David Asher, "A Dangerous Nexus: Terrorism, Crime and Corruption," testimony before the House Committee on Financial Services, Task Force to Investigate Terrorism Financing, U.S. House of Representatives, May 21, 2015.

[112] Isikoff, "Bank Accused of Laundering Drug Money for Hezbollah."

[113] Devin Barrett, "U.S. Intensifies Bid to Defund Hezbollah," *Wall Street Journal*, December 16, 2015, available at <http://www.wsj.com/articles/u-s-intensifies-bid-to-defund-hezbollah-1450312498>.

[114] U.S. Strategy to Combat Transnational Organized Crime, Addressing the Threats to National Security, *White House*, July 2011

8

Convergence in Criminalized States: The New Paradigm

Douglas Farah

The reality of convergence of different terrorist and other threat groups with transnational organized crime (TOC) networks is still much debated in U.S. policy circles. Field research and open-source judicial cases where different types of convergence can be observed and documented are dismissed as anecdotal by skeptics. The intelligence and law enforcement communities dedicate few of their resources to exploring how and if groups join forces, under what circumstances, and the conditions necessary for this to happen, stunting a conversation critical to our national strategic interests.

Those skeptical of the reality of convergence of multiple illicit actors fail to recognize how dramatically and profoundly the world has changed over the past 15 years. There has been a deep fracturing of the Westphalian consensus in many regions of the world, including Latin America, where there was at least lip service paid to a set of shared values and where definitional consensus was the norm in recent decades. This fracturing has a direct relationship to new and dangerous forms of convergence among transnational criminal actors, criminalized states, and terrorist groups.

This chapter focuses on Latin America, where for almost 50 years a broad consensus on democratic governance and rule of law was expressed through the Organization of American States (OAS). Now the OAS is far less relevant, replaced by a multiplicity of voices and forums that espouse the authoritarian values of the radical populist governments led by Venezuela, where political dissent is criminalized and the rule of law is replaced by presidential fiat.[1]

In Latin America, as in many other parts of the world, there is deep conflict and disagreement over the very definitions of the rule of law, democracy, freedom of expression, private enterprise, illicit action, and many other fundamentals of governance. For example, the eight members of the Bolivarian Alliance for the Peoples of Our America (ALBA), led by Venezuela, espouse as doctrine the replacement of representative democracy with "participatory and protagonist" democracy modeled on the Cuban system and "permanent confrontation between the Latin American and Caribbean peoples and imperialism" embodied by the United States.[2] The leaders have a radical populist caudillo corollary: the leaders of these governments perceive themselves to be the embodiment of the collective will of the people. Therefore their legitimacy cannot be challenged or their rule ended without violating that collective will and any opposition—in the political and judicial arenas, the media or civil society—must be eradicated. The bloc's economic model is "21st Century Socialism" and the abolition of the "neoliberal" free market model.

On the other side are nations that are more traditional liberal, representative democracies: Colombia, Chile, Uruguay, Peru, and Brazil; countries that, while suffering setbacks, have worked hard to establish a system, whether socialist or conservative, that embraces the concepts of alternation of power, an independent judiciary, rule of law, freedom of expression even when critical of the government, and open and free political debate, both in the political arena and the media. The two sides have little common ground on which to build a consensus, greatly limiting the number of countries where the United States can build military, political, and economic alliances with trusted partners.

It is important to note that criminalization is not used to define a traditional left/right divide, but rather to illustrate one that pits radical populist leaders with a desire to cling to power in perpetuity against more traditional liberal democratic principles of free and fair elections, rule of law, and the willingness to leave power when elections are lost. Numerous socialist governments in Latin America function within the liberal democratic norms (e.g., Uruguay, Brazil, and Chile), operating under the rule of law and eschewing the siren song of unlimited terms in office and unbridled power grabs. Nor can one excuse the behavior of many of the Bolivarian opponents when in power, many of whom presided over government-sanctioned corruption and mismanagement. Yet, they all left power when the rules of the game dictated they must.

But the sharp split, rather than being a disagreement over the rules of the game among groups playing on the same game on the same field, shows there is no longer any consensus on what game is being played.

It is the deep rending of this consensus that has opened the possibility of new forms of convergence that are real, yet little understood. In the ALBA bloc, one sees the emergence of a group of "criminalized states"—meaning states that use TOC groups and terrorist proxies as instruments of statecraft—has been a fundamental driver and facilitator of new and innovative dynamics of convergence.[3] This presents significant new policy challenges to traditional statecraft because there is currently little discussion of this divide or how to respond when a country or group of countries decides to play by a separate and incompatible set of rules.

By and large the criminal actions or support for terrorist organizations by leaders of governments from Russia to Venezuela are viewed by policymakers as anomalous actions of a few rogue actors within a state apparatus. In reality, I argue, in a growing number of nations these actions are policies of the state. The failure to distinguish rogue actors from criminalized states is a fundamental misunderstanding of the new dynamics of much of the world.

The Role of Criminalized States

A clear example, discussed in detail below, is the Venezuelan state's embrace of both major drug trafficking organizations and groups designated as terrorist entities as instruments of statecraft. Other countries in the ALBA bloc have taken similar actions.

In March 2015, Director of National Intelligence James R. Clapper testified to Congress that in Russia "the nexus among organized crime, state actors and business blurs the distinction between state policy and private gain."[4] This is true in much of Latin America now as well, where over the past decade the ALBA bloc of nations, joined by Argentina under the administrations of Cristina Fernández de Kirchner (2007-2015), have opted for a model that shares many of the same characteristics.

The ALBA bloc—the core of the Community of Latin American and Caribbean States (CELAC) and the Union of South American Nations (UNASUR)—embraces, as a policy of the state, alliances with TOC groups and terrorist groups such as the Revolutionary Armed Forces of Colombia (FARC)· Hezbollah, the Spanish ETA separatists, and others.[5]

Alliances of the ALBA states at the highest level with the FARC, ETA, Hezbollah, and drug cartels that move cocaine abroad are presented internally as part of a broad struggle against the United States, imperialism, and neoliberalism, all of which require resources.[6] Alliances with Iran and Hezbollah are portrayed as an extension of that struggle on a global scale. The embrace of Russia (both the state and state-affiliated TOC groups) and China as extra-regional actors corresponds to a set of shared values, both in terms of geopolitical interest and governance models. This is shown in the multiple visits of the heads of the Bolivarian states to Russia and China, where expressions of solidarity are a staple, as well as the constant visits by Russian and Chinese leaders to the Bolivarian states to strengthen military, political, and economic ties.[7]

In the construct of the new rules they are writing for their game, none of the state-sanctioned or state-sponsored activities with TOC groups or terrorist groups are illegal or questionable—they are revolutionary tools to obtain a strategic objective.

The cases of Walid Makled in Venezuela and many others demonstrate how, with state sponsorship of criminal activities and enormous amounts of personal and institutional corruption surrounding the revolutionary movements, vast amounts of money are diverted in the system for personal gain.[8] The well-being of the individual official is equated with the well-being of the state, and as Clapper noted regarding Russia, the lines among personal, state, and criminal are largely erased. The FARC's cocaine trafficking activities may fund revolutionary activities in several countries, but they also enrich many at the top of the revolutionary movement and provide lucrative employment to revolutionary cadres.

I have written in the past of the model of shadow facilitators, super fixers, and fixers, all of which are necessary in criminalized states as well. The illicit activities under consideration need a structure in which to operate and involve the movement of high-value commodities, and these commodity chains often span significant geographic space and require multiple steps, in multiple countries, to be successfully completed.

It is seldom possible for one individual, or even one criminal and/or terrorist group, to have the capacity to operate throughout this complex landscape, so they rely on specialized individuals, often primarily motivated by economic incentives rather than ideology, who can navigate specific links in that chain. These individuals are the crucial links or bridges between different worlds that do not often otherwise overlap.[9]

In the criminalized state model, senior officials, acting on behalf of the state, are the crucial links or bridges between different worlds that do not otherwise overlap. This removes all risk from the equation for the illicit actor, and with the removal of risk, new horizons open.

This is a crucial point in the new convergence model. TOC groups, terrorist organizations, and legitimate businessmen all crave stability, predictability, and risk minimization. When the state is a partner in an enterprise, whether licit or illicit, the primary benefit is the creation of an ideal business environment that enhances all three elements.

The clearest example is the support of Venezuela, both under Chávez and current President Nicolás Maduro, for the FARC, Hezbollah, Iran, ETA, and major drug trafficking organizations as a matter of state policy. FARC documents captured by the Colombian military in 2008 show that the Chávez government, with the direct participation of the president, loaned the FARC $300 million for new weapons and other equipment, money the FARC agreed to repay in cocaine shipments. In addition, the documents show, the discussions of the loan and other vital strategic support, including weapons shipments and the creation of front groups, took place in Fuerte Tiuna, the headquarters of both the military and intelligence structures in Caracas.[10] It would be difficult to have more direct evidence of direct state sponsorship than this.

There is also significant evidence from the same documents and other research that clearly shows the support of the governments of Bolivia, Ecuador, and Nicaragua for the same groups, and of the strong overall support the Bolivarian project enjoyed from the government of Fernández de Kirchner in Argentina.[11] This has opened up a new world of possibilities for both state and nonstate actors with shared business interests, ideological outlooks, and common enemies to meet in a stable, secure environment.

The Trust-Based ALBA Model

This chapter will focus on the Venezuelan government's well-documented ties to TOC groups and terrorist groups, and how those are leveraged with other complicit states into a sophisticated multifaceted structure for moving hundreds of millions of dollars in funds that have no traceable legitimate source.

Since 2008, different U.S. government entities have named more than a dozen senior Venezuelan officials as major drug traffickers, supporters of the FARC, supporters of Hezbollah, and for sanctions-busting on behalf of Iran. For example, in 2008, the U.S. Treasury Department's Office of Foreign Assets Control (OFAC) sanctioned Hugo Armando Carvajal, then Director of Venezuela's Military Intelligence Directorate, for protecting cocaine shipments, providing weapons to the FARC, and issuing Venezuelan identification cards to FARC members. It sanctioned Henry de Jesus Rangel, Director of the Venezuelan Intelligence Directorate, for cocaine trafficking and pushing "for greater cooperation between the Venezuelan government and the FARC." Also sanctioned was Ramón Emilio Rodriguez, then Venezuela's Minister of Interior, as the Venezuelan government's "main weapons contact for the FARC."[12]

In 2008, OFAC designated senior Venezuelan diplomats for facilitating the funding of Hezbollah. One of those designated, Ghazi Nasr al-Din, served as the *charge d'affaires* of the Venezuelan Embassy in Damascus, and then served in the Venezuelan Embassy in London. According to the OFAC statement in late January 2008, al-Din facilitated the travel of two Hezbollah representatives of the Lebanese Parliament to solicit donations and announce the opening of a Hezbollah-sponsored community center and office in Venezuela. The second individual, Fawzi Kan'an, is described as a Venezuela-based Hezbollah supporter and a "significant provider of financial support to Hezbollah." He met with senior Hezbollah officials in Lebanon to discuss operational issues, including possible kidnappings and terrorist attacks.[13]

This is not to say that everyone in the illicit world is allied with everyone. Rather, they have the opportunity, under the protection of different states, to find mutually beneficial relationships based on shared objectives, be they economic, political, or both.

At its core the ALBA alliance is possible because the basis for the alliance is a long shared history of armed insurrection that has built in a level of trust very difficult to replicate under other circumstances. Senior members of the Sandinista National Liberation Front (FSLN) in Nicaragua fought alongside the Farabundo Martí National Liberation Front (FMLN) insurgents in El Salvador; at the end of the Central American conflicts in the early 1990s, both groups provided significant support to the FARC in Colombia and developed a close relationship with Chávez and his inner circle in Venezuela.

Chávez in turn used that trust with this group and the trust born of his friendship with former Iranian President Ahmadinejad to bring Iran and Hezbollah into the inner circle, making the introductions to regional leaders based on their trust relationship with him. Daniel Ortega in Nicaragua, who built close ties to the ETA, the Red Brigade, the FARC, and Libyan terrorist groups during his revolutionary rule as president of Nicaragua from 1980 to 1990, brought different terrorist groups into the Bolivarian circle based on the trust relationships he had established. Ortega was the key link in bringing Iran, Hezbollah, and the Spanish ETA into relationships with Chávez.[14]

These long personal histories of shared revolutionary experiences, shared combat, and shared political vision make such organizations very difficult for outsiders to penetrate, be it for business purposes or intelligence gathering. It is this trust that is the glue of the alliance of the Bolivarian criminalized states and their nonstate allies.

An imperfect but useful analogy is that Venezuela and its allies operate as a singles bar or brothel, where different groups who share a common understanding of the outside world can meet, pick each other up, have a fling, share lessons learned, develop a relationship, and network with groups from across the globe.

The Role of PDVSA

One of the state vehicles where the convergence of state, criminal, and terrorist interests is visible is *Petróleos de Venezuela* (PDVSA), the Venezuelan state oil company that provides more than 95 percent of Venezuela's hard currency earnings. The president and his trusted inner circle directly control PDVSA. The U.S. government has sanctioned the company numerous times for money laundering, violating international sanctions on Iran, and supporting the FARC, a designated terrorist group.

A recent and revealing insight came to light in March 2015, when the U.S. Treasury Department's Financial Crimes Enforcement Network (FinCEN) designated the Banca Privada D'Andorra (BPA) a bank of "primary money laundering concern." The statement noted multiple billions of dollars' worth of money laundering activity in the bank, benefiting a host of actors, including Russian and Chinese organized crime and PDVSA. The Treasury notice reported that BPA and PDVSA set up shell companies and "complex financial products to siphon off funds from…PDVSA. BPA processed approximately $2 billion in…this money-laundering scheme."[15]

The figure of $2 billion siphoned off from PDVSA—which FinCEN documented in only a two-year period and is likely much higher—is stunning, given that the country is in an economic free-fall. PDVSA's oil production has fallen sharply in recent years due to lack of maintenance and aging equipment, while the price of crude oil has plummeted. In addition, the Venezuelan government has borrowed more than $49 billion from the Chinese government to stay afloat, and the loans are being paid back in oil. As a result, more half of Venezuela's oil exports generate no revenue, and only go to pay off the debt.[16] This indicates either a massive money laundering operation of funds that were not really originated in PDVSA's oil sales, or a significant looting of government funds at a time of deep economic crisis.

Yet, despite the company's failing economic state some of its main subsidiaries are exhibiting remarkable and irrational growth that sources involved in the subsidiaries say is directly tied to the large flow of illicit funds from the FARC, other cocaine cartels, and the personal fortunes of Venezuelan government officials.

According to sources with direct knowledge of events, one part of the PDVSA's illicit money movement structure centers around largely fictitious oil exports to the allied governments of Nicaragua and El Salvador and their respective oil companies, Albanisa and ALBA Petróleos, both directly controlled by senior government officials. In Nicaragua, this includes President Daniel Ortega, his son Laureano, and Francisco López, who is both the treasurer of the ruling Sandinista Party and deputy director of the oil company.

In El Salvador, the key architect is José Luis Merino, more commonly known by his *nom de guerre* Ramiro Vásquez, a former Communist Party urban commando who carried out a series of high-profile kidnappings for profit in order to raise money for his organization after the Salvadoran civil war ended in 1992. He also built a strong relationship with the FARC leadership and Chávez in Venezuela. For more than a decade, usually working through the Chávez government, Merino supplied the Colombian insurgents with sophisticated weapons, financial services, international solidarity events, and strategic advice. Documents captured by the Colombian military in May 2008 publicly exposed Merino's close relationship with the insurgency.[17]

According to the founding documents of Albanisa and ALBA Petróleos, PDVSA is to supply oil to each company under the umbrella of Petrcocaribe, an organization founded by Chávez in 2005 to deliver cheap oil to his allies in the Caribbean region. Under the standard PDVSA agreement with each country, half of the delivery is to be paid for at time of delivery, and half is to be repaid at concessionary prices with very low interest (1 to 2 percent) over 22 years.[18] The companies can sell the subsidized oil at market rates and in turn were to use the profits on education, health care, and other public benefits "to remake the fabric of society that was destroyed by neoliberalism."[19]

But there are multiple anomalies that indicate the structures are not what they appear to be on paper. PDVSA is the majority owner (60 percent) of both companies where it is illegal for a foreign company to invest in a state enterprise. Both Albanisa and ALBA Petróleos control sprawling business empires, investing heavily, at least on paper, in food production, financial institutions, airlines, think tanks, alternative energy projects, land acquisitions, bottled water, gasoline stations, and other activities. In both cases, interlocking networks of party stalwarts, senior government officials and PDVSA officials run the subsidiary agencies. And in both cases, many of the projects the oil companies claim to fund do not appear to exist.

The biggest anomaly however, is the true origin of the funds. In a rational world, the drastic cutbacks in Venezuelan cheap oil sales due to falling production and internal economic collapse, coupled with the fact that the price of oil remains below $50 a barrel, would lead to a significant drop in the budgets of Albanisa and ALBA Petróleos, as it has in the other Petrocaribe nations.

Barclay's Bank, in March 2015, estimated that Venezuelan oil going to Petrocaribe, including El Salvador and Nicaragua, had dropped by 50 percent from 2012 through 2014, from a total of 400,000 barrels per day (BPD) to 200,000 BPD. The bank expects the oil flow to be further reduced in 2015, from 200,000 BPD to 80,000.[20] Yet, both Central American oil companies have undergone remarkable and inexplicable economic booms that continue. During that period, ALBA Petróleos' revenues grew by 50 percent.

Ortega has publicly stated that Albanisa generates some $400 to $500 million that do not pass through any sort of government or private accountability, but basically operate as his own personal slush fund. The total budget of the Nicaraguan state is about $2.5 billion a year. This means that a sum totaling between 16 and 20 percent of the national budget is spent at the sole discretion of the president, with no accountability or oversight, and with few actual oil shipments to explain the origin of the funds.[21] The digital newspaper *Confidencial* in 2011 ran an exhaustive investigative series documenting how, over the course of five years, some $3 billion went to Ortega rather than the development projects the money was supposed to finance.[22]

ALBA Petróleos presents an even more dramatic growth in times of scarcity.

According to its own tax filings and public statements by company leaders, ALBA Petróleos started with $1 million in 2007, and its revenues grew to $862 million by the close of 2013. In 2014, according to Merino's public statements, the revenues were projected to exceed $1 billion because there are "investors from all over the world" lined up to invest with the anomalous and opaque company.[23]

In 2013 Merino acknowledged the company's unusual growth by stating that, "There are some who are afraid because ALBA Petróleos was born six or seven years ago with $1 million and it now has $400 million. Let me correct myself, $800 million, and we are trying to change the lives of Salvadorans."[24]

Figure 8.1. Alba Petróleos Income 2007-2014

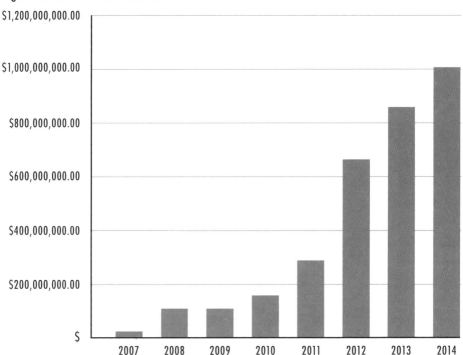

An analysis of possible oil revenues for ALBA Petróleos done by IBI Consultants found that, with oil at a price of $100 and all of the promised Venezuelan oil arriving on time and being sold at optimal prices, the company could generate about $220 million in revenues. Given the steady drop of oil prices as well as the steady decline of Venezuelan oil production, the ALBA Petróleos growth defies rational economic behavior.

ALBA Petróleos, in turn, has set up a sprawling network of companies dealing in everything from solar panels to think tanks, airlines to agriculture businesses, and financial institutions to petroleum refining. The legally required public filings on these companies have largely disappeared from the official registry since the election of Salvador Sánchez Cerén, a former guerrilla commander and close friend of Merino, in June 2014.

Figure 8.2. Price of a Barrel of Oil 2007 to August 2015

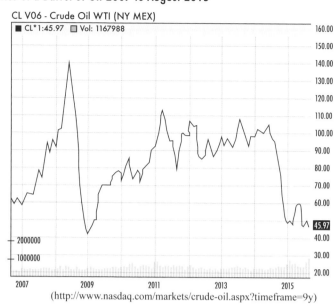

(http://www.nasdaq.com/markets/crude-oil.aspx?timeframe=9y)

This economically irrational behavior is generally seen when illicit money is being laundered into financial systems in order to justify its origin. In instances where the initial investments of the money can be traced, the businesses seldom generate profits and many are not operational. Rather, the projects exist on paper, hundreds of millions of dollars are legitimized as investments and the money can enter the financial stream to be repurposed for other uses.

For example, one of the signature programs of Albanisa in Nicaragua was to have been the construction of a large oil refinery to refine PDVSA crude. Then President Chávez and President Ortega jointly laid the ceremonial stone in 2007, announcing that the total cost of the project would be $6.6 billion.[25] Named the "Supreme Dream of Bolivar" (*Sueño Supremo de Bolívar*), the refinery received $32 million in start-up funding in 2008 and an additional $60 million over the following three years. In 2012, the program received an additional $141.2 million, and in 2013, an additional $200 million.[26]

Yet, when I visited the site in March 2015, all that was visible of the $437.2 million investment were a few abandoned oil tanks, a sagging chain-link fence and a few wooden watchtowers on the verge of toppling over. The only person at the site was a guard at the main gate.

ALBA Petróleos uses the same model of announcing mega projects that are never carried out, as a vehicle for justifying the flow of large sums of money. For example, in June 2013, Merino and his associates announced their participation in a $462 million renewable energy joint venture in El Salvador, an enterprise that has since produced no visible results.[27] Despite announcing a $25 million investment in a solar panel project in

October 2013 and promising production within 45 days, no start-up activity has taken place.[28] A promised $3 million for a water bottling business has produced no bottled water.

Some projects do come to fruition. ALBA Petróleos invested $60 million in a regional airline that flies three times a week to Guatemala and Costa Rica, as well as charter services used by government officials to fly to Cuba and Venezuela.[29] What is interesting is that the airline is several years behind schedule on expanding its routes and, according to local and regional analysts, is not likely to come close to making the ALBA Petróleos investment profitable.

The relationships of the leaders of ALBA Petróleos and Albanisa to PDVSA's opaque money movements, linked by law enforcement and intelligence officials to money laundering activities by the FARC, ETA, Hezbollah, and Russian organized crime, are not new. Both the Ortega structure and the Merino organization have documented, long-standing ties to these and other organizations, dating back, in some cases, to the Cold War. In that sense, both served as more traditional convergence centers for more than a decade.

But it is their control of state levers that has catapulted these networks into the ranks of important criminal structures and centers of convergence for TOC groups, terrorist organizations, and criminalized states. Now that these organizations operate with end-to-end state protection, they have access not only to government protection, but also to immunity from any form of public scrutiny or accountability for their actions.

In this model, with all sides agreeing that the movement of drug money or support for a terrorist organization is useful and not subject to sanction, such movement and support is legitimate and therefore, not illegal to those participating. This is the new game, with new rules, that is being played by criminalized states and their allies and is the fundamental chasm with the traditional models of state comportment.

Conclusion

The emergence of a bloc of states that has a fundamentally different understanding of the role of government and rule of law than those espousing a liberal democratic model has altered and accelerated the convergence of TOC groups and some terrorist entities. The members of these alternative blocs of nations, in Latin America and elsewhere, are not just states with significant corruption but governments that view partnership with TOC and terrorist groups as an important policy instrument, useful for both raising money and countering a common enemy—the United States. The means to the policy objective cannot be viewed internally as illegal if the state using those means views them as legitimate and even necessary.

The shattering of the once-shared Westphalian consensus has created a new and dangerous dynamic where groups traditionally viewed as illicit actors (e.g., the FARC, Hezbollah, and Russian organized crime groups) can seek not only sanctuary but state sponsorship for their ideological and economic objectives. The opportunities for convergence among groups in that alliance are far greater than when the groups operated independently and without the security of state protection.

The ALBA bloc relationships are trust-based relationships, especially those forged in conflict, and this shared history offers multiple advantages to those involved but may not always transfer to others. In the case of the ALBA alliance there has been significant fraying of the relationship between Merino's group and the Maduro government in Venezuela since the death of Chávez, because the relationships are less tested and resources are far scarcer.

Chávez was a trusted entity while Maduro is viewed as less so, but more than just mutual trust is required for the criminalized state model to function. There must be a political agenda that overlaps at least minimally; in this case, the view of the United States as the most dangerous threat they face because of unbridled American imperialism.

Finally, the relationship must be mutually beneficial. The ALBA states do not support the FARC solely out of ideological kinship. The state reaps enormous profits from the drug trade—law enforcement and intelligence officials in the United States, Colombia, and Central America believe that at least part of the unexplained flood of cash to the enterprises discussed above is FARC money—and keeps an armed ally or proxy force on call.[30] The FARC, in turn, operates its cocaine production with stability, protection, and enhanced profits, while also maintaining secure supply lines and logistical support. Other countries such as Ecuador, Bolivia, and Nicaragua have issued passports and other identification documents to members of the FARC, Hezbollah, and Iranian intelligence agents, providing a significant benefit in the form of state protection.

When these conditions are met—trust among actors, minimal common political agenda, and mutual benefit—the new convergence with the criminalized state as a key actor can take place.

This new reality makes the development of policies that recognize the distinction between rogue actors and criminalized states imperative. The policy must also acknowledge that the convergence among TOC groups and terror actors is real, and the strategic challenges these different convergences bring vary depending on the groups, their objectives, and their ties to states. A state that has chosen to play a new game with its own rules is quite different from a state playing by the Westphalian system but struggling with endemic violence, corruption, or governance issues. The first likely represents strategic challenges to the United States, while the second likely represents a potential ally. In an era of constrained resources and crisis, such distinctions are crucial.

Notes

[1] Among the most important of these organizations are: the Bolivarian Alliance for the Peoples of Our America (ALBA); the *Comunidad de Estados Latinoamericanos y Caribeños* (CELAC), founded in 2010 by then Venezuelan President Hugo Chávez, incorporating Latin American countries but expressly excluding the United States and Canada; and the *Unión de Naciones Sudamericanas* (UNASUR), founded in 2008.

[2] ALBA (*Alianza Bolivariana Para los Pueblos de Nuestra América* in Spanish) was founded in 2004 by Cuba and Venezuela, and currently has 11 members, including: Venezuela, Cuba, Nicaragua, Ecuador, Bolivia, Suriname, and several small Caribbean island nations. In keeping with its strong anti-U.S. stance, its two foreign observer nations are Iran and Syria. See Joel Hirst, "A Guide to ALBA," *Americas Quarterly, Council of the Americas*, available at <http://www.americasquarterly.org/hirst/article>.

[3] For a full discussion of criminalized states and their importance, see Douglas Farah, *Transnational Organized Crime, Terrorism, and Criminalized States in Latin America: An Emerging Tier-One National Security*

Priority (Carlisle, PA: Strategic Studies Institute, U.S. Army War College, August 2012), available at <http://www.strategicstudiesinstitute.army.mil/pubs/display.cfm?pubID=1117>.

[4] Statement of James R. Clapper, Director of National Intelligence, "Statement for the Record: Worldwide Threat Assessment of the U.S. Intelligence Community," before the House Appropriations Subcommittee on Defense, March 25, 2015, available at <http://docs.house.gov/meetings/AP/AP02/20150325/103200/HHRG-114-AP02-Wstate-ClapperJ-20150325.pdf>.

[5] The FARC, first designated a terrorist entity by the United States in 1997, is one of three groups in the world designated as both a major drug trafficking organization and terrorist group. The other two are the Taliban and *Sendero Luminoso* (Shining Path) in Peru. The FARC has also been designated a terrorist entity by the European Union since 2001. The Lebanese-based Hezbollah was designated a terrorist organization by the United States in 1997 and by the European Union in 2013. ETA was designated a terrorist organization by the United States in 1997. See: State Department, "Foreign Terrorist Organizations," *Bureau of Counterterrorism*, available at <http://www.state.gov/j/ct/rls/other/des/123085.htm>.

[6] Douglas Farah, *Transnational Organized Crime, Terrorism, and Criminalized States in Latin America: An Emerging Tier-One National Security Priority* (Carlisle, PA: Strategic Studies Institute, U.S. Army War College, August 2012).

[7] For the documentation of the significant Russian visits to the region and the increased rate for the visits see R. Evan Ellis, "Russian Engagement in Latin American and the Caribbean: Return to the 'Strategic Game' in a Complex-Interdependent Post-Cold War World?" *Strategic Studies Institute, U.S. Army War College*, April 24, 2015, available at <https://strategicstudiesinstitute.army.mil/index.cfm/articles/Russian-Engagement-in-Latin-America/2015/04/24>.

[8] Walid Makled is a Venezuelan who was formally designated a drug kingpin by the U.S. government. Arrested by Colombian police after he fled Venezuela, Makled was eventually extradited back to Venezuela. Preet Bharara, U.S. Attorney for the Southern District of New York, dubbed Makeld, also known as "The Turk," a "king among kingpins." While in Colombian custody, Makled gave multiple interviews and showed documents that he claimed showed he acquired control of one of Venezuela's main ports, as well as an airline used for cocaine trafficking, by paying millions of dollars in bribes to senior Venezuelan officials. According the U.S. indictment against him, Makled exported at least 10 tons of cocaine a month to the United States by keeping more than 40 Venezuelan generals and senior government officials on his payroll. "All my business associates are generals. The highest," Makled said. "I am telling you, we dispatched 300,000 kilos of coke. I couldn't have done it without the top of the government." Among the documents he presented in prison were checks of his cashed by senior generals and government officials and videos of what appear to be senior government officials in his home discussing cash transactions. For details of the case see José de Córdoba and Darcy Crowe, "U.S. Losing Big Drug Catch," *The Wall Street Journal*, April 1, 2011, available at <http://www.wsj.com/articles/SB10001424052748704471904576229001472736780>; U.S. Attorney Southern District of New York Public Affairs Office, "Manhattan U.S. Attorney Announces Indictment of one of World's Most Significant Narcotics Kingpins," press release, November 4, 2010, available at <https://www.justice.gov/archive/usao/nys/pressreleases/November10/makledwalidindictmentpr.pdf>; "Makled: Tengo suficientes pruebas sobre corrupción y narcotráfico para que intervengan a Venezuela," *NTN24 TV* (Colombia), April 11, 2011, available at <http://www.elestenoticias.com/makled-tengo-suficientes-pruebas-sobre-corrupcion-y-narcotrafico-para-que-intervengan-a-venezuela/>.

[9] For a fuller discussion of the Shadow Facilitator-Super-Fixer-Fixer model, see Douglas Farah, "Fixers, Super Fixers, and Shadow Facilitators: How Networks Connect," in *Convergence: Illicit Networks and National Security in the Age of Globalization*, ed. Michael Miklaucic and Jacqueline Brewer (Washington, DC: National Defense University Press, 2013).

[10] For an extensive look at the support of the FARC by Chávez, and a full explanation of captured FARC documents following the death of FARC Commander Raúl Reyes, see James L. Smith, "The FARC Files: Venezuela, Ecuador and the Secret Archive of 'Raul Reyes,'" *International Institute for Strategic Studies*, May 10, 2011. See also Farah, *Transnational Organized Crime, Terrorism, and Criminalized States in Latin America.*

[11] For an extensive look at the support of the FARC by Chávez, Ecuadoran President Rafael Correa, Bolivian President Evo Morales, and Nicaraguan President Daniel Ortega, see James L. Smith, "The FARC Files: Venezuela, Ecuador and the Secret Archives of 'Raul Reyes.'" See also Douglas Farah, "Into the Abyss: Bolivia Under Evo Morales and the MAS," *International Assessment and Strategy Center*, February 2009, available at <http://www.strategycenter.net/docLib/20090618_IASCIntoTheAbyss061709.pdf>; Douglas Farah and Glenn Simpson, "Ecuador at Risk: Drugs, Thugs, Guerrillas and the Citizens' Revolution," *International Assessment and Strategy Center*, February 2010, available at <http://www.strategycenter.net/docLib/20101214_EcuadorFINAL.pdf>.

[12] U.S. Treasury Department Press Center, "Treasury Targets Venezuelan Government Officials Support of the FARC," press release, September 12, 2008, available at <http://www.treasury.gov/press-center/press-releases/Pages/hp1132.aspx>._

[13] U.S. Treasury Department Press Center, "Treasury Targets Hizbullah in Venezuela," press release, June 18, 2008, available at <http://www.treasury.gov/press-center/press-releases/Pages/hp1036.aspx>.

[14] Farah, *Transnational Organized Crime, Terrorism, and Criminalized States in Latin America.*

[15] "FinCEN Names Banca Privada d'Andorra a Foreign Financial Institution of Primary Money Laundering Concern," *FinCEN*, March 10, 2015, available at <http://www.fincen.gov/news_room/nr/html/20150310.html>.

[16] Corina Pons and Alexandra Ulmer, "Source: China to Lend Venezuela $10 billion in coming months," *Reuters*, March 19, 2015, available at <http://www.reuters.com/article/venezuela-china-idUS-C2N0T700V20150319>.

[17] For a comprehensive look at captured FARC documents showing Merino's involvement see James L. Smith, "The FARC Files: Venezuela, Ecuador and the Secret Archives of 'Raul Reyes.'" See also Jose de Cordoba, "Chavez Ally May Have Aided Colombian Guerrillas: Emails Seem to Tie El Salvador Figure to Weapons Deal," *The Wall Street Journal*, August 28, 2008, available at <http://www.wsj.com/articles/SB121988527305078325>.

[18] Kaia Lai-Coha, "PetroCaribe: Chávez's Venturesome Solution to the Caribbean Oil Crisis," *Venzuelanalysis*, January 31, 2006, available at <http://venezuelanalysis.com/analysis/1592>.

[19] "José Luis Merino defiende a Alba Petróleos por ataques de ANEP," *Verdad Digital* (El Salvador) October 31, 2013, available at <http://verdaddigital.com/archivo/index.php/44-nacional/7093-merino-defiende-alba-petroleos-ante-cuestionamientos-de-anep>.

[20] Alejandro Arreaza and Alejandro Grisanti, "Venezuela: Reducing Generosity," *Barclay's Emerging Market Research*, March 25, 2015.

[21] For the most complete look at Albanisa, the Nicaraguan company run by Ortega and his family, see Carlos F. Chamorro and Carlos Salinas Maldonado, "Las cuentas secretas de Albanisa," *Confidencial* (Nicaragua), March 5, 2011, available at <http://www.confidencial.com.ni/articulo/3388/las-cuentas-secretas-de-albanisa>.

[22] See the series, Carlos Fernando Chamorro, "Los Petrodólares de Venezuela: Desvio de más de tres mil millones de dólares de la cooperación Venezolana a las arcas de Daniel Ortega," *Atavist*, available at <https://confidencial.atavist.com/los-petrodlares-de-venezuela918v4>.

[23] "José Luis Merino habla de las futuras inversions en ALBA," YouTube video, 2:04, posted by El Noticiero de Canal 6 (El Salvador), January 17, 2014, available at <https://www.youtube.com/watch?v=dxz0JkeJx6U>.

[24] "José Luis Merino defiende a Alba Petróleos por ataques de ANEP," *Verdad Digital* (El Salvador) October 31, 2013, available at <http://verdaddigital.com/archivo/index.php/44-nacional/7093-merino-defiende-alba-petroleos-ante-cuestionamientos-de-anep>.

[25] Gisella Canales Ewest, "Sueño Supremo de Bolivar costar US$6.620 millones," *La Prensa*, September 9, 2012.

[26] For a more comprehensive look at the refinery project and interesting graphics see José Denis Curz, "El Supremo Sueño de Bolívar no avanza," *La Prensa* (Nicaragua), March 25, 2013, available at <http://www.laprensa.com.ni/2013/03/25/lptv/1076870-el-supremo-sueno-de-bolivar-no-avanza>.

[27] "Anuncian acuerdo para transformar basura en energía," *Central America Data*, June 27, 2013, available at <http://www.centralamericadata.com/es/article/home/Anuncian_acuerdo_para_transformar_basura_en_energa>.

[28] Juan José Morales, "Alba iniciará producción de paneles solares con $25 mlls.," *El Diario de Hoy* (El Salvador), October 10, 2013, available at <http://www.elsalvador.com/articulo/negocios/alba-iniciara-produccion-paneles-solares-con-mlls-46214>.

[29] Alberto López, "Despega VECA Airlines con inyección de US$60 millones de ALBA petroleos de El Salvador," *Estrategia y Negocios*, April 9, 2015, available at <http://www.estrategiaynegocios.net/inicio/829348-330/despega-veca-airlines-con-inyecci%C3%B3n-de-us60-millones-de-alba-petr%C3%B3leos-el>.

[30] Law enforcement and intelligence officials, interviews by author, January to August 2015.

9

ISIL and the Goal of Organizational Survival

Jessica Stern

The Islamic State, or ISIL, has emerged as a manifestation of two trends: the convergence between transnational terrorist and criminal organizations, and the erosion of the post-Westphalian world order. The conventional wisdom has long held that criminal organizations are driven by the venal pursuit of wealth, while terrorist organizations are driven exclusively by ideological motives, and would presumably be repelled by the materialism of ordinary criminals.[1] ISIL, however, from its inception, has represented a merging of criminal activity for profit and terrorism. ISIL was founded as al-Qaeda in Iraq in 2004 by an infamous Jordanian thug known by his *nom de guerre*, Abu Mus'ab al-Zarqawi. Since its creation, the organization has changed names several times, but it has retained and expanded upon many of the innovations put in place by its founder, who used his experience as a gangster to create an unusually wealthy, vicious, and crude criminal/terrorist organization that also claims to be a state.

Abu Mus'ab al-Zarqawi was a high school dropout, known around his hometown of Zarqa, Jordan, as a boozer and a brawler, certainly not as a pious man, let alone a fundamentalist. He was reportedly arrested 37 times for his involvement in violent crime and drug dealing, and was well known to the local police. His mother encouraged him to study Islam, hoping to rescue her son from a life of crime, but studying religion did not help Zarqawi find peace. The Islam that he discovered was an unusually violent one. His jihad had nothing to do with elevating himself spiritually, and everything to do with justifying his preferred lifestyle of burglary and brutality.

Like many terrorist operatives, Zarqawi's jihadi views were solidified in prison. He was incarcerated in Swaqa Prison from 1993 to 1999, during which time he studied Islam under the tutelage of his fellow inmate and mentor, Sheikh Abu Muhammad al-Maqdisi.[2] Maqdisi is one of the architects of jihadi Salafism, an ideology based on the principle that any government that does not rule through a strict interpretation of Shariah is an infidel regime that must be violently opposed.[3]

Prison was transformative for Zarqawi. It was there that he made the transition from a small-time criminal to terrorist mastermind and leader.[4] After he was released from prison in 1999, he traveled first to Pakistan and from there to Afghanistan, where he met Osama bin Laden.[5] In the days prior to September 11, 2001, bin Laden repeatedly sought *bayah* from Zarqawi, a religiously binding oath of allegiance. Until 2004, the former gangster refused.[6]

Ironically, it was the invasion of Iraq that pushed Zarqawi into an alliance with bin Laden and led to al-Qaeda's enduring presence in Iraq, and ultimately, the rise of ISIL.[7] Soon after allied forces overthrew Saddam Hussein in 2003, the Coalition Provisional Authority (CPA) disbanded Saddam Hussein's military, security, and intelligence forces.[8] More than 100,000 Sunni Baathists were removed from the government and military, leaving them unemployed, angry, and, at least for the military personnel, armed.[9] Lieutenant General Jay Garner warned that the policy rendered a large number of educated and experienced Iraqis "potential recruits for the nascent insurgency."[10]

One particularly important function impacted by the purge was the Iraqi border patrol. The weakened force provided little resistance to the flow of foreign fighters into the country.[11] Those weak borders also facilitated crime. Thus, the predecessor organizations to ISIL were both fomenting and exploiting weaknesses in the state system.

Some Baathist insurgents ended up in U.S.-run detention facilities, such as Camp Bucca, where the jihadists and former military personnel forged closer ties. Louise Shelley deemed prisons "corporate headquarters for crime-terror interactions," citing that of the 25 members of ISIL's senior leadership, 17 were incarcerated in American facilities between 2004 and 2011.[12]

An estimated one-third of Abu Bakr al-Baghdadi's deputies served in the Iraqi military under Saddam.[13] For example, Haji Bakr (a *nom de guerre* for Samir Abd Muhammad al-Khlifawi), a former Baathist military leader, became a top strategist for ISIL. During his 2006 to 2008 detainment in Camp Bucca and Abu Ghraib, Bakr established connections with other former high-ranking members of the military and intelligence sector, with whom he developed ISIL's master plan for the takeover of Iraq and Syria.[14]

In June 2006, Abu Musab al-Zarqawi was killed in a U.S. air strike. ISIL acquired expertise, knowledge, and inspiration from this erstwhile gangster, leading it to form a hybrid criminal organization, proto-state, and apocalyptic cult that flaunts its brutality over social media. Soon after Zarqawi was killed, al-Qaeda in Iraq (AQI) changed its name to the Islamic State of Iraq (ISI). Thus, in 2006, the organization was already beginning to see itself as an alternative to a state. In 2010, Abu Bakr al-Baghdadi (a *nom de guerre* for Ibrahim Awwad Ibrahim Ali al-Badri al-Samarrai) took over the leadership of ISI, which would eventually become known as ISIL, and then the Islamic State. In the remainder of this chapter, I will use the acronym ISIL.

The Rise of ISIL

In early June 2014, ISIL captured Mosul, a city of 1.5 million people and the site of Iraq's largest dam.[15] Because it was so dangerous for journalists and other noncombatants to operate in areas afflicted by insurgency, the victory seemed to come out of nowhere. ISIL was reported to have stolen $429 million from Mosul's central bank and millions more from other banks in the city.[16] These reports were later denied by the Iraqi government. But the denials—sourced to Iraqi bankers and officials whose own businesses rested on their ability to secure funds and the country's economy—were not any more credible than the original reports.[17]

As a result of its criminal activities, ISIL quickly became the richest terrorist organization in the world, worth an estimated $2 billion by mid-2014.[18] Unlike al-Qaeda and many other terrorist groups, which rely on external sources of funding, including "charitable" donations, much of ISIL's revenue was generated internally from taxes on local populations, looting, the sale of antiquities, and oil smuggling.[19] By June 2015, U.S. officials ranked extortion as ISIL's primary source of income.[20]

Human Trafficking

By early August 2014, ISIL had advanced into the northern Iraqi town of Sinjar, which had a large population of minorities, including a Kurdish-speaking population known as the Yazidis.[21] ISIL believes the Yazidis are devil worshippers and constructed a religious justification to kill all the men and enslave the women and children.[22] ISIL hunted and then surrounded the Yazidis as they fled to Iraq's Mount Sinjar with no food and no water. The United States, United Kingdom, and France made emergency airdrops of food and water to the Yazidi refugees to forestall what the UN referred to as a threatened genocide.[23] Sinjar was retaken in November 2015.[24]

Only later would it become clear that ISIL's aim was to set up a sex trafficking operation. Thousands of Yazidis were abducted. Matthew Barber, a scholar of Yazidi history at the University of Chicago, estimates that as many as 7,000 women were taken captive in August 2014.[25] "The offensive on the mountain was as much a sexual conquest as it was for territorial gain," Barber said.[26] The operation to enslave the girls was entirely preplanned, according to both Human Rights Watch and Amnesty International.[27] The girls are sold to ISIL operatives as sex slaves, and have also been sold in Jordan and the United Arab Emirates.[28]

The sexual enslavement of Yazidi girls would appear to fulfill a number of ISIL's needs. Premarital sex is forbidden by Islamic law. In societies where young men cannot afford to marry, the availability of sex, even with an unwilling victim, is part of ISIL's recruitment drive, enabling ISIL to outcompete other jihadi groups. According to ISIL's "scholars," the renunciation of slavery created a problem for Muslim men. They argue that in the absence of sex slaves, "the shari'a [shariah] alternative to marriage is not available, so a man who cannot afford marriage to a free woman finds himself surrounded by temptation towards sin."[29] Thus, by ISIL's logic, sexual slavery is a way to avoid the sin of premarital sex or adultery, since premarital or extramarital rape of a slave does not count as sex. One of ISIL's essays on sexual slavery, "The Revival of Slavery Before the Hour," explains that polytheist and pagan women can and should be enslaved. Indeed, their enslavement is one of the "signs of the hour as well as one of the causes of al Malhalah al Kubra," the Final Battle that will take place in Dabiq.[30]

ISIL is also abducting children to use as child soldiers. Children of ethnic minorities, particularly the Kurds and Yazidis, have been kidnapped and forced to join ISIL. According to the Syrian Observatory for Human Rights, in one case, more than 600 Kurdish students were kidnapped on their way home from taking exams in Aleppo. Their captors gave the boys an Islamic "education," encouraging the children to join the jihad, showing them videos of beheadings and suicide attacks.[31]

Smuggling both Syrian and African migrants from Libya to Europe has become a big business. ISIL, along with other militias active in Libya, has been taxing this human trafficking business, which at the time of this writing generated an estimated $320 million per year. These funds have reportedly helped to finance ISIL's provinces in Libya, called Barqa, Tarabulus, and Fezzan.[32] The Global Initiative Against Transnational Organized Crime estimated in May 2015 that the value of migrant smuggling in Libya had risen from $8 to $20 million in 2010 to $255 to $323 million in 2014.[33] In recent years, Syrian refugees en route to Europe have shifted their routes through Africa into the Sinai and Libya, where ISIL has a foothold and can levy a lucrative tax on their transport.

ISIL has capitalized on a business model previously employed by Congolese rebels, in that it creates a refugee crisis and subsequently services the refugee camps.[34] By attacking civilians in Iraq, Syria, and in Jordanian and Lebanese refugee camps, ISIL forces them to flee, thus facilitating the migrant-trade system. This criminal scheme finances ISIL's operations while aiding the expansion of its so-called "Caliphate" throughout North Africa.

Kidnapping for Ransom

In 2014 alone, ISIL raised around $20 million by ransoming Western hostages, according to David Cohen, who served as the U.S. Treasury's Under Secretary for Terrorism and Financial Intelligence from 2011 to 2015.[35] While the United States and the United Kingdom have government policies that forbid paying ransoms, many other countries, including some in Europe, have paid to have hostages released.[36]

Oil Sales

Between June and late 2014, when the coalition bombing campaign against ISIL began to impact its oil production, ISIL was earning around $1 million a day from oil production alone, according to Cohen. Some of the oil was refined using makeshift facilities, and some was sold as crude. Matthew Levitt reported that facilities located in ISIL-controlled territory were producing an estimated 80,000 barrels per day (BPD).[37] In late 2014, Louise Shelley reported that ISIL was selling crude at a discount (around $20 to $35 per barrel) to truckers or middlemen. Smugglers were paying about $5,000 in bribes at checkpoints to export the crude out of ISIL-controlled territory.[38]

In May 2015, Celina Realuyo testified before the House Committee on Financial Services that ISIL's oil production had gone significantly down, but was still estimated at 48,000 BPD (44,000 BPD from Syrian wells and 4,000 BPD from Iraqi ones). Realuyo reported that ISIL's putative enemies were rumored to be its customers—the Assad regime, Turks, and Iraqi Kurds.[39] But Charles Lister points out that ISIL does not just trade in oil; it relies on it for its own needs—not only to run its generators, vehicles, and bakeries, but also to establish a system of dependency in which civilians under its governance rely on its ability to provide inexpensive oil.[40]

After the coalition started striking ISIL's oil production facilities, the U.S. Treasury Department stated that ISIL began to shift its operations away from oil smuggling in favor of extortion.[41] Some reports challenge the assertion that coalition air strikes have significantly disrupted ISIL's oil production. In September 2015, some 50 intelligence officers claimed that their intelligence had been politicized, and that U.S. bombing of ISIL's oil facilities had damaged the economy less than had been reported by CENTCOM.[42] In October 2015, oil traders and engineers on the ground approximated that crude production in militant-held territory amounted to 34,000 to 40,000 BPD, generating revenues of $1.5 million per day.[43]

The coalition was initially wary of attacking oil transport vehicles for fear of inflicting civilian casualties.[44] But by mid-November 2015, the United States intensified its efforts to thwart ISIL's oil sales, targeting the tanker trucks used to smuggle crude produced in Syria, warning the drivers to leave their vehicles behind in an effort to minimize collateral damage. In the first week of the new campaign, U.S. warplanes struck nearly 500 tanker trucks, reportedly destroying half of ISIL's transport vehicles.[45] Truckers consequently became reluctant to approach the oil fields, which reportedly resulted in a significant decrease in trade.[46]

Theft of Equipment

ISIL's military conquests have allowed it to acquire territory, infrastructure, and military equipment.[47] Prime Minister Haider al-Abadi admitted in May 2015 that the Iraqi Armed Forces had lost U.S.-made weapons to ISIL on multiple occasions, reporting that they "lost 2,300 Humvees in Mosul alone," worth over $1 billion.[48]

Antiquities Trafficking

ISIL has dismayed the international community with its destruction and pillaging of ancient heritage sites throughout Iraq and Syria. In late 2014, an archaeologist from Iraq's Department of Antiquities estimated that ISIL controlled more than one-third of the country's 12,000 historical sites.[49] The illicit sale of artifacts is a relatively low-risk crime, as European art markets are not typically closely monitored by police. Even for offenders who are caught, the likelihood of serious punishment is low. In addition to being low-risk, it is also highly lucrative.[50]

"The Taliban learned to finance their terror through opium. They don't have opium in the Middle East. What they do have is antiquities. It's the cash crop," Matthew Bogdanos, a U.S. Marine colonel and assistant district attorney in New York City with experience investigating the illegal antiquities trade in Iraq, explains.[51] ISIL turns a profit both by selling stolen artifacts and by imposing taxes on traffickers who move the items through militant-controlled territory.[52] U.S. officials estimate that the value of ISIL's illicit antiquities trade exceeds $100 million per year.[53] ISIL's artifact trafficking is aided by criminally inclined middlemen with years of experience stealing, looting, and smuggling under Saddam's regime.[54] ISIL has tapped into long-standing black markets and smuggling routes, making traditional instruments for fighting terrorist financing far less useful.[55]

Stolen pieces often end up in the hands of wealthy art collectors in the West, many of whom are willing to turn a blind eye to the unlawful origins of their purchases.[56] The value of declared artifacts imported from Syria into the United States increased by 134 percent between 2013 and early 2015, and as of February 2015, almost 100 Syrian antiquities had been smuggled into England, some of them worth hundreds of thousands of dollars.[57]

The widespread pillaging of the cradle of civilization provides ISIL with a valuable opportunity to demonstrate its opposition to the worshipping of idols through various propaganda outlets. However, ISIL's methodical system of organized looting and marketing of stolen goods suggests that the group's goals may revolve around income rather than ideology.[58] On camera, ISIL destroys relics with sledgehammers and drills, but behind the scenes the group has reportedly hired professional contractors to execute its excavations as efficiently as possible.[59] At some sites ISIL outsources digging jobs to locals and collects a 20 to 50 percent Islamic *khums* tax, historically levied on the spoils of war, on artifacts that are excavated.[60]

In his late-2014 testimony before the House Committee on Financial Services, Matthew Levitt named the trafficking of looted artifacts as ISIL's second-highest source of income and a testament to the organization's perpetually diversifying criminal economy.[61] Levitt reported that despite the decline in oil profits, "ISIL is not likely to just sit there and watch itself wither on the vine. It has shown it can adapt...for a long time they were not focusing on [antiquities], and now it's become very important to them."[62]

The supply of antiquities in ISIL-controlled territory is finite, and, though lucrative now, ISIL's illicit trafficking is by no means a sustainable business model. Daveed Gartenstein-Ross of the Foundation for Defense of Democracies has proposed that cutting out the former Baathist middlemen is the key to curtailing ISIL's black market antiquities trade. He posits that cracking down on these middlemen by arresting or incarcerating them would result in a dearth of facilitators for extortion, theft, and looting, thereby cutting off a large portion of ISIL's cash flow.[63]

Drug Smuggling

In Syria, production of the synthetic stimulant Captagon has generated hundreds of millions of dollars per year in revenues.[64] The pills are produced cheaply in Syria and smuggled via Lebanon into the Gulf, where they are sold for up to $20 each.[65] Syrian fighters have admitted to taking Captagon tablets before battle to energize themselves for long periods of strenuous fighting.[66]

Smaller-Scale Crime

While the majority of ISIL's funds come from organized operations such as human trafficking, kidnapping for ransom, oil sales, and theft, the group has managed to finance itself through smaller-scale crime as well.

Cigarette smuggling has proven to be a low-risk and profitable business venture for ISIL. A cigarette smuggler in Raqqa who was apprehended by militants reported, "[ISIL] never beat me at all, they never tortured me. They just saw us as a profitable catch. They didn't even try to learn how we smuggled in the cigarettes or who the big traders were."[67]

ISIL banned smoking in areas under its control and has punished offenders with large fines, beatings, and in some cases, executions.[68] Militants posted billboards in Raqqa suggesting that as an alternative to smoking, citizens should chew on branches from the *arak* tree, in imitation of the Prophet Mohammad's reported practice.[69] However, there is mounting evidence to suggest that ISIL's cigarette ban has more to do with money than morality. Iraqi civilians in Mosul stated in June 2015 that ISIL actually controls the black market, publicly outlawing smoking in order to partner secretly with smugglers, who can then sell illegal cigarettes at an inflated price.[70] A pack of cigarettes reportedly costs three times as much in ISIL-controlled territory as elsewhere in Syria, and tobacco for flavored water pipes costs almost seven times as much.[71] Smugglers import cigarettes from Turkey to Syria to Iraq, paying 20 cents per pack and working with corrupt ISIL officials to resell them at quadruple the price on the black market in militant-held territory.[72]

ISIL has reportedly advertised car and real estate auctions in Iraq and Syria selling property seized from "apostates and disbelievers."[73] ISIL recruits also fundraise through the sale of passports. Before crossing the border from Turkey to Syria, foreign fighters en route to join ISIL often sell their travel documents for thousands of dollars, using the profits to finance their jihadi ventures.[74] Financing major attacks through petty crime is not exclusive to ISIL. According to Louise Shelley, the planners of the January attack on the *Charlie Hebdo* offices in Paris funded their operation with criminal operations such as the sale of counterfeit Nikes.[75]

ISIL's Staying Power

In war-torn Iraq and Syria, ISIL provides citizens with jobs and services and, through its harsh law enforcement, has projected a sense of order after years of lawlessness. In this sense, ISIL fills an important void for civilians. ISIL is clearly hoping to stay in the communities it occupies. For example, it indoctrinates children and their teachers by crafting extremist curricula intended to secure its popularity among the next generation of citizens.[76] It also provides employment to its supporters as a way of economically incentivizing capitulation among inhabitants of ISIL-controlled territory. A laborer from Raqqa employed by ISIL described living under the militant group's rule; "As a way of life, people got used to it. It is not our life, all the violence and fighting and death, but they got rid of the tyranny of the Arab rulers."[77]

ISIL enhances its staying power through a practice that Hassan Abu Hanieh, an expert on Islamism, calls "geographic cleansing," in which minorities and enemies of the militant group either emigrate or are killed, leaving behind mostly Sunni Arabs willing to live under ISIL's harsh but semi-stable rule.[78]

An anti-ISIL activist criticized the group's manipulative tactics, stating that "their policy is to make people hungry while they pay their fighters so that becoming one of them is the only way to live and eat."[79] ISIL systematically controls every aspect of the economy in areas under its governance, taking over local businesses and making it impossible for nonmembers to find jobs, thus blocking off all alternative avenues of income.[80] ISIL enforces this economic subjugation while raising the prices of necessities and taxing local populations heavily on basic services such as water and electricity. "Only their people or those who swear allegiance to them have a good life…only the air people breathe is not taxed," reported a Syrian man who fled from militant-held Deir al-Zor.[81] With ISIL traders controlling the prices of goods in the markets, the cost of food has reportedly skyrocketed by as much as 1,000 percent in parts of Syria that ISIL controls.[82]

While bankrupting the population stuck in militant-controlled territory, ISIL pays its employees a living wage. Activists have deemed ISIL's economic warfare a tactic to coerce civilians to pledge allegiance to the militants. Faced with a choice between going hungry and submitting to the extremists, many have joined ISIL's ranks. It is estimated that within the first three months following ISIL's takeover in Palmyra, 1,200 young men were recruited by the militant group.[83] ISIL's wages have become particularly important in the wake of coalition air strikes, which have disrupted the oil-based economy upon which many Syrian livelihoods depend.[84]

In a May 2015 statement encouraging Muslims around the world to make *hijrah* (migration) to the Islamic State, Baghdadi claimed that Muslims living outside of ISIL-held territory were "homeless" and "humiliated," while assuring that inhabitants of the so-called caliphate lived "with might and honor, secure by God's bounty alone."[85]

Thus, ISIL aims to create a post-Westphalian world. In June 2014, the organization announced that it had destroyed the international border between Iraq and Syria. It announced its accomplishment with great fanfare, including a slick film accompanied by *nasheed*, the haunting jihadi music ISIL often uses as a soundtrack for its propaganda films, and established a Twitter hashtag #SykesPicotOver. With this accomplishment, ISIL was taunting the West that it was undoing the Sykes-Picot Agreement, the Anglo-French agreement of 1916 to carve up the Middle East into spheres of influence, which ultimately led to the new Arab nation-states of Iraq and Syria. As of this writing, ISIL claims dominion over 10 provinces in 8 states, proving itself particularly adept at exploiting areas where states are failing or weak.

How are we to understand how an ideologically driven organization—one that holds itself out as practicing a pure version of Islam—justifies moneymaking operations that are forbidden by Islam? After all, the Quran condemns bribery, theft, and rape.[86]

Some jihadists have come up with a way to rationalize their crimes in religious terms. For example, Anwar al-Awlaki ruled that it is permissible to "dispossess" the "disbelievers'" wealth by any means possible, including theft, embezzlement, and seizure of property. "All of our scholars agree on the permissibility of taking away the wealth of the disbelievers in *dar al-harb* [the territory of war] whether by means of force or by means of theft or deception," he writes.[87] He was writing about the West, but his ruling could potentially apply to all "disbelievers."

More importantly, terrorist groups are not just mission-driven organizations. Terrorist organizations are also organizations *qua* organizations, the goal of which is often first and foremost, to survive. James Q. Wilson argued a quarter-century ago, "Organizations tend to persist. This is the most important thing to know about them."[88] Interestingly, ISIL's slogan of *baqiyah* (to remain) appears to be one of its principal goals.[89] In the remainder of this chapter, I will assess ISIL's survival strategy from an organizational perspective, an approach I developed in an earlier book and in several articles.[90]

The organizational approach allows us to explore terrorism as a definable and distinctive product rendered by terrorist "firms." It allows us to analyze the production of terrorism, the evolution of groups, and the types of behaviors, shapes, and attributes that make groups more or less effective and resilient. Here we will focus most concretely on ISIL's use of criminal activity to promote resilience.

The organizational approach is most akin to that of Martha Crenshaw, who, in several articles published in the 1980s, emphasizes the internal politics of terrorist groups, concluding that "terrorist behavior represents the outcome of the internal dynamics of the organization rather than strategic action." She argues further that some terrorist groups can best be understood as self-sustaining organizations whose fundamental purpose is to survive.[91] The organizational approach described herein assumes that making progress—or at least appearing to make progress—towards stated goals (the "mission," defined below) is an important consideration for terrorists, but rejects the assumption that the terrorists' principal aim is necessarily to achieve those goals. This approach is to be distinguished from an instrumental paradigm, which assumes that terrorists select strategies to achieve a set of political objectives.[92]

Organizational theorists distinguish between a rational system paradigm, which asserts that organizations seek specific, stated collective goals, and a natural system paradigm (now dominant in the literature), which asserts that participants in organizations pursue multiple interests, both disparate and common, instrumental and expressive, with the perpetuation of the organization as the single most important objective. Natural system theorists emphasize that even when the organization is pursuing its stated goals, it will also promote the needs of the individuals who belong to the group. These maintenance goals, required to secure the capital and labor needed to keep the organization in business, often "absorb much energy, and in the extreme (but perhaps not rare) case, become ends in themselves."[93] Over time, groups that survive tend to evolve from the cause-maximizing end of the spectrum to the incentive-maximizing end.

To survive, organizations need to secure capital and labor. They also need to develop a brand. Very few terrorist groups get past identifying their collective cause or mission, to successfully secure the capital, labor, and brand that are required to persist. Indeed, according to a study by political scientist David Rapoport, 90 percent of terrorist organizations survive less than a single year; of those that manage to survive beyond the first year, more than half disappear within the following decade.[94]

ISIL is much better at securing capital and labor than previous terrorist organizations, so much so that its claim to be a "state" is partly credible. It is also significantly more skilled at branding itself, producing films with unprecedentedly high production values and spreading its message over social media.[95] It clearly values branding above all else, paying its filmmakers and social media gurus more than its fighters.[96]

What is ISIL trying to achieve? Abu Muhammad al-Adnani, the spokesperson for ISIL, describes the purpose of jihad as recovering the lost honor of the Muslim Ummah:

> The time has come for those generations that were drowning in oceans of disgrace, being nursed on the milk of humiliation, and being ruled by the vilest of all people, after their long slumber in the darkness of neglect.... The time has come for the Ummah of Muhammad (sallallahu 'alayhi wa sallam) to wake up from its sleep, to remove the garments of dishonor, and shake off the dust of humiliation and disgrace, for the era of lamenting and moaning has gone and the dawn of honor has emerged anew. The sun of jihad has risen.[97]

But it is hard to measure honor, or describe the inputs to its creation in tangible terms. I will refer to these intangible, collective goals as the terrorist mission, and the terrorist product as violence (or the credible threat of violence) in the service of that mission.

Capital includes such things as weapons, camps, equipment, factories, and the physical plants for any businesses run by the terrorist firm. Labor includes the terrorist leaders plus all the personnel employed by the organization, including managers, killers, marketers, financiers, public relations officers, etc. Like other nonprofit firms, terrorist groups are distinct from for-profit firms in that they are not maximizing profits for shareholders, but maximizing the appearance of achieving their mission. Another distinguishing feature is that terrorist groups claim to be producing a service needed by oppressed peoples, and they often attract donors. Over time, like other NGOs, they may evolve to service the needs of their donors and recruits more than the people whose needs they claim to be addressing.[98]

The mission serves many functions. It distinguishes the organization from a purely profit-driven criminal organization, it helps the group raise funds from donors, it provides a *raison d'être* for action, and it provides a narrative about collective identity. It also enables the group to attract (but not necessarily retain) recruits. The mission of the terrorist group can be expressive (to communicate something), instrumental (to achieve something), or both. ISIL has both expressive and instrumental objectives. ISIL engages in expressive acts of extra-lethal violence, such as beheading or crucifying "apostates." Before invading Mosul, ISIL deliberately spread videos of its brutal executions in order to intimidate the Iraqi Army into retreating from the city.[99] ISIL's instrumental goals include spreading its caliphate throughout the Middle East and eventually, globally. ISIL boasts that its "soldiers continue to hope for Allah's further support and the conquest of Constantinople and Rome."[100]

There are several things to note about terrorist missions. First, making progress toward achieving the mission does not guarantee organizational survival. Indeed, the two objectives—mission achievement and organizational survival—are quite distinct. This is different from the for-profit world, where mission achievement (or value maximization), financial performance, and organizational survival are aligned. Maximizing profits is the firm's long-term goal, and the production of goods and services is a means to that end.

Second, terrorist organizations must make two calculations instead of one. They must attend to the financial performance of the enterprise and its long-term survival, and they must ensure that the enterprise is promoting, or at least appears to be promoting, its social objectives. For nonprofit firms, including terrorist ones, revenues are not the objective but the means to the desired end of achieving the organization's collective goals.[101] (It is important to point out that some terrorist organizations evolve into organized criminal groups with no apparent mission other than generating profits. For example, the Revolutionary Armed Forces of Colombia [FARC] appears to be moving in this direction.)

Terrorist leaders encourage participation by offering two broad types of incentives to individual members: material incentives in the form of physical protection, housing, food, and cash; and nonmaterial incentives in the form of spiritual and emotional rewards. Transgressions are discouraged with negative incentives, both material and emotional. ISIL advertises these positive incentives to potential fighters, boasting about a "five-star jihad," that includes free housing, "[health care] in the Khilafah," schooling for fighters' children, taking care of orphans, and the opportunity for individuals who cannot afford a wife to acquire sexual slaves or concubines.[102] For example, two German recruits who escaped from ISIL and were then tried upon their return said that they had been recruited in Germany by a "false preacher" who emphasized religion more than the requirement to join in the fighting. He promised that they "would drive the most expensive sports cars and have many wives" and that they could leave whenever they wished.[103] None of these claims were true.

There are also negative incentives to discourage defection.[104] One defector who managed to get out told the *BBC* that he feared not only for his life, but also the life of his family, whom he had left behind in ISIL-controlled territory. He summarized ISIL's approach to securing cooperation: "If you're against me, then you'll be killed. If you're with me, you work with me. You submit to my will and obey me, under my power in all matters."[105]

In the spiritual realm, ISIL offers its followers the opportunity to live in the only place on earth "where the shari'a [shariah] of Allah is implemented and the rule is entirely for Allah."[106] It provides adventure, camaraderie, and, most importantly, a collective identity with honor for its followers. In an ISIL member's words, "[T]he Islamic State, by the grace of its Lord alone, brought out the Islamic punishments and rulings of the shari'a [shariah] from the darkness of books and papers, and we truly lived them after they were buried for centuries."[107] ISIL also spins an "end-time" narrative, capitalizing on the growing apocalyptic mood among Sunni Arabs.[108]

The goals of terrorism thus fall on two continua: from the purely instrumental (aiming to achieve something, such as expanding the "caliphate") to the purely expressive (aiming to communicate something, such as terrorizing enemies) and from promoting a mission to promoting the wealth or personal power, identity, or enjoyment of the participants.[109]

Changes over Time

Weber first observed the tendency for organizations to shift their mission from achieving their objectives to promoting their own survival. When spontaneous movements create bureaucratic structures, he argues, the organization's ends are inevitably distorted, with the substitution of self-preservation for the objectives it was formed to promote. Thus, terrorist organizations, like other nonprofits, can shift their missions in two principal ways—from one stated objective to another stated objective or from their stated objectives to no objective other than organizational survival or personal enrichment.

As terrorist organizations become more concerned with self-preservation and securing benefits for laborers and less concerned about their ostensible *raison d'être*, the mission becomes a marketing tool for securing organizational survival or a source of social identity. When a group becomes too focused on maintenance goals, it may shift into pure crime, and may lose some of its supporters. Thus, in the words of one jihadist I interviewed in Pakistan, "I feel they are running a business. They are…suppliers of human beings." He added, "They use poor and illiterate boys for their own private cause, and call it 'jihad.' This 'jihad' has nothing to do with religion," he said. Asked how his organization receives its funds, he said that:

> the…real methods for raising funds is smuggling of goods through Afghanistan, Iran, and India. This includes drug trafficking, in some cases to India. Mujahedeen cross the borders and carry drugs, delivering them to the Indian underworld mafia. Similarly, the mujahedeen bring with them many smuggled items such as cosmetics and…electronic goods from Afghanistan to Pakistan to raise funds.[110]

The Competition for Resources

When multiple groups purportedly promote similar goals, competition for scarce resources such as donors, government assistance, and personnel can make them fierce competitors. This leads to a dynamic in which marketing becomes increasingly important.[111] The group's competitors can become their actual enemies, rather than those enemies that they identify publicly. Thus, ISIL aims to fight its enemies, the governments of Iraq and Syria, as well as the members of the 65-nation coalition engaged in fighting the organization. But it also needs to fight the many groups it is competing with, among them the Syrian al-Qaeda group Jabhat al-Nusra and Ahrar al-Sham. Jabhat al-Nusra is internally divided about whether to sever its relations with al-Qaeda and change its name, in order to secure more funding from the Gulf.[112] A Qatari official confirmed to *Reuters* that his government had pledged "more support, i.e.[,] money, supplies" if they cut their relations with al-Qaeda.[113]

When groups are competing with one another, violence can become a critical marketing tool, directed at political and financial backers rather than at the "target audience" (potential victims) that is usually described in the literature. The day-to-day task of political combat becomes one of competition for supporters and financial backers. Neutralization of the competition can become the group's most important goal.[114]

Conclusion

ISIL is unusual in that from the very beginning, it has focused on securing capital and labor more attentively than other terrorist organizations. It has acquired capital mainly through criminal activities including theft and selling oil, counterfeit goods, and antiquities on the black market. ISIL has illustrated the means by which terrorist organizations can leverage the resources available by interacting with criminal networks to produce a resilient, well-resourced organization. Part and parcel with counterterrorism efforts must be efforts to stem the sort of criminal networks that have padded the coffers of ISIL.

To acquire labor, ISIL offers a wide variety of incentives, including offering fighters wives, salaries, and camaraderie. ISIL is also unusually skilled at marketing these incentives—both positive ones for joining and negative ones for defecting. The group's innovative use of social media and online networks presents a great challenge, as it has quickly gained a transnational audience and the ability to attract volunteers from around the globe.

ISIL presents a new type of threat to the international community. On the one hand, it is in control of territories within two states recognized by the United Nations: Iraq and Syria. It boasts in its online magazine, *Dabiq*, that Western scholars have described it not as a terrorist organization, but as a state.[115] On the other hand, it rejects the post-Westphalian system of states. It claims to have reestablished the caliphate. The entire self-image and propaganda narrative of the Islamic State is based on emulating the early leaders of Islam, in particular the Prophet Muhammad and the four "rightly guided caliphs" who led Muslims from Muhammad's death in 632 until 661. ISIL claims to control provinces within the territory of eight states around the globe. While it has referred to itself by many names since it was first established in 2003, among them the "Islamic State of Iraq" and the "Islamic State of Iraq and the Levant," in June 2014, it changed its name to reflect its grandiose ambition to become a new kind of state that transcends state borders. As Henry Kissinger warns in his 2014 book, *World Order*, "[t]he state itself—as well as the regional system based on it—is in jeopardy, assaulted by ideologies rejecting its constraints as illegitimate and by terrorist militias that, in several countries, are stronger than the armed forces of the government."[116] Further, Kissinger warns that:

> [t]he world has become accustomed to calls from the Middle East urging the overthrow of regional and world order in the service of a universal vision. A profusion of prophetic absolutisms has been the hallmark of a region suspended between a dream of its former glory and its contemporary inability to unify around common principles of domestic or international legitimacy. Nowhere is the challenge of international order more complex—in terms of both organizing regional order and ensuring the compatibility of that order with peace and stability in the rest of the world.[117]

In many ways, ISIL has appropriated the mechanism of interstate relations and legitimacy to advance its cause—to establish a shariah-based borderless caliphate that eventually encompasses the entire world. While there is no risk that it will achieve this goal, it nonetheless presents a serious threat to weak states in the region where governance is poor.

Notes

[1] Michael Miklaucic, "World Order or Disorder: The SOF Contribution," in SOF Role in Combating Transnational Organized Crime, ed. William Mendel and Dr. Peter McCabe (MacDill AFB, FL: Joint Special Operations University Press, 2016).

[2] Mary Anne Weaver, "The Short, Violent Life of Abu Musab al-Zarqawi," The Atlantic, July/August 2006, available at <http://www.theatlantic.com/magazine/archive/2006/07/the-short-violent-life-of-abu-musab-al-zarqawi/304983/>.

[3] Assaf Moghadam, "The Salafi-Jihad as a Religious Ideology," CTC Sentinel 1, no. 3 (February 2008), <https://www.ctc.usma.edu/posts/the-salafi-jihad-as-a-religious-ideology>.

[4] Jessica Stern and J.M. Berger, ISIS: The State of Terror (New York, NY: Harper Collins Publishers, 2015).

[5] Jean-Charles Brisard, Zarqawi: The New Face of Al-Qaeda (New York, NY: Other Press, 2004).

[6] Weaver, "The Short, Violent Life of Abu Musab al-Zarqawi," The Atlantic; Bruce Reidel, The Search for Al-Qaeda: Its Leadership, Ideology, and Future (Washington, DC: Brookings Institution Press, 2010), 94.

[7] R. Jeffrey Smith, "Hussein's Prewar Ties to Al-Qaeda Discounted: Pentagon Reports Says Contacts Were Limited," Washington Post, April 6, 2007, available at <http://www.washingtonpost.com/wp-dyn/content/article/2007/04/05/AR2007040502263.html>.

[8] The Coalition Provisional Authority (CPA) was the transitional government of Iraq established by the U.S.-led coalition, which held executive, legislative, and judicial authority over the Iraqi government from April 2003 to June 2004.

[9] Sharon Otterman, "Iraq: Debaathification," Council on Foreign Relations, April 7, 2005, available at <http://www.cfr.org/iraq/iraq-debaathification/p7853#p3>.

[10] Sarah Childress, Evan Wexler, and Michelle Mizner, "Iraq: How Did We Get Here?" PBS Frontline, July 29, 2014, available at <http://www.pbs.org/wgbh/pages/frontline/iraq-war-on-terror/iraq-decade/iraq-how-did-we-get-here/>; Lee Hudson Teslik, "Profile: Abu Musab al-Zarqawi," Council on Foreign Relations, June 8, 2006, available at <http://www.cfr.org/iraq/profile-abu-musab-al-zarqawi/p9866#p4>; Lieutenant General Jay Garner was appointed in 2003 as Director of the Office for Reconstruction and Humanitarian Assistance (ORHA) for Iraq. Less than a month later, he was dismissed and succeeded by L. Paul Bremer, and the ORHA was dissolved and replaced by the CPA.

[11] Bruce R. Pirnie and Edward O'Connell, Counterinsurgency in Iraq (2003–2006): RAND Counterinsurgency Study—Volume 2 (Santa Monica, CA: RAND Corporation, 2008), 50-51; email interview with Charles Lister, Visiting Fellow, Foreign Policy, Brookings Institution Doha Center.

[12] Louise Shelley, "Criminal Minds," American Legion, June 1, 2015, available at <http://www.legion.org/magazine/227319/criminal-minds>.

[13] Louise Shelley, "Blood Money: How ISIS Makes Bank," Foreign Affairs, November 30, 2014, available at <https://www.foreignaffairs.com/articles/iraq/2014-11-30/blood-money>.

[14] Christopher Reuter, "The Terrorist Strategist: Secret Files Reveal the Structure of the Islamic State," Der Spiegel, April 18, 2015, available at <http://www.spiegel.de/international/world/islamic-state-files-show-structure-of-islamist-terror-group-a-1029274.html>.

[15] Liz Sly and Ahmed Ramadan, "Insurgents Seize Iraqi city of Mosul as Security Forces Free," The Washington Post, June 10, 2014, available at <http://www.washingtonpost.com/world/insurgents-seize-iraqi-city-of-mosul-as -troops-f lee/2014/06/10/21061e87-8fcd-4ed3-bc94-0e309af0a674_story.html>.

[16] "ISIS Steals $429m from Central Bank After Capturing Mosul," Al Arabiya, June 13, 2014, available at <http://english.alarabiya.net/en/News/middle-east/2014/06/13/Report-ISIS-steals-429mn-in-Mosul-capture.html>; Statement of Richard Barrett, "A Dangerous Nexus: Terrorism, Crime, and Corruption" testimony before the Task Force to Investigate Terrorist Financing, Committee on Financial Services, U.S. House of Representatives, May 21, 2015, available at <http://financialservices.house.gov/uploadedfiles/hhrg-114-ba00-wstate-rbarrett-20150521.pdf>.

[17] Jeremy Bender, "Iraqi Bankers Say ISIS Never Stole $430 Million from Mosul Banks," Business Insider, July 17, 2014, available at <http://www.businessinsider.com/isis-never-stole-430-million-from-banks-2014-7>.

[18] Martin Chulov, "How an Arrest in Iraq Revealed ISIS's $2bn Jihadist Network," The Guardian, June 15, 2014, available at <http://www.theguardian.com/world/2014/jun/15/iraq-isis-arrest-jihadists-wealth-power>.

[19] Statement of Matt Levitt, "Terrorist Financing and the Islamic State," before the House Financial Services Committee, Terrorist Financing and the Islamic State, Hearing, November 13, 2014, available at <http://www.washingtoninstitute.org/uploads/Documents/testimony/LevittTestimony20141113.pdf>.

[20] Jay Solomon, "Territorial Gains Could Bolster ISIS's Balance Sheet," The Wall Street Journal, June 1, 2015, available at <http://blogs.wsj.com/washwire/2015/06/01/territorial-gains-could-bolster-isiss-balance-sheet>. A 2015 investigation by analysts from the Defense Intelligence Agency concluded that coalition

airstrikes targeting ISIL-held refineries did not disrupt terrorist oil revenues to the extent that CENTCOM previously reported. It is unclear how significantly the airstrikes have damaged ISIL's black market oil sales.

[21] Bill Roggio, "Islamic State Takes Control of Sinjar, Mosul Dam in Northern Iraq," Long War Journal, August 3, 2014, available at <http://www.longwarjournal.org/archives/2014/08/islamic_state_takes.php>.

[22] "Who, What, Why: Who Are the Yazidis?" BBC Magazine Monitor, August 7, 2014, available at <http://www.bbc.com/news/blogs-magazine-monitor-28686607>.

[23] Jane Arraf, "Islamic State Persecution of Yazidi Minority Amounts to Genocide, UN Says," Christian Science Monitor, August 7, 2014, available at <http://www.csmonitor.com /World/Middle-East/2014/0807/ Islamic-State -persecution-of-Yazidi -minority-amounts-to-genocide-UN-says-video>.

[24] Michael Gordon and Rukmini Callimachi, "Kurdish Fighters Retake Iraqi City of Sinjar from ISIS," The New York Times, November 13, 2015, available at <http://www.nytimes.com/2015/11/14/world/middleeast/sinjar-iraq-islamic-state.html>.

[25] Rukmini Callimachi, "ISIS Enshrines a Theology of Rape," The New York Times, August 13, 2015, available at <http://www.nytimes.com/2015/08/14/world/middleeast/isis-enshrines-a-theology-of-rape.html?_r=0>.

[26] Ibid.

[27] "Iraq: Yezidi Women and Girls Face Harrowing Sexual Violence," Amnesty International, December 23, 2014, available at <https://www.amnesty.org/en/latest/news/2014/12/iraq-yezidi-women-and-girls-face-harrowing-sexual-violence>;

"Iraq: ISIS Escapees Describe Systematic Rape," Human Rights Watch, April 14, 2015, available at <https://www.hrw.org/news/2015/04/14/iraq-isis-escapees-describe-systematic-rape>.

[28] Emma Batha, "Iraqi Women Trafficked Into Sexual Slavery – Rights Group," Reuters, February 17, 2015, available at <http://www.reuters.com/article/2015/02/17/us-iraq-trafficking-women-idUSKBN0L-L1U220150217>.

[29] "The Revival of Slavery Before the Hour," Dabiq 4, July 2014, available at <http://media.clarionproject.org/files/islamic-state/islamic-state-isis-magazine-Issue-4-the-failed-crusade.pdf>.

[30] Charles Lister, "Assessing Syria's Jihad," Survival: Global Politics and Strategy 56, no. 6 (2014): 87-112.

[31] Olivia Becker, "ISIS Is Radicalizing Kidnapped Kurdish Students," Vice, June 23, 2014, available at <https://news.vice.com/article/isis-is-radicalizing-kidnapped-kurdish-students>; Salma Abdelaziz, "Syrian Radicals 'Brainwash' Kidnapped Kurdish School Children," CNNWorld, June 26, 2014, available at <http://www.cnn.com/2014/06/25/world/meast/syria-isis-schoolboys/>.

[32] Orlando Crowcroft, "ISIS: People Trafficking, Smuggling, and Punitive Taxes Boost Islamic State Economy," International Business Times, June 16, 2015, available at <http://www.ibtimes.co.uk/isis-people-trafficking-smuggling-punitive-taxes-boost-islamic-state-economy-1506473>.

[33] "Libya: A Growing Hub for Criminal Economies and Terrorist Financing in the Trans-Sahara," The Global Initiative Against Transnational Organized Crime, May 11, 2015, available at <http://www.globalinitiative.net/download/global-initiative/Libya%20Criminal%20Economies%20in%20the%20trans-Sahara%20-%20May%202015.pdf>.

[34] Ibid.

[35] Statement of David Cohen, "The Islamic State and Terrorist Financing," hearing before the House Financial Services Committee, November 13, 2014, available at <http://financialservices.house.gov/uploadedfiles/hhrg-113-ba00-wstate-dcohen-20141113.pdf>.

[36] Peter Bergen, "Should Western Nations Just Pay ISIS Ransom?" CNN Opinion, August 22, 2014, <http://www.cnn.com/2014/08/22/opinion/bergen-schneider-isis-ransom/>.

[37] Statement of Matt Levitt, "Terrorist Financing and the Islamic State."

[38] Shelley, "Blood Money."

[39] Celina Realuyo, "Leveraging the Financial Instrument of National Power to Counter Illicit Networks," Congressional Hearing entitled "A Dangerous Nexus: Terrorism, Crime, and Corruption," before the Task Force to Investigate Terrorist Financing, Committee on Financial Services, U.S. House of Representatives, May 21, 2015, available at <http://financialservices.house.gov/uploadedfiles/hhrg-114-ba00-wstate-crealuyo-20150521.pdf>.

[40] Charles Lister, "Cutting Off ISIS' Cash Flow," Brookings Institution, October 24, 2014, available at <http://www.brookings.edu/blogs/markaz/posts/2014/10/24-lister-cutting-off-isis-jabhat-al-nusra-cash-flow>.

[41] Statement of David Cohen, "The Islamic State and Terrorist Financing."

[42] Mark Mazzetti and Matt Apuzzo, "Analysts Detail Claims that Reports on ISIS Were Distorted," The New York Times, September 15, 2015, available at <http://www.nytimes.com/2015/09/16/us/politics/analysts-said-to-provide-evidence-of-distorted-reports-on-isis.html>.

[43] Erika Solomon, Guy Chazan, and Sam Jones, "ISIS Inc: How Oil Fuels the Jihadi Terrorists," Financial Times, October 14, 2015, available at <http://www.ft.com/intl/cms/s/2/b8234932-719b-11e5-ad6d-f4ed76f0900a.html#axzz3sQJx2O19>.

[44] Michael Gordon, "U.S. Warplanes Strike ISIS Oil Trucks in Syria," The New York Times, November 16, 2015, available at <http://www.nytimes.com/2015/11/17/world/middleeast/us-strikes-syria-oil.html?_r=0&module=ArrowsNav&contentCollection=Middle%20East&action=keypress&region=FixedLeft&pgtype=article>.

[45] David Martin, "U.S. Airstrikes Against ISIS Target Oil Tanker Trucks," CBS News, November 23, 2015, available at <http://www.cbsnews.com/news/u-s-airstrikes-against-isis-target-oil-tanker-trucks/>.

[46] Erika Solomon, "Upsurge in Air Strikes Threatens ISIS Oil Production," Financial Times, November 18, 2015, available at <http://www.ft.com/intl/cms/s/0/54dd32ee-8df1-11e5-a549-b89a1dfede9b.html#axzz3sQJx2O19>.

[47] Sarah Almukhtar, "ISIS Finances Are Strong," The New York Times, May 19, 2015, available at <http://www.nytimes.com/interactive/2015/05/19/world/middleeast/isis-finances.html?_r=0>.

[48] Angelo Young, "ISIS Has $1B Worth of US Humvee Armored Vehicles; One Was Used in Monday's Suicide Bombing Near Baghdad," International Business Times, June 1, 2015, available at <http://www.ibtimes.com/isis-has-1b-worth-us-humvee-armored-vehicles-one-was-used-mondays-suicide-bombing-1946521>.

[49] Janine Di Giovanni, Leah McGrath Goodman, and Damien Sharkov, "How Does ISIS Fund Its Reign of Terror?" Newsweek, November 6, 2014, available at <http://www.newsweek.com/2014/11/14/how-does-isis-fund-its-reign-terror-282607.html>.

[50] Louise Shelley, Dirty Entanglements (New York, NY: Cambridge University Press, 2014).

[51] Reid Wilson, "The Illegal Antiquities Trade Funded the Iraqi Insurgency. Now It's Funding the Islamic State," The Washington Post, March 9, 2015, available at <https://www.washingtonpost.com/posteverything/wp/2015/03/09/how-shady-art-dealers-help-fund-the-islamic-states-violent-insurgency>.

[52] "FATF Report: Financing of the Terrorist Organisation Islamic State in Iraq and the Levant," Financial Action Task Force, February 2015, available at <http://www.fatf-gafi.org/media/fatf/documents/reports/Financing-of-the-terrorist-organisation-ISIL.pdf>.

[53] Joe Parkinson, Ayla Albayrak, and Duncan Mavin, "Culture Brigade: Syrian 'Monuments Men' Race to Protect Antiquities as Looting Bankrolls Terror," The Wall Street Journal, February 10, 2015, available at <http://www.wsj.com/articles/syrian-monuments-men-race-to-protect-antiquities-as-looting-bankrolls-terror-1423615241>.

[54] Yasmeen Sami Alamiri, "ISIS's Wealth: Stronger Than Ever or In Decline?" Al Arabiya, May 28, 2015, available at <http://english.alarabiya.net/en/perspective/analysis/2015/05/28/ISIS-s-wealth-Stronger-than-ever-or-in-decline html>.

[55] David Cohen, "Attacking ISIL's Financial Foundation" (remarks presented at Carnegie Endowment for International Peace, October 23, 2014). Juan Zarate, who served in the U.S. government from 2005-2009, helped to develop laws and regulations to target sources of funding for al-Qaeda and other terrorist groups. See Juan Zarate, Treasury's War (New York, NY: Public Affairs, 2013). According to Zarate, Cohen, and Levitt, these instruments are far less useful in regard to ISIL because its main sources of funding are criminal operations, and there is less money laundering to target.

[56] Parkinson, Albayrak, and Mavin, "Culture Brigade;" Daniela Dae, "Islamic State is selling looted Syrian art in London to fund its fight," The Washington Post, February 25, 2015, available at <https://www.washingtonpost.com/world/is-looted-syrian-art-showing-up-in-london-to-fund-activities/2015/02/25/785ab630-bcd0-11e4-b274-e5209a3bc9a9_story.html>.

[57] Ibid.

[58] David Kohn, "ISIS's Looting Campaign," The New Yorker, October 14, 2014, available at <http://www.newyorker.com/tech/elements/isis-looting-campaign-iraq-syria>.

[59] Ibid.; Anne Barnard, "ISIS Engulfs Assyrian Christians as Militants Destroy Ancient Art," The New York Times, February 26, 2015, available at <http://www.nytimes.com/2015/02/27/world/middleeast/more-assyrian-christians-captured-as-isis-attacks-villages-in-syria.html>.

[60] Amr Al-Azm, Salam Al-Kuntar, and Brian Daniels, "ISIS' Antiquities Sideline," The New York Times, September 2, 2014, available at <http://www.nytimes.com/2014/09/03/opinion/isis-antiquities-sideline.html>.

[61] Statement of Matt Levitt, "Terrorist Financing and the Islamic State."

[62] Howard LaFranchi, "What Syrian Antiquities Reveal About Islamic State's Billion-dollar Economy," Christian Science Monitor, August 25, 2015, available at <http://www.csmonitor.com/USA/Foreign-Policy/2015/0825/What-Syrian-antiquities-reveal-about-Islamic-State-s-billion-dollar-economy>.

[63] Alamiri, "ISIS's Wealth: Stronger Than Ever or In Decline?"

[64] Stephen Kalin, "War Turns Syria Into Major Amphetamines Producer, Consumer," Reuters, January 13, 2014, available at <http://uk.reuters.com/article/2014/01/13/uk-syria-crisis-drugs-idUKBRE-A0B04K20140113>.

[65] Aryn Baker, "Syria's Breaking Bad: Are Amphetamines Funding the War?," TIME, October 28, 2013, available at <http://world.time.com/2013/10/28/syrias-breaking-bad-are-amphetamines-funding-the-war/>.

[66] Kalin, "War Turns Syria Into Major Amphetamines Producer, Consumer."

[67] Erika Solomon, "Syrian Smugglers Shun Weapons and Turn to Cigarettes for Profits," Financial

Times, May 15. 2015, available at <http://www.ft.com/intl/cms/s/0/081071ee-f975-11e4-ae65-00144feab7de.html#axzz3kb9n7afL>.

[68] Morgan Windsor, "ISIS Beheads Cigarette Smokers: Islamic State Deems Smoking 'Slow Suicide' Under Sharia Law," International Business Times, February 12, 2015, available at <http://www.ibtimes.com/isis-beheads-cigarette-smokers-islamic-state-deems-smoking-slow-suicide-under-sharia-1815192>.

[69] Solomon, "Syrian Smugglers Shun Weapons and Turn to Cigarettes for Profits."

[70] Vivian Salama and Bram Janssen, "What It's Like to be a Cigarette Smuggler in ISIS-Controlled Areas," Business Insider, June 18, 2015, available at <http://www.businessinsider.com/what-its-like-to-be-a-cigarette-smuggle-in-isis-controlled-areas-2015-6>.

[71] Solomon, "Syrian Smugglers Shun Weapons and Turn to Cigarettes for Profits."

[72] Vivian Salama and Bram Janssen, "What It's Like to be a Cigarette Smuggler in ISIS-Controlled Areas."

[73] Nawzat Shamdeen, "The Islamic State's New Money-Making Business: Auction Off Homes It Stole," Your Middle East, October 10, 2014, available at <http://www.yourmiddleeast.com/business/the-islamic-states-new-moneymaking-business-auction-off-homes-it-stole_27152>.

[74] Shelley, "Blood Money."

[75] Louise Shelley, "Smuggling of Oil is 'Only One of IS Group's Illicit Activities,'" interview by Armen Georgian, France 24, March 12, 2015, available at <http://www.france24.com/en/20150311-interview-louise-shelley-terrorism-financing-islamic-state-group-smuggling>.

[76] Ben Hubbard, "Offering Services, ISIS Digs in Deeper in Seized Territories," The New York Times, June 16, 2015, available at <http://www.nytimes.com/2015/06/17/world/middleeast/offering-services-isis-ensconces-itself-in-seized-territories.html>.

[77] Ibid.

[78] Ibid.

[79] Ibid.

[80] Joanna Paraszczuk, "The ISIS Economy: Crushing Taxes and High Unemployment," The Atlantic, September 2, 2015, available at <http://www.theatlantic.com/international/archive/2015/09/isis-territory-taxes-recruitment-syria/403426/>.

[81] Ibid.

[82] Ibid.

[83] Ibid.

[84] Hassan Hassan, "In Syria, Many Families Face a Terrible Dilemma," The National, September 20, 2015, available at <http://www.thenational.ae/opinion/comment/in-syria-many-families-face-a-terrible-dilemma#full>.

[85] "In New Audio Speech, Islamic State (ISIS) Leader Al-Baghdadi Issues Call to Arms to All Muslims," MEMRI Jihad and Terrorism Threat Monitor, May 14, 2015, available at <http://www.memrijttm.org/in-new-audio-speech-islamic-state-isis-leader-al-baghdadi-issues-call-to-arms-to-all-muslims.html>.

[86] On rape, the Quran says, "And those who launch a charge against chaste women, and produce not four witnesses (to support their allegations) – flog them with eighty stripes; and reject their evidence ever after: for such men are wicked transgressors." [Quran 24:13] On theft, the Quran says, "As for the thief, both male and female, cut off their hands. It is the reward of their own deeds, an exemplary punishment from Allah." [Quran 5:38] On bribery, the Quran says, "And do not consume one another's wealth unjustly or send it [in bribery] to the rulers in order that [they might aid] you [to] consume a portion of the wealth of the people in sin, while you know [it is unlawful]." [Quran 2:188]

[87] Anwar al-Awlaki, "The Ruling on Dispossessing the Disbelievers' Wealth in dar al-harb," Inspire, November 2010, 55, available at <https://azelin.files.wordpress.com/2011/01/inspire-magazine-4.pdf>.

[88] J.Q. Wilson, Political Organizations (Princeton, NJ: Princeton University Press, 1995), 30.

[89] Dabiq 5, November 2014, 22-33.

[90] Jessica Stern, Terror in the Name of God: Why Religious Militants Kill (New York, NY: Harper Perennial, 2004); Jessica Stern, with Amit Modi, "Organizational Forms of Terrorism," in Countering the Financing of Terrorism, ed. Thomas Biersteker and Sue Eckert (New York, NY: Routledge, 2008).

[91] Martha Crenshaw, "The Logic of Terrorism: Terrorist Behavior as a Product of Choice," Terrorism and Counter Terrorism 2, no. 1 (1998): 54-64; Martha Crenshaw, "An Organizational Approach to the Analysis of Political Terrorism," Orbis 29, no. 3 (Autumn 1985): 473-487; Martha Crenshaw, "Theories of terrorism: Instrumental and organizational approaches," Journal of Strategic Studies 10, no. 4 (1987). In the first cited work, Crenshaw outlines an instrumental approach in response to works by a group of psychologists. In the second, she distinguishes an instrumental approach from an organizational one, but does not specify what the organization would entail.

[92] The instrumental paradigm is based on the work of Thomas Schelling and his successors. See, for example, Thomas C. Schelling, The Strategy of Conflict (Cambridge, MA: Harvard University Press, 1960);

Thomas C. Schelling, Arms and Influence (Westport, CT: Praeger, 1976). When Thomas Schelling wryly observes that "despite the high ratio of damage and grief to the resources required for a terrorist act, terrorism has proved to be a remarkably ineffectual means to accomplishing anything," he is assuming that terrorists have substantively rational objectives and that their main purpose is to achieve those objectives. Thomas Schelling, "What Purposes can 'International Terrorism' Serve?" in Violence, Terrorism, and Justice, ed. R.G. Frey and Christopher W. Morris (Cambridge, MA: Cambridge University Press, 1991), 20-21. See also Sun-Ki Chai, "An Organizational Economics Theory of Antigovernment Violence," Comparative Politics 26, no. 1 (September 1993): 99-110.

[93] W.R. Scott, Organizations: Rational, Natural, and Open Systems, Second Edition (Englewood Cliffs, NJ: Prentice-Hall Inc., 1981), 52.

[94] David C. Rapoport, "Terrorism," in Routledge Encyclopedia of Government and Politics, Volume 2, ed. Mary Hawkesworth and Maurice Kogan (London: Routledge, 1992), 1067.

[95] Stern and Berger, ISIS: The State of Terror.

[96] Greg Miller and Souad Mekhennet, "Inside the Surreal World of the Islamic State's Propaganda Machine," The Washington Post, November 20, 2015, available at <https://www.washingtonpost.com/world/national-security/inside-the-islamic-states-propaganda-machine/2015/11/20/051e997a-8ce6-11e5-acff-673ae-92ddd2b_story.html>; Gina Scott Ligon, Mackenzie Harms, and Douglas C. Derrick, "Lethal Brands: How VEOs Build Reputations," Journal of Strategic Security 8, no. 1-2 (2015): 27-42.

[97] "The Return of the Khilafah," Dabiq 1, 9, available at <http://media.clarionproject.org/files/09-2014/isis-isil-islamic-state-magazine-Issue-1-the-return-of-khilafah.pdf>; Isa Ibn Sa'd Al Ushan, "Advice to the Mujahidin: Listen and Obey," Dabiq 12, 2015, 9-10.

[98] For examples, see Stern, Terror in the Name of God.

[99] William McCants, "The Believer," Brookings Institution, September 1, 2015, available at <http://www.brookings.edu/research/essays/2015/thebeliever>.

[100] Dabiq 10, July 2015, 51.

[101] Mark Moore, conversations with author, 2002-2003.

[102] Jessica Stern and J.M. Berger, "Thugs Wanted – Bring Your Own Boots," The Guardian, March 9, 2015, available at <http://www.theguardian.com/world/2015/mar/09/how-isis-attracts-foreign-fighters-the-state-of-terror-book>; Dabiq 9, May 2015, 24-27; Dabiq 1, July 2014, 4.

[103] "'Jail in Germany is Preferable to Freedom in Syria' – German ISIS Defector," RT, August 6, 2015, available at <https://www.rt.com/news/311795-jail-better-than-ISIS/>.

[104] Simon Tomlinson, "Barbaric Turf War Escalates in Afghanistan as ISIS Executes Three 'Defectors' Days After the Taliban Warned Terror Group to Stay Out of the Country," Daily Mail, June 19, 2015, available at <http://www.dailymail.co.uk/news/article-3131248/Barbaric-turf-war-escalates-Afghanistan-ISIS-execute-three-defectors-days-Taliban-warned-terror-group-stay-country.html>; Deborah Amos, "Islamic State Defector: 'If You Turn Against ISIS, They Will Kill You,'" NPR, September 25, 2014, available at <http://www.npr.org/sections/parallels/2014/09/25/351436894/islamic-state-defector-if-you-turn-against-isis-they-will-kill-you>; Morgan Winsor, "ISIS Executes 100 Foreign Fighters For Trying to Flee Syria," International Business Times, December 20, 2014, available at <http://www.ibtimes.com/isis-executes-100-foreign-fighters-trying-flee-syria-1764018>.

[105] Paul Wood, "ISIS Defector Speaks of Life Inside Brutal Jihadist Group," BBC News, July 14, 2014, available at <http://www.bbc.com/news/world-middle-east-28269596>.

[106] Dabiq 10, July 2015, 51.

[107] Dabiq 5, November 2014, 45.

[108] Jessica Stern, "ISIS's Apocalyptic Vision," Hoover Institution, February 25, 2015, available at <http://www.hoover.org/research/isiss-apocalyptic-vision>.

[109] Jessica Stern, "Obama and Terrorism," Foreign Affairs, September/October 2015, available at <https://www.foreignaffairs.com/articles/obama-and-terrorism>.

[110] Like the other interviewees for the Harvard project on leadership, this interviewee's identity cannot be identified. Interviews from Stern, Terror in the Name of God.

[111] John Fawcett, telephone and email interview with author, July 15, 2001. Among humanitarian relief organizations, the drive to beat out other organizations in the competition for funds can replace the drive to provide humanitarian relief where it matters most. John Fawcett argues that donor agencies wanted the NGOs active in the former Yugoslavia to carve up the problem and coordinate their activities. But the aid agencies are actually competing for funding, he says. They did not want to coordinate; they wanted to outbid each other. Each aid agency feared the others would steal its ideas.

[112] Jack Moore, "Al-Qaeda Losing 'Deadly Competition' With ISIS," Newsweek, March 4, 2015, available at <http://europe.newsweek.com/al-Qaeda-decline-syria-affiliate-nusra-front-considers-split-311348>.

[113] Ibid.

[114] Philip Selznick, The Organizational Weapon: A Study of Bolshevik Strategy and Tactics (Santa Monica, CA: RAND Corporation, 1952), 229.

[115] "Paradigm Shift Part II," Dabiq 12, November 2015, 48.
[116] Henry Kissinger, World Order (New York, NY: Penguin Press, 2014), 7.
[117] Ibid., 96.

I would like to thank my research assistant, Jaclyn Roache, and the editors of this volume for their comments.

III. Pandora

10

Virtually Illicit: The Use of Social Media in a Hyper-Connected World

Tuesday Reitano and Andrew Trabulsi

With her customized pink AK-47 cocked on her shiny latex-covered hip, Claudia Ochoa Felix, the woman reported to be the new matriarch of "Los Antrax," the killing squad of Mexico's Sinaloa drug trafficking cartel, personifies the spirit of narco-narcissism. Ochoa Felix' moniker as the "Kim Kardashian of the crime world" refers not only to her scarlet pucker, contoured cheeks, and pert glutes, but also to her obsessive use of social media to flaunt the wealth and status that accompany her fearsome reputation. The 20-something mother-of-three's Instagram and Twitter accounts are replete with images of luxury cars, designer clothes, beautiful people, and fabulous parties. One now infamous image shows her toddler lying on a bed blanketed in dollar bills. Despite her success, she is far from alone in successfully cultivating such an image on social media. From the street gangs in the United States to the mafia in Palermo, social media is becoming increasingly ostentatious and ubiquitous. Where organized crime was once something that previously lived furtively in the shadows, with secrecy being a premium commodity, in today's hyper-connected world, social media has become the showcase of organized criminal groups to demonstrate their power and profits with impunity.

By virtue of its prodigious use of social media, the Islamic State of Iraq and the Levant (ISIL) has managed to develop a large number of globally dispersed supporters, which has enabled ISIL to exert an outsized impact on how it is perceived by the world. Violent achievements such as the beheadings of Western journalists and aid workers are trumpeted on Twitter, YouTube, and Instagram, raising the group's profile, mobilizing funds, and recruiting new fighters.[1] As of January 2015, ISIL had recruited an estimated 4,500 foreign terrorist fighters from Western countries, many of whom have been lured by an aggressive social media outreach strategy that preys on the isolated, marginalized, or vulnerable.[2] Regardless of the means of recruitment, their physical journey to join ISIL will be coordinated by countless interactions with facilitators; corruption brokers offering fraudulent documents or safe passage through an accommodating border post; and faithful allies offering safe houses, secure transport, and fraudulent documents, all via a series of messages on ever-changing Twitter accounts, messaging apps, and encrypted platforms.

Society, identity, and connectivity are increasingly being defined in the social media space, rather than by the physical borders of geographic states. Facebook, for example, has

1.3 billion active users across the globe; only 2 of the world's countries can boast a larger population. This has created an open forum for the connection of individuals and exchange of ideas, goods, and services through which criminal organizations can ideologically thrive. It has become one of the primary means by which groups identify the like-minded, engage with them, and garner their support. This process is harmless when applied to gardening groups or fan pages dedicated to icons of pop culture, but infinitely more damaging when used to further criminal or terrorist agendas. The potency of social media comes in its unique ability to broadcast en masse, whilst at the same time delivering messages that seem intimate, allowing individuals to respond. As this chapter will explore, a myriad of deviant groups is using social media to shape opinion and elicit respect, fear, and terror. They are exploiting the functionalities of social media along all parts of their enterprise chain, from identification of allies and victims, to executing operational capacities such as logistics and fundraising. In doing so, social media is blurring definitional categorization between criminal groups, terrorists, political activists, and insurgents; perhaps more importantly, it is increasingly conferring legitimacy on their acts by drawing average citizens into this spectrum.

The challenges of responding to this growing convergence are significant, and the majority of efforts deployed by states at counter-messaging and/or identifying and disabling key nodes in social media networks have proven largely fruitless, sometimes laughable, and often counterproductive. These efforts inflame, rather than subdue, the calls to arms that such social media campaigns represent. Traditional law enforcement tools have struggled to gain traction in this landscape: intelligence gathering techniques are challenged to embed themselves effectively to understand transnationally spread groups and operations, with highly contextualized roots but a compelling shared ideology; anti-money laundering (AML) and countering the financing of terrorism (CFT) protocols remain focused on formal economy transactions and lack the data or instruments to engage in cash economies and through informal value transfer systems. Much is spoken of the capacity of big data to offer solutions, but efforts are still highly experimental and require technical knowledge, systems, and advanced contextual analyses that are often missing from state capacity. Efforts to offset the threat by mobilizing the power of social media for the public good have equally fallen short of the mark, with campaigns "going viral" and creating a short-lived public pressure that might result in rapid and flamboyant mobilization but rarely sustained commitment. Instead, the private sector and motivated vigilantes are most often seen standing at the frontier of combating social media security threats and their manifestations, while the primary onus on preventing radicalization and criminalization lies in the hands of the individual. Social media's potency comes from its authenticity of voice, and its ability to create a community of shared vision and understanding. This is best done organically, and in that case, arguably the correct role for the international community is to serve as incubator and defender of those brave voices that continue to champion freedom from fear and insecurity.

Individuals, Identity, and Ideology

In his 2006 book *Terror on the Internet*, Gabriel Weimann outlined the major uses of the internet by extremist groups: communication among members, transfer of information such as instructions on bomb-making or training videos, research of potential targets, and cyberterrorism.[3] Since then, however, with new technology and greater experience, the uses of the internet and social media by groups with a violent extremist or terrorist agenda have become manifold and prolific. Terrorist groups now use the internet, social media, and messaging applications for recruitment, fundraising, spreading ideology, influencing public opinion, and calling a diverse audience to arms in support of their cause. This is a nuanced agenda that plays on individuals' fears, isolation, and desire for empowerment, and as a consequence is arguably more potent and potentially damaging than was previously understood.

An area of social media usage that has risen to particular prominence with the ascendency of ISIL is the capacity of social media to be used as a tool for recruiting foreign fighters. Foreign fighters add not only considerable cachet to a jihadist agenda, but they may also bring to the table sets of skills and local knowledge that increase the capacity to manipulate Westphalian narratives and inspire terror. For many, however, the concern is less the possibility for recruitment of foreign fighters, which will always only reach out to a very tiny margin of the population. Instead, it is the narrative power that ISIL and other terrorist groups gain by using social media as a broadcasting tool. Like all propaganda, public perception can be partially engineered with these tools, and ISIL does this remarkably well. Highlighting their violence and their successes, even if the narrative is overplayed, influences the thoughts of the general public, which in turn becomes an obstacle that must be addressed by both policymakers and the intelligence community.

The messaging used by terrorist groups on social media involves appealing to three motives: humanitarian or moral imperative, ideology, and identity.[4] In the first instance, they speak to abuses by the opposing force and express moral outrage in order bestow legitimacy on their struggles and to portray themselves as fighting injustice. Foreign fighters in Syria regularly post pictures of young children allegedly killed by the Assad regime, and Jabhat al-Nusra has posted videos of their fighters rescuing civilians. Throughout the war in Afghanistan, the Taliban regularly used Twitter to document war atrocities committed by the International Security Assistance Force (ISAF) troops and spoke aggrievedly of the "occupiers" and their illegitimate wars.[5]

Terrorist groups also broadcast their ideology through social media as a call to arms, to raise local morale and attract supporters further afield. Somalia's al-Shabaab has long had quite active campaigns on Twitter and Facebook, and spends considerable energy on propagating its radical ideology through slick propaganda, targeted media campaigns, and clever use of media networks—and their use of social media has evolved over time. In 2011 and 2012, al-Shabaab created its first Twitter accounts, and used it to provoke the Kenya Defence Force (KDF), and later the African Union Mission in Somalia (AMISOM), into long "Twitter duels," where they traded insults and military successes with a surprising degree of engagement and wit. One entertaining interchange followed a post by the KDF spokesman, Major Emmanuel Chirchir (@MajorEChirchir), threatening to bomb concentrations of donkeys that might be moving weapons for the insurgents. Al-Shabaab (@HSMPress) responded, "Your eccentric battle strategy has got animal rights groups quite concerned, Major."[6] More recently, either inspired by (or in competition with) ISIL, the group appears to be putting more effort into using social media platforms to attract international attention and call followers to action. On February 22, 2015, al-Shabaab released a video on YouTube in which they called for an attack on London's Oxford Street, as well as Westfield and White City Malls. The video compared such attacks to the attack by al-Shabaab on Westgate Mall in Kenya in 2013. Other malls, like the Mall of Americas in the United States, were also mentioned. These locations are home to the largest Somali diaspora communities, which is apparently the target audience of the appeal, rather than foreign fighters, which were the ambition of ISIL strategies. In the video, an al-Shabaab member, speaking with a strong English accent, called upon jihadists living in the West to "answer the call of Allah and target disbelievers wherever they are."[7]

Finally, and perhaps their most potent weapon is that of identity, appealing to aspirations and vulnerabilities of their target group, to their sense of sense of self, pride, and family.[8] Conveyed across multiple social media channels, Twitter, YouTube, Facebook, Instagram, the messaging of ISIL manages to swell youths with a (false) sense of empowerment. "Access to weaponry and the ability to intimidate represent an antidote to the feelings of marginalization, alienation and powerlessness that young men felt in their former environment, whether a slum in a European or Middle Eastern city, or in a village as a peasant trying to help his family make ends meet."[9] They offer a sense of purpose, suggest that "war is cool," that they will find friendship and acceptance, both amongst brothers and with women. Much of the propaganda aimed at young men parallels popular culture: they use GoPro cameras to share the excitement of conflict, with set-up skirmishes that are self-labelled as being like popular video games.[10] ISIL has been the most successful group when it comes to recruiting women to their cause. Here, they target the socially isolated, appealing to their empathy, a sense of grievance, and message around acceptance, belonging, and the possibility of finding a mate.[11] Jihadist propaganda aimed at women tends to be counter-feminist, emphasizing that women are valued not as sexual objects, but as mothers to the next generation and guardians of the ISIL ideology, and they are encouraged to anticipate and think fondly of the husbands that they will have. Interestingly, it also frequently emphasizes the potential transience of such a union, as fighters give their lives for jihad.[12] Far from being exclusively "jihadi brides," female migrants into ISIL play crucial roles in the development of the group's internal social structure, from taking care of men who have returned from battle to vaccinating children. There is no singular profile for Western women in ISIL. Thus, the use of social media for female recruitment operations is varied. While some women romanticize their lives in the Caliphate on social media, others express hardship and sacrifice in making their lives inside it.

Around these primary sending accounts, groups create a bevy of followers that amplify, glorify, and reinforce the central message and its senders.[13] A 2014 census of

ISIL-supporting Twitter accounts, for example, found that the core of the activity emanated from a relatively small group of hyperactive users, numbering between 500 and 2,000 accounts that tweeted at concentrated bursts of high volume. However, around that central traffic were 46,000 to 70,000 accounts which would play a secondary supporting role, creating the vibrant and self-reinforcing perception of an active and inclusive community, and sending as many as 100,000 tweets per day.[14] New members of the dialogue are quickly identified and encouraged, with efforts made to then shift conversations from the public forums to private mediums, using internet-based messaging apps such as Skype, Facebook Messenger, Surespot, Telegram, Kik, or WhatsApp, where the relationship can be cultivated on an individualized basis, using textbook methods of attitude modification and indoctrination used in many faith-based groups, with creating a shared identity being central.[15]

The emphasis on identity used by terrorist groups on social media is also used by organized crime groups. Two main strategies long used by organized crime groups to ensure the loyalty of their subordinates are: first, through violence or the threat of violence; and second, by creating and emphasizing a sense of identity and belonging, and communicating this widely.[16] This is often because organized crime groups have grown out of a population that is marginalized, disenfranchised, or actively persecuted, and have created violent wings for self-protection. As the external threat has diminished, those providing protection have slowly morphed into a source of insecurity, demanding fees for their "protection," also known as extortion.[17] Identity becomes an important factor, as it creates a unity of purpose and reduces the likelihood of betrayal and competition. There are many markers used to affirm these identity—initiation rites, for example, or a strong emphasis on family. The use of tattoos is almost synonymous with criminality, used from the Asian groups like the Triads or the yakuza, to the Russian mafia prison groups and the street gangs of the Americas. This indelible brand of belonging simultaneously serves to unify cultures and represent distance from the established regime.[18] Similarly, symbolism, graffiti, code words, and signature moves have been used to create criminal subcultures, claim ownership of territory or violent acts, and to intimidate rivals.

The communication of power and the capacity to commit violence has always been an important part of the strategy of criminal groups, and social media has significantly amplified that. In the hyper-connected world, groups are taking these two distinguishing strategies online.[19] Social media has become the new means by which to display potency and initiate conflict. Graffiti "tags" have evolved into hashtags, and the same signs of prowess and success play out on social media sites, with the same intent as the symbols of the previous generation. In the typology of groups active along a spectrum of criminality, the ability to communicate and project their potency differs. The United Nations Office on Drugs and Crime (UNODC) has developed a working typology of five types of criminal groups, ranging from highly organized, hierarchical mafia-type groups which dominate some markets to more loosely organized networks or gangs which are active in others, and their communication requirements are different.[20] Arguably, groups with a wide membership, who are very secure in their support, are the most likely to broadcast widely the symbols of their influence, whereas highly localized crime groups are least likely. For example, for the highly geographically dispersed Mara Salvatrucha (MS-13) group, social media (and violence) has been instrumental in maintaining allegiance across multiple countries and territories. Unlike many of the traditional hierarchical mafia groups, the MS-13 operates with a hierarchy, "but it is a 'hierarchy of influence' where 'respect' and loyalties are expressed through a networked structure," and the evidence of influence, respect, and authority are displayed online. Social media mapping work undertaken by the SecDev Foundation has demonstrated that social media has facilitated this networked structure to engage in transnational crime, with certain cities (and prisons) serving as communication hubs across multiple regions.[21] Robert Muggah, director of Brazil's Igarapé Institute and the leader of SecDev's analysis, observed, "They use it to tag (mark territory), they use it to coerce, they use it to recruit, they use it to move product, they use it to communicate directives."[22]

Groups posture and self-promote online. The Sinaloa cartel, arguably the most powerful crime group in Mexico, has a Twitter account (@carteidsinaloa) with more than 57,000 followers. The alleged account of their leader, El Chapo Guzman, has more than half a million followers (@elchap0guzman), and his bravado has extended to insults and threats to Mexican President Peña Nieto and U.S. presidential candidate Donald Trump.[23] In the highly violent drug wars of the Americas, in a practice known as "cyber-banging," groups trade threats and insults over Facebook and Twitter, tagging rivals as the next in line for hits. This may cause spikes in homicides in concentrated and unpredictable bursts that confound the standard intelligence and analytical tools of law enforcement that do not systematically monitor social media nor have the ability to predict which insults or threats might prompt the next bout of violence.[24]

Using data analytics software, we can find this behavior evident in ISIL's use of social media as well.[25] Data visualization software from San Francisco-based analytics firm Quid illustrates that over half of all tweets from ISIL's top Twitter handles focus on conflict reporting and promoting new Twitter accounts. Quid's software uses natural language processing and mathematical physics to identify and visualize patterns within structured and unstructured datasets. Further, Quid's analysis demonstrates that new Twitter account promotion and security precautions drive major spikes in conversation amongst ISIL's digital communities, fostering both support and enhancing operations. In Figure 10.1 below, each node represents one tweet, and the proximity of nodes designates the degree of semantic similarity each tweet has with others in the network.

Figure 10.1. Network of Tweets in English-Speaking ISIL Network from August 6 to 26, 2015 (conflict reporting outlined in black and account promotion in blue; n = 878 tweets)

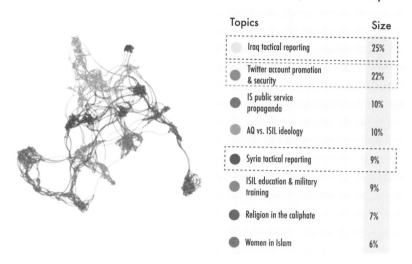

Topics	Size
Iraq tactical reporting	25%
Twitter account promotion & security	22%
IS public service propaganda	10%
AQ vs. ISIL ideology	10%
Syria tactical reporting	9%
ISIL education & military training	9%
Religion in the caliphate	7%
Women in Islam	6%

Social Media as a Service Provider for Underground Operations

Not only is social media helping deviant groups win hearts and minds and intimidate their opposition, it is also proving a more practical, tangible instrument in executing operations. Criminal groups and terrorists are exploiting the functionalities of social media along all parts of their enterprise chains, from the identification of allies and victims, to the execution of operational capacities such as logistics and fundraising, the procurement of services, and the development of their technical capacity. In the Americas, there has been a rise in kidnappings of software engineers and programmers as organized crime groups seek to reinforce their digital capabilities to ensure their technological dominance.[26]

One of the primary ways in which social media aids criminal and terrorist organizations is in identification and recruitment of victims, clients, service providers, and allies. Social networking sites contain reams of personal information about their users, which criminal groups use as intelligence gathering for possible targets. When soliciting new recruits, deviant groups can gather information about the friends, family, and habits of their victims. This helps them to customize their grooming strategy and to apply pressure on new recruits to conform. Individuals trying to exit from criminal or terrorist groups find themselves intimidated by threats, blackmailed and extorted until they find themselves forced to continue engaging with the groups. In one case of a teen courier working the Mexican-U.S. border, documented by *CNN*, he found himself locked into the cartel:

> "I told him that I didn't want to work for Los Zetas anymore and that I would rather be killed than to continue to work under him," Cesar said he declared after one night of constant text orders. The new boss called at 5 a.m. and told Cesar he was coming to pick him up to kill him. They met outside Cesar's house. The man pointed a gun at Cesar's face. Then, he started laughing. He told Cesar he wasn't going to kill him. But that he

Figure 10.2. Timeline of Tweets in English-Speaking ISIL Network from August 6 to 26, 2015 (n = 878 tweets)

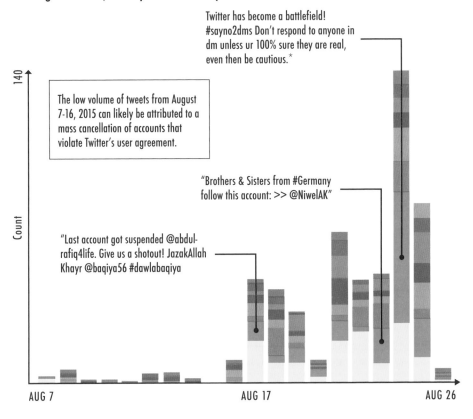

Twitter has become a battlefield! #sayno2dms Don't respond to anyone in dm unless ur 100% sure they are real, even then be cautious.*

The low volume of tweets from August 7-16, 2015 can likely be attributed to a mass cancellation of accounts that violate Twitter's user agreement.

"Brothers & Sisters from #Germany follow this account: >> @NiwelAK"

"Last account got suspended @abdul-rafiq4life. Give us a shoutout! JazakAllah Khayr @baqiya56 #dawlabaqiya

Topics		Size			
Iraq tactical reporting	25%		Syria tactical reporting	9%
Twitter account promotion & security	22%		ISIL education & military training	9%
IS public service propaganda	10%		Religion in the caliphate	7%
AQ vs. ISIL ideology	10%		Women in Islam	6%

would keep working or his family would be killed. He had been working with the cartel for too long. He would have to work again that morning.[27]

In a similar vein, one of the most standard practices and lucrative earners for criminal and terrorist groups alike comes in the form of fraud, blackmail, and extortion. In these crimes, which range from 419 cyber-scams to "virtual kidnapping," social media again allows the capacity to profile victims, and can provide kidnappers with the exact location of a victim through photos, live feeds, and apps with geolocation features, which can be embedded in posts and pictures without owners even realizing it.[28]

In addition to supporting the planning and execution of international terrorism, secure social messaging apps like Kik, Wickr, and Surespot, which encrypt or destroy communications data, play a critical role in ISIL's bride-recruiting and kidnapping operations. According to the U.S. Homeland Security Committee, an estimated 550 Western women have traveled to the conflict zone to marry ISIL fighters, as of October 2015. Conversations that begin on social networking platforms like Twitter or ASKfm migrate to secure channels, where men and women affiliated with ISIL lure potential brides to Syria, to marry jihadists living within ISIL-controlled territories. A 2015 report from the Institute for Strategic Dialogue (ISD) and the International Centre for the Study of Radicalization (ICSR) found that within three months of arrival, women who join ISIL are expected to marry a fighter and start producing children. Under strict shariah law, women are prohibited from combat and combative activities, but those who are social media savvy may themselves become part of the recruitment effort.[29]

The same tools are being used to facilitate kidnappings and allow troops to communicate on the battlefield. The September 2015 issue of *Dabiq*, the ISIL magazine, lists the sale of newly captured hostages as advertisements. Each solicitation includes each man's purported occupation, date of birth, and home address, as well as a Telegram number—a mobile social networking app that encrypts communication—for "whoever would like to pay the ransom for his release and transfer." Similar encryption-based messaging apps allow fighters to communicate both inside and outside of ISIL, even while on the battlefield. Use of such applications generates additional layers of exclusivity and security for ISIL members and their evangelists. These secure spaces allow fighters on the frontlines to connect with others through social media to promote messages, warn of impending security risks, and facilitate transactions that keep operations humming within the operations of ISIL itself. Such technological advances are wholly new and represent a transformation in the ways that global jihadist movements, and more broadly criminal enterprises, can conduct business and expand their presence.

Social media also serves as a marketplace for the exchange of illicit goods and deviant services. Websites where document forgers, hackers, heavies, and hit men all proffer their services have been discovered. While these have migrated mainly to the Dark Web, onto sites such as the Silk Road or variations thereof—they also have been found on normal social media sites like Facebook.[30] On one social media site found in Mexico, while others boasted of their exploits, aspiring contract killers sought mentors and training: "I want to learn how to be a hit man," wrote one on Los Pulpos' wall, "Someone train me. I am capable of killing whoever [sic]."[31] Criminals operating online can be surprisingly service-oriented, offering help desks, customer service, and translations to support non-native speakers in their effort to communicate with potential victims.[32] In the migrant smuggling trade, smugglers advertise their illicit migration packages overtly to potential migrants, not only openly offering different classes of travel at varying prices, but also fraudulent documents and "relocation consultants" who will help illegal migrants successfully seek asylum and maximize their benefit packages.[33]

Social media is enabling a new category of professional "terrorist financiers" within terrorist groups, as they rely more heavily on social media to solicit donations and communicate with both donors and recipient radicals.[34] The U.S. National Terrorist Financing Risk Assessment identified nine terrorist financing cases that involved personal fundraising online or through social media, and in August 2014, the United States designated an al-Nusra Front financial facilitator who regularly solicited funds over social media.[35] Such social media accounts, known as *Ansar* accounts to ISIL—literally translating into "helpers," in reference to Arabian tribesmen who gave shelter and a home to the Prophet Muhammad and Islam "at the most critical point of the Prophet's mission,"—provide channels beyond media distribution and recruiting, to support operational coordination and broader development goals.[36] Social media serve as the primary interface for a number of terrorist financing transactions that take place in the informal economy, through *hawala* dealers, contributions to front nongovernmental organizations (NGOs), or even fake crowdfunding activities.[37] This presents an extraordinary challenge to law enforcement, as the majority of informal financial transfer systems have their foundations in trust-based relationships, rather than actual physical or electronic transactions. The majority of *hawala* transactions, an estimated 80 percent, occur only through local cash transactions, with a periodic "net settlement" of funds between *hawala* operators. Thus, the ability to ensure secure channels of communication are critical to ensuring that trust can be built and that the system functions effectively.[38]

Social media also facilitates information sharing of privileged material, both discretely as well as more broadly. An ISIL-affiliated British-born hacker and propagandist under the *nom de guerre* Hussain al-Britani, for example, was linked to fomenting lone-wolf terrorist attacks internationally, including a shooting at a Garland, Texas event featuring cartoon images of the Prophet Muhammad. Before his death via drone strike in August 2015, the 21-year-old disseminated a kill list of 1,351 American government and military personnel—obtained from a Kosovo national—via social media, threatening to "strike at your neck in your own lands."[39]

Tools that allow for encryption of communication make the job of intelligence officials all the more difficult. As the Federal Bureau of Investigation's top counterterrorism official, Michael B. Steinbach said in response to new uses of social media technologies by global criminal organizations, "We're past going dark in certain instances. We are dark."[40] In addition, *Ansar* accounts continuously monitor supporting social media communities for both terrorist groups and drug cartels. For example, in response to hacks allegedly stemming from Western intelligence communities, an ISIL sympathizer (@Moha5er) recently tweeted, "Warning significant Ansar accounts compromised today. Beware of opening any link without verifying the integrity of the author." Use of such methods seeks to delegitimize efforts taken by Western governments to stem the growth of such criminal organization online.

Evaluating and Strengthening Responses to Social Media Capture

Recognizing and acknowledging the extent to which illicit markets and deviant groups are enabled by social media is the first obstacle, and a considerable one. Finding effective and appropriate responses is proving an even greater challenge. The threat presented by terrorist and criminal activities on social media is clearly vast. Awareness and initiatives are growing in number, but are still highly fragmented and experimental. Along with task forces developed by states and social media companies themselves, online vigilantes—from collectives tied to the online activist group "Anonymous" to individual white hat hackers—are taking on criminal organizations, reporting their activity and working to mitigate their outreach.

If use of social media is propaganda in the modern day, and an instrumental tool in the hybrid warfare of the 21st-century battlefield, the insurgents are far outstripping states in their efficacy of use. The U.S. Department of State's counter-narrative effort is a campaign known as "Think Again, Turn Away" (@ThinkAgain_DOS), which has a mere 23,000 followers and lacks engagement, personalization, or creativity. In Europe, the Western region hardest hit by the foreign terrorist fighter phenomenon, the number of social media counter-messaging and outreach efforts have proliferated exponentially, but have had minimal impact. For example, a British group has started a video campaign known as #NotAnotherBrother, which targets young Muslim men in the UK who are considering going to fight for ISIL, highlighting the damage to families, but the video has been heavily criticized for its failure to resonate with the priorities of its target demographic.[41] States' institutions and mechanisms are, arguably, poorly suited to achieve the level of individualized attention required to successfully counter-message either within the framework of violent radicalization or gang identity. States necessarily work from the viewpoint of the protection needs of their citizens and gain their legitimacy precisely from the mass majority from which those attracted to deviant groups feel a sense of alienation or grievance. Their clumsy efforts to counter-message read more like propaganda than compelling rhetoric, and efforts to engage in dialogue with opposing groups have come across as defensive, often reinforcing terrorist arguments rather than breaking them down.[42] As a recent leak of a confidential U.S. State Department assessment of ISIL strategy soberly concluded, "When it comes to the external message, our narrative is being trumped by ISIL's. We are reactive—we think about 'counter-narratives' not 'our narrative.'"[43]

Experts have suggested that states' efforts would be better directed at development interventions targeting root causes, or classic law enforcement investigations to disrupt operations, rather than trying to battle extremist groups and criminal enterprises in the battle for hearts and minds. Fundamentally, states will rarely have the kind of manpower to engage in the long-term cultivation of relationships on an individual basis that ISIL recruiters have shown, for example, nor would it be their priority to do so. It is only in the case of high-value targets that such investment of resources would be justifiable, but in those cases the result is more likely to be sting operations or indictments, rather than ideological challenges, though these strategies are not without their concerns.[44] As criminals and terrorists increasingly take to social media, law enforcement agencies have developed new investigative strategies, and had some notable successes. One self-promoting, web-addicted drug lord, Rodrigo Arechiga Gamboa (known as El Chino and rumored to be the lover of Claudia Ochoa Felix) was arrested by U.S. authorities in December 2013, as he arrived at Schiphol Airport in Amsterdam. Authorities identified him not by his face, which was always obscured, but by the skull-shaped diamond ring, which he used as his signature in the pictures on his multiple social media accounts of his sports cars, yachts, weapons, jewelry, money, and extravagant parties.[45] Eighty percent of U.S. law enforcement professionals profess to actively use social media in investigations, but this remains an organic and informal process; 75 percent of those officials were self-taught, as less than half of these agencies have a formal procedure for the use of social media in criminal investigations. They have employed social media for everything from collecting evidence, seeking witnesses, identifying criminals and crime scenes, and mapping criminal networks. "My biggest use for social media has been to locate and identify criminals. I have started to utilize it to piece together local drug networks," said a police officer quoted in a LexisNexis survey.[46]

Tracking and monitoring social media interactions through social network analysis (SNA) tools allows the visual documentation of the relationships between actors, highlighting strong connections that could span a global domain. While use of SNA-style investigations are embryonic in the United States, they are even scarcer elsewhere, especially in countries where escalating crime and violence are pressing social problems. Fundamentally, moving towards SNA-style analysis is a big shift for law enforcement, which has traditionally focused on specific crime types (e.g., drug trafficking, human trafficking, arms dealing) or even individual incidents, each in its own silo. This is a nonsensical restriction when organized crime groups are increasingly poly-crime syndicates working in a versatile manner across criminal markets. Furthermore, national law enforcement agencies are predominantly constrained within their national boundaries, with knowledge concentrated within their local environment. For example, even while awareness has grown about the links between West African and Latin American groups trafficking cocaine to Europe, cooperation is minimal and there is still precious little information or common knowledge of their connections and interactions.[47] Instead, when SNA enabled by technology is included within the analysis, it can help to identify pivotal "nodes" in the criminal economy, or actors that function as a bridge between the criminal and the legitimate economy (e.g., lawyers, bankers, businessmen, corrupt state officials) and which are often points of vulnerability to be seized by law enforcement investigations.[48]

The need to update tools to counter the financing of terrorism through social media and informal economies is becoming increasingly urgent. The Financial Action Task Force (FATF)—the principle custodian of AML, CFT policy, and best practices—remains unable to offer better solutions, but flags the issue as being of primary concern. Reducing access to funding will have a potent impact on the ability of both criminal and terrorist groups to conduct their activities and meet their objectives, and warrants considerable extra analysis and attention.[49] Continuing to invest in and rely upon traditional AML approaches may give the impression of action in the traditional sense of technical cooperation (i.e., passing legislation, building capacity, etc.), but is largely ineffectual. Symbolic prosecutions and asset seizure in the case of criminal groups in particular, given that the ability to communicate power and influence is so central to their identities and efficacy, can prove a powerful counter-message and undermine their authority.

In environments of chronic insecurity, whether due to terrorist conflict or criminally motivated gang violence, a further impact has been to compromise the capacity of state institutions and the media to play their role as a bulwark against insecurity and the erosion of civil liberties. A combination of violence and corruption has had a highly detrimental effect on public protest and media freedom in a number of theaters, as crime groups and terrorists have targeted state officials, law enforcement, and journalists. In the war of words, ideas, and influence, the role of those that are the principle broadcasters have become particularly central. ISIL and al-Shabaab both target journalists as a way of sending a message to the broader population; as a consequence, Somalia is consistently ranked in the top three most dangerous places in the world for journalists.[50] A Reporters Without Borders inquiry report in 2013 found that organized crime has become a "fearsome predator for journalists in many parts of the world," identifying Honduras, Guatemala, Brazil, and Paraguay in Latin America, as well as Afghanistan, Pakistan, China, Kyrgyzstan, and the Balkans, as countries where the beat is most dangerous. They found that 141 media workers and journalists had been killed during the decade of the 2000s, in attacks and reprisals blamed on criminal groups.[51] Increasingly, these killings are preceded by threats, frequently delivered through social media. The Committee to Protect Journalists estimated that almost a third of the murdered journalists were either taken captive or tortured before their deaths, with the intention to send "a chilling message to the entire news media."[52] The result has been a self-censoring effect on news media across a number of parts of the world, and a devastating impact on one of the primary means by which to fight organized crime.[53]

Similarly, states are making efforts to silence the voices of those promoting criminal acts or terrorist agendas. As states have struggled with issues of capacity and resources, much of the emphasis has also been placed on the potential role and responsibility of the private sector to engage in the fight against deviant networks and behavior. States have put pressure on internet firms and social media sites to be more vigilant in monitoring content, filtering, and closing accounts that are deemed to be a security risk, and in some cases have heavily criticized platforms for failing to proactively alert law enforcement to potential risks. For example, the UK's Intelligence and Security Committee (ISC) investigation into the murder of a British soldier, Lee Rigby, by two Islamic extremists concluded that the only "decisive" possibility for preventing the attack involved cooperation and proactive engagement by Facebook. The social media platform had committed to communicate the graphic intent to murder a soldier, but the company failed to either identify or pass on the threat.[54] The UK's subsequent introduction of the Counter Terrorism and Security Act in 2015, grants intelligence agencies the power to conduct mass surveillance and store data from emails and other internet data from social networking sites and messaging services.[55] Similarly, the U.S. Senate Intelligence Committee approved a bill in June 2015, which would require social media companies to alert federal authorities when they become aware of terrorist-related content on their sites. Repressive states that make moves to limit the access of their citizens to social platforms come under heavy censure from the international community (consider, for example, the recent efforts by the Turkish government to ban Twitter and YouTube), but according to watchdog Freedom House, in one year alone between May 2013 and May 2014, 41 countries (20 percent of the world's states) passed or proposed legislation to penalize legitimate forms of speech online, increase government powers to control content, or expand government surveillance capabilities, including many Westphalian states that pride themselves on their democratic tradition and freedom of speech, as the fight against crime and terrorism increasingly come into conflict with the right to privacy.[56]

Consequently, the various platforms have struggled with their obligations in this war of ideas and identity. Most have developed policies forbidding the posting of violent, extremist, or offensive (hate crime) material on their platforms, but have found it difficult to control the rapid exchange of content and ideas, especially as accounts can be opened and closed in a manner of minutes. But at a more fundamental level, both the internet and social media were precisely created—and have since thrived—to allow the free exchange of ideas and information. The social media companies, while broadly supportive of efforts to prevent terror and compliant with national laws for disclosure, have, at the same time, been reticent to infringe on their users' rights to privacy, to impact on the platforms' ability to promote free speech, or to deny historical fact, regardless of its propaganda value.[57] Subsequently, they have shown even greater resistance to overtures from states to be drawn into direct efforts to create explicit counter-messaging strategies and content.[58] In their own words, "The internet is fundamentally a connective technology, and as it continues to grow, it has enabled illicit actors to better connect and coordinate complex actions, better manipulate and launder money, and better map and understand data in realtime [sic]. It behooves us in the technology community to ensure that innovative tools are also being used to disrupt illicit networks, and that on balance technology is a force for good. This is a mission we welcome."[59] And thus, their engagement has predominantly been restricted to providing some basic capacity building for public officials on how to make messaging impactful, funding civil society efforts to enhance security through social media, and offering space for discussion and debate. For example, the first Google Ideas conference in 2011 took the theme of "Terrorist Networks," the second in 2012 was titled "Illicit Networks," and the follow-up was to create a secure platform for "formers"—former terrorist combatants—to contribute to efforts to counter violent extremism.[60]

With the state rendered ineffective, traditional media silenced, and the private sector reluctant to actively engage, countering the negative influences of social media has, thus, increasingly fallen into the domain of civil society and nonstate actors. There are a number of ways that social media in the hands of civil society can and have been used to enhance civilian security and counter the pernicious influence of deviant groups. These have ranged from addressing local conflicts and creating community-based solutions, to internationally focused advocacy and awareness raising.

On a number of occasions, social media have been instrumental in campaigns to galvanize the international community to engage in ongoing crisis or conflict resolution efforts, calling attention to violence or injustice. The risk in these large-scale campaigns is that while they can be incredibly powerful in capturing attention and mobilizing public opinion in the short term, they have a tendency to very quickly fall out of vogue. As opposed to genuine activism, this has resulted in a phenomenon known as "clicktivism" that uses the power of social media to exert pressure on political actors. However, because of its nature, it tends to result in a specific style of politically expedient response, one that is typically high in visibility but lacking in long-term commitment and sustainable momentum required to have a genuine impact on the challenge. There are numerous examples. The high-profile #BringBackOurGirls campaign, which was launched in response to the kidnapping of more than 275 schoolgirls by the Nigerian terrorist group Boko Haram in April 2014, resulted in widespread and high-profile outrage, but little in terms of results—the majority of those girls are still missing over 2 years later, and Amnesty International estimates that Boko Haram has subsequently abducted more than 2,000 women with little fanfare from the international community.[61] What the campaign did mobilize, however, was the necessary justification for the Nigerian government to mount a highly militarized and violent campaign against Boko Haram, consisting of a security crackdown that certainly exacerbated tensions between the group and the state and made a negotiated solution more remote.

A similar example of clicktivism at work can be found in the sudden global awakening to the crisis of Syrian migration, responding to images of #AlyanKurdi, a three-year-old child drowned on the shores of Turkey while trying to get to Europe. That single image brought home the severe consequences of a crisis that has been building for years, and triggered a response that the thousands who died in boats in the Mediterranean, or in the brutal Sahara, had failed to mobilize. But what was that response? Celebrity singles and touching artwork, campaign funding to NGOs to support humanitarian work, and migrant protection; all of these are useful, but ultimately insufficient, including a promise by European leaders to "get tough on smugglers" which materialized as warships in the Mediterranean and a meaningless Security Council Resolution, a response that two UN Special Envoys said in an open letter, "misses the mark."[62] As with the Boko Haram case, a clicktivist response comes with the flamboyant use of the military, without offering legal channels for migration or addressing the long-term root causes of displacement and mobility. The damaging consequence of such a response is that not only does it not respond to the genuine drivers of the crisis, but it is far more likely to result in pushing migrants more deeply underground and further from protection, entrenching the role of smuggling rings and increasing the risks of human rights violations. Furthermore, in less than a month, searches for the word "refugee" or "migrant" have already declined by half, as the refugee crisis fades again from public consciousness in favor of a new *cause célèbre.* The seeds for longer-term social, economic, and justice challenges have not been sown.[63]

There is little doubt that social media activism has a tendency to create lazy, feel-good solutions and short-term outpourings of aid, as opposed to genuine, long-term advocacy for reform. Where it can have value, however, is when it is properly curated within a national audience, aligned to a genuine political movement, and targeted to apply systematic pressure on political systems to change, or to create a grassroots campaign to enhance information, transparency, and democratic reform. In this way, social media can be used to make states more sensitive to audience costs (that is, the benefits and drawbacks that it could accrue from lying or telling the truth) as it allows citizens to engage with their governments and with others in civil society in ways that were not possible in the past. Much has been attributed to the role of social media in initiating the Arab Spring, for example, though its subsequent sustainability has clearly been brought into question. This has also been used to good effect in the case of the #Iguala, where student protestors in the Mexican city of Iguala were handed to organized crime groups by local governors to have them "disappear." Sustained momentum on the part of social activists has seriously damaged the credibility of the Peña Nieto administration and ensured that the impressive economic performance has failed to offset his government's inadequacy in addressing human rights abuses, criminal impunity, and rising violence. At a less transformative level, social media can also lead to a greater degree of clarity or veracity in reporting about events and provide alternatives to state-controlled media.

At the local level, digital activists are self-organizing in virtual communities and are using their networks to curate and disseminate information to protect themselves.[64] Social media platforms have helped to reduce civilian casualties by serving as early warning systems, helping citizens stay connected to rapid response humanitarian organizations or security providers, and by providing information to citizens during and in the aftermath of crimes. A Kenyan village chief claims to have drastically reduced crime rates in his community by sending out tweets instructing citizens what to do in the aftermath of insecurity.[65] In Mexico, Twitter has been used to create a real-time "security monitor" that allows average citizens to report crimes as they happen, allowing a mapping of flare-ups of violence or areas of particular insecurity, and research suggests that about 1.5 percent of all Mexicans have tweeted about the drug war, which amounts to almost 5 percent of the country's online population.[66] The costs of doing so, however, have become increasingly lethal. Mexican bloggers and online activists are finding themselves violently targeted in parallel ways to the fear tactics used to silence the country's traditional media reporters. In October 2014, the site administrator for one such security monitoring site, *Valor por Tamaulipas*, a community site which has more than 100,000 followers on Twitter and over half a million on Facebook, had her execution at the hands of local drug trafficking groups broadcast via her own Twitter feed. This middle-aged mother, a physician and concerned citizen, had used the site to build her community and urged its citizens to speak out against gang violence.[67]

Whether social media ultimately will prove itself to be a tool of greater pacification or belligerence remains to be seen; that it certainly serves a powerful lever capable of conveying advantages to whichever side in a conflict wields it most strategically is beyond question. Fundamentally, free societies have always relied upon their citizenry and their values to be their last line of defense, and as nascent lessons from state-led efforts have demonstrated, authenticity of voice is extremely important. Seeking better solutions to the criminalization of social media will require creating safe spaces for courageous voices to share information and communicate without fear and intimidation. Doing this well is a design challenge that poses further questions about free speech, rights of expression, and personal protection in the global digital ecosystem, but it is an urgent and necessary priority.[68] It will also require all of the online community recognizing the power and choices that it makes when engaging with social media. Clicktivism has demonstrated the ease of mobilization on social media, but it has also made us lazy and immune to consequences. Clicking on something—whether to watch, "like," or share—amplifies its message, whether that is one of social good or fear, and as citizens within a global social media community we need to learn to click responsibly. In the social media space, where identity, ideas, and ideology have the greatest currency, a multitude of voices and sustained commitment are required to have an impact.

Notes

[1] J.M. Berger and Jonathon Morgan, *The ISIS Twitter Census Defining and describing the population of ISIS supporters on Twitter* (Washington, DC: Brookings Institution, 2015), available at <http://www.brookings.edu/~/media/research/files/papers/2015/03/isis-twitter-census-berger-morgan/isis_twitter_census_berger_morgan.pdf>.

[2] Peter Bergen, Courtney Schuster, and David Sternman, "ISIS in the West," *New America Foundation,* November 16, 2015, available at <https://www.newamerica.org/new-america/isis-in-the-west-2/>; Rachel Briggs and Tanya Silverman, *Western Foreign Fighters Innovations in Responding to the Threat* (London: Institute for Strategic Dialogue, 2015), available at <http://www.strategicdialogue.org/ISDJ2784_Western_foreign_fighters_V7_WEB.pdf>; Rukmini Callimachi, "ISIS and the Lonely Young American," *The New York Times,* June 27, 2015, available at <http://www.nytimes.com/2015/06/28/world/americas/isis-online-recruiting-american.html?_r=0>.

[3] Gabriel Weimann, *Terror on the Internet: The New Arena, the New Challenges* (Washington, DC: U.S. Institute of Peace Press, 2006).

[4] Briggs and Silverman, *Western Foreign Fighters Innovations.*

[5] Rory Medcalf, "War Tweets: War propaganda has finally moved onto Twitter," *American Review,* available at <http://americanreviewmag.com/opinions/War-tweets>.

[6] Duncan Omanga and Pamela Chepngetich-Omanga, "Twitter and Africa's 'War on Terror': News Framing and Convergence in Kenya's Operation Linda Nchi," in *New Media Influence on Social and Political Change in Africa,* ed. Anthony Olorunnisola and Aziz Douia (Hershey, PA: IGI Global, 2013), 241-256; David Smith, "Al-Shabaab in war of words with Kenyan army on Twitter," *The Guardian*, December 13, 2011, available at <http://www.theguardian.com/world/2011/dec/13/al-shabaab-war-words-twitter>.

[7] Darren Boyle, "Somali terror group Al-Shabaab calls for 'Westgate-style' shopping centre attack on London's Oxford Street and Westfield malls in Stratford and White City in chilling new video fronted by 'English' Jihadi," *DailyMail*, February 23, 2015, available at <http://www.dailymail.co.uk/news/article-2964218/Somali-terror-group-Al-Shabaab-calls-Westgate-style-shopping-centre-attack-London-s-Oxford-Street-chilling-new-video.html>.

[8] Briggs and Silverman, *Western Foreign Fighters Innovations.*

[9] Eric Davis, "ISIS's Strategic Threat: Ideology, Recruitment, Political Economy," *The New Middle East,* August 17, 2014, available at <http://new-middle-east.blogspot.com/2014/08/isiss-strategic-threat-ideology.html>.

[10] Briggs and Silverman, *Western Foreign Fighters Innovations.*

[11] Erin Saltman and Melanie Smith, *Till Martyrdom Do Us Part: Gender and the ISIS Phenomenon* (London: Institute for Strategic Dialogue, 2015).

[12] Carolyn Hoyle, Alexandra Bradford, and Ross Frenett, *Becoming Mulan? Female Western Migrants to ISIS* (London: Institute for Strategic Dialogue, 2015), available at <http://www.strategicdialogue.org/ISDJ2969_Becoming_Mulan_01.15_WEB.PDF>. The phenomenon is beautifully described in the outstanding investigation in Callimachi, "ISIS and the Lonely Young American."

[13] Briggs and Silverman, *Western Foreign Fighters Innovations.*

[14] Berger and Morgan, *The ISIS Twitter Census Defining and describing the population of ISIS supporters.*

[15] Hoyle, Bradford, and Frenett, *Becoming Mulan?*

[16] Gambetta, *Codes of the Underworld.*

[17] Reitano and Hunter, *Contests and Compacts.*

[18] Wahlstedt, *Tattoos and Criminality.*

[19] Gambetta, *Codes of the Underworld.*

[20] UNODC, *Results of a pilot survey of forty organized criminal groups in sixteen countries* (Vienna: United Nations, 2002).

[21] John P. Sullivan and Samuel Logan, "MS-13 Leadership: Networks of Influence," *The Counter Terrorist*, August 2010, available at <http://digital.ipcprintservices.com/display_article.php?id=428186>.

[22] Ciara Byrne, "Drugs, Guns and Selfies: Gangs on Social Media," *Fast Company*, February 15, 2015, available at <http://www.fastcompany.com/3041479/drugs-guns-and-selfies-gangs-on-social-media>.

[23] Hasani Gittens, "El Chapo's 'Official' Twitter Takes On Trump, Mexican President," *NBC News,* July 14, 2015, available at <http://www.nbcnews.com/news/latino/el-chapos-official-twitter-takes-trump-mexican-president-n391411>.

[24] Sandy Banks, "'Cyber banging' drives new generation of gang violence," *LA Times,* October 3, 2015, available at <http://www.latimes.com/local/crime/la-me-1003-banks-lapd-gang-shootings-20151003-column.html>.

[25] For more information on Quid, the data analytics software used in the chapter, please visit http://www.quid.com.

[26] Robert Muggah, "The rising threat of organised crime on social media," *World Economic Forum,* July 27, 2015, available at <http://www.weforum.org/agenda/2015/07/social-media-violence/>.

[27] Evelio Contreras, "Inside the life of a drug trafficking teen," *CNN*, August 15, 2015, available at <http://www.cnn.com/2015/08/12/us/inside-the-life-of-a-drug-trafficking-teen/>.

[28] News Team, "Use of Social Media in Kidnapping and Extortion," *KR Magazine*, October 2, 2015, available at <http://www.krmagazine.com/2015/10/02/analysis-olive-group-use-of-social-media-in-kidnapping-and-extortion/>.

[29] Saltman and Smith, *Till Martyrdom Do Us Part.*

[30] Raj Samani, "Cybercrime as a Service," *McAfee Research*, 2014, available at <http://www.mcafee.com/us/resources/white-papers/wp-cybercrime-exposed.pdf>.

[31] Miriam Wells, "Facebook Hitman Highlights Organized Crime's Online Presence," *InSight Crime,* June 21, 2013, available at <http://www.insightcrime.org/news-analysis/facebook-hitman-highlights-organized-crimes-online-presence>.

[32] Samani, "Cybercrime as a Service."

[33] Based on an extensive series of interviews by the Global Initiative with law enforcement in Europe and North Africa.

[34] Samuel Rubenfeld, "Social Media Emerges as Terrorism Fundraising Tool," *Wall Street Journal,* August 11, 2014, available at <http://blogs.wsj.com/riskandcompliance/2014/08/11/social-media-emerges-as-terrorism-fundraising-tool/>.

[35] Financial Action Task Force (FATF), *Emerging Terrorist Financing Risks* (Paris: FATF, 2008), available at <http://www.fatf-gafi.org/media/fatf/documents/reports/Emerging-Terrorist-Financing-Risks.pdf>.

[36] S.H.M. Ja'fri, *The Origins and Early Development of Shi'a Islam* (Oxford: Oxford University Press, 2002).

[37] FATF, *Emerging Terrorist Financing Risks.*

[38] FATF, *The role of hawala and other similar service providers in money laundering and terrorist financing,* (Paris: FATF, 2013), available at <http://www.fatf-gafi.org/publications/methodsandtrends/documents/role-hawalas-in-ml-tf.html>.

[39] Ja'fri, *The Origins and Early Development of Shi'a Islam.*

[40] Brian Bennett, "With Islamic State using instant messaging apps, FBI seeks access to data," *LA Times,* June 8, 2015, available at <http://www.latimes.com/world/middleeast/la-fg-terror-messaging-20150608-story.html>.

[41] Trevor Gundy, "In Britain, glut of anti-terrorism campaigns draws fans and critics," *USA Today,* August 15, 2015, available at <http://www.usatoday.com/story/news/world/2015/08/15/britain-glut-anti-terrorism-campaigns-draws-fans-and-critics/31773737/>.

[42] Briggs and Silverman, *Western Foreign Fighters Innovations*.

[43] Richard Stengel, "State Department Memo on the Islamic State Group," *New York Times,* June 9, 2015, available at <http://www.nytimes.com/interactive/2015/06/12/world/middleeast/document-state-department-memo-on-the-islamic-state-group.html>.

[44] Vanda Felbab-Brown, *Despite its siren song, high-value targeting doesn't fit all: matching interdiction patters to specific narcoterrorism and organised crime contexts* (Washington, DC: Brookings Institution, 2013), available at <http://www.brookings.edu/research/papers/2013/10/01-matching-interdiction-patterns-narcoterrorism-organized-crime-contexts-felbabbrown>.

[45] See: Tom McKay, "These 28 Instagram Pictures Just Busted One of the Biggest Mexican Drug Lords," *News Mic,* January 23, 2014, available at <http://mic.com/articles/79841/these-28-instagram-pictures-just-busted-one-of-the-biggest-mexican-drug-lords#.E3CCtDA50>; Tom Porter, "Mexican Cartel Members Face Arrest After Sharing Crimes on Facebook and Twitter," *International Business Times,* January 5, 2014, available at <http://www.ibtimes.co.uk/mexican-cartel-members-face-arrest-after-sharing-crimes-facebook-twitter-1431081>.

[46] LexisNexis Risk Solutions, *Survey of Law Enforcement Personnel and Their Use of Social Media*, 2014, available at <www.lexisnexis.com/investigations>.

[47] Margaret Shaw, "Illicit Narcotics Transiting West Africa: Actors, Finances and Impact," in *Illicit Financial Flows: The economy of illicit trade in West Africa,* OECD/AfDB (Paris: OECD Publishing, forthcoming).

[48] Anine Kriegler, "Using Social Network Analysis to Profile Organised Crime," *Institute for Security Studies*, 2014, available at <https://www.issafrica.org/uploads/PolBrief57.pdf>.

[49] FATF, *The role of hawala and other similar service providers.*

[50] Reporters Without Borders, "Non-State Groups: Tyrants of Information," 2015, available at <https://www.reporter-ohne-grenzen.de/fileadmin/Redaktion/Presse/Downloads/Ranglisten/Rangliste_2015/150211_Nichtstaatliche_Gruppen_EN.pdf>.

[51] Benoit Hervieu, "Organized Crime Muscling in on the Media," 2014, available at <https://www.reporter-ohne-grenzen.de/fileadmin/Redaktion/Presse/Downloads/Berichte_und_Dokumente/2011/110224_RSF-Bericht_Organisierte_Kriminalitaet_Pressefreiheit.pdf>.

[52] Elisabeth Witchel, "Getting Away with Murder," *Committee to Protect Journalists,* April 16, 2014, available at <https://cpj.org/reports/2014/04/impunity-index-getting-away-with-murder.php>.

[53] Global Initiative, "Dangerbeat: Journalists Under Fire for Investigating Organized Crime," *Global Initaitive against Transnational Organized Crime*, August 14, 2014, available at <http://www.globalinitiative.net/dangerbeat-journalists-under-fire-for-investigating-organized-crime/>.

[54] Natasha Lomas, "U.K. Government Points Finger Of Blame At Web Firms For Counter-Terror Failures," *Tech Crunch,* November 26, 2014, available at <http://techcrunch.com/2014/11/26/surveillance-scapegoats/>.

[55] *Counter-Terrorism and Security Act 2015,* London, 2015, available at <http://www.legislation.gov.uk/ukpga/2015/6/contents/enacted>.

[56] Eliana Dockterman, "Turkey Bans Twitter," *TIME,* March 20, 2014, available at <http://time.com/32864/turkey-bans-twitter/>; "Freedom on the Net 2014," *Freedom House,* 2015, available at <https://freedomhouse.org/report/freedom-net/freedom-net-2014#.Vp5QW_nR_IU>.

[57] Scott Higham, and Ellen Nakashima, "Why the Islamic State leaves tech companies torn between free speech and security," *The Washington Post,* July 16, 2015, available at <https://www.washingtonpost.com/world/national-security/islamic-states-embrace-of-social-media-puts-tech-companies-in-a-bind/2015/07/15/0e5624c4-169c-11e5-89f3-61410da94eb1_story.html>.

[58] Nancy Scola, "Obama's anti-ISIL push falls flat on social media," *Politico,* August 12, 2015, available at <http://www.politico.com/story/2015/08/obamas-anti-isil-push-falls-flat-on-social-media-121301>.

[59] Jared Cohen, Director, Google Ideas, as quoted in Neal Ungerleider, "How Google Fights Terrorists and Human Traffickers," *Fast Company,* July 17, 2012, available at <http://www.fastcompany.com/1842993/how-google-fights-terrorists-and-human-traffickers>.

[60] Google Ideas, *Summit against Violent Extremism,* June 26-29 2011, available at <https://www.google.com/ideas/events/save-2011/formers/>.

[61] See: Maeve Shearlaw, "Did the #bringbackourgirls campaign make a difference in Nigeria?" *The Guardian,* April 14, 2015, available at <http://www.theguardian.com/world/2015/apr/14/nigeria-bringbackourgirls-campaign-one-year-on>; Amnesty International, "Our job is to shoot, slaughter and kill," *Boko Haram's reign of terror in North East Nigeria*, 2015, available at <http://www.coalitionfortheicc.org/documents/AFR4413602015ENGLISH.PDF>.

[62] François Crépeau and Francisco Carrión Mena, "Statement by the Special Rapporteur on the human rights of migrants, François Crépeau, and the Chair of the UN Committee on the Protection of the Rights of Migrant Workers and Members of Their Families, Francisco Carrión Mena," *United Nations Human Rights,* October 23, 2015, available at <http://www.ohchr.org/EN/NewsEvents/Pages/DisplayNews.aspx?NewsID=16641&LangID=E>.

[63] Kim Ghattas, "The Sad Fading Away of the Refugee Crisis Story," *Foreign Policy,* October 19, 2015, available at <http://foreignpolicy.com/2015/10/19/the-sad-fading-away-of-the-refugee-crisis-story/>.

[64] Muggah, "The rising threat of organised crime on social media."

[65] Gabrielle Ramaiah and Jason Warner, "Four ways social media could transform conflict in Africa," *CNN,* July 16, 2012, available at <http://globalpublicsquare.blogs.cnn.com/2012/07/16/fours-ways-social-media-could-transform-african-conflicts/>.

[66] Andres Monroy-Hernández, Emre Kiciman, Danah Boyd, and Scott Counts, "Narcotweets: Social Media in Wartime," *Microsoft Research*, 2012, available at <http://research.microsoft.com/pubs/160480/IC-WSM12-093.pdf>; Muggah, "The rising threat of organised crime on social media."

[67] Jason McGahan, "She Tweeted Against the Mexican Cartels, They Tweeted Her Murder," *The Daily Beast*, October 21, 2014, available at <http://www.thedailybeast.com/articles/2014/10/21/she-tweeted-against-the-mexican-cartels-they-tweeted-her-murder.html>.

[68] Muggah, "The rising threat of organised crime on social media."

11

"We Pay, You Pay": Protection Economies, Financial Flows, and Violence

Mark Shaw

Across the globe illicit flows have not only continued to sustain themselves; they have expanded into new sectors, with a series of new criminal markets now emerging. What is remarkable about this development is not so much the growth, although in some sectors this has been truly astonishing, but that the policy solutions to the phenomenon have been so underwhelming.[1] After decades of focus on drugs markets, most analysts have conceded that law enforcement and associated development policies have failed to curb the power and expansion of illicit networks. In conflict zones, notably in Africa, where illicit trafficking has now become a standard—albeit understudied—feature, a series of tentative programmatic experiments appears not to have achieved the envisaged results.[2]

As their depth and breadth have grown, "organized criminal networks" and "illicit trafficking" are often now identified as key drivers of conflict. Yet, exactly how illicit markets contribute to conflict is not well understood. More often than not, the connection between organized crime and conflict is assumed rather than explained. But understanding the connection between illicit economies, their funding, and the promotion of violence is crucial for policymakers. This applies to several conflict zones, but methodologies for doing so have been less developed in the specific context of Africa, where countries are experiencing new forms of warfare, often on state peripheries, across several states, and involving multiple protagonists.[3] The intent of this chapter is to provide a conceptual framing through which violence or the threat thereof can be linked to illicit economies; by doing so, the chapter will provide a better understanding of the financial transactions and flows that provide the stimulus for violence. This approach underlines that not all illicit markets generate conflict and the nature and types of violence differ between illicit economies. This distinction is critical for policymakers constrained by limited resources who need to prioritize their interventions.

This chapter argues that a useful analytical device to understand the linkage between local power relations and evolving conflict dynamics is that of "protection economies." Such economies, and the transactions they involve, provide a means to structure the relationship between criminal entrepreneurs, armed groups, state structures, and the use of violence. This analysis takes its inspiration from a wider body of literature, often referred to as "protection theory," which draws on the formation of mafias and other groups and their use of violence to obtain tribute or to extort payment, and has sought to understand how violence, or the threat thereof, becomes a marketable commodity in itself.[4] In other words, protection emerges as a key service in an economy where there is a high demand for it and where the state does not or cannot provide that service, or does not do so in an even-handed way. The buying and selling of protection of the transport of illicit goods creates dependent relationships between smugglers and those capable of protecting the lines of transport; the existence of this relationship provides valuable opportunities for policymakers to tailor interventions to stem the flow of illicit goods by targeting the market for protection.

Assumptions and Methodology

Protection economies provide a useful lens through which to examine the linkage between organized crime and conflict, precisely because they mirror two important developments in the current global political economy: the dramatic growth of illicit flows and their channelling through weak or conflict ridden states; and, a proliferation of armed groups in such zones of instability which do not necessarily have external funding sources.

Though to build this framework, a number of key assumptions should be laid out clearly at the outset. The first is an acceptance that organized crime or illicit trafficking may be key drivers of conflicts, but they are not the only ones. How organized crime or illicit markets affect conflict is often context-specific, including how such resources are absorbed and used by criminal and/or political groups. The second is a dismissal of the idea that only nonstate actors engage in cooperative efforts with organized crime and illicit trafficking during conflict. State actors, too, make these linkages. Importantly, how state and nonstate actors participate in protection economies needs to be accommodated in the same model for it to be analytically useful. An important theme of this chapter then is "to bring the state back in" to our discussion of the vectors and beneficiaries of illicit flows and markets. Third is a concession that the flows of resources between different actors (what will be termed "transactions" in the analysis that follows) are difficult, if not impossible, to document accurately. So, assumptions are made as to how resources have flowed, based on interviews in the field and the presentation of a series of case studies. Finally, the term "organized crime" is a somewhat clumsy conception when applied to complex criminal markets and their connections to conflict, particularly in Africa.[5] For this reason, the chapter avoids the theoretical and conceptual debate around this issue by sticking to a narrower terminology around criminal flows or markets. The latter are seen as the demand or supply of goods or services (including violence itself) that are broadly considered to be illegal, "uncivil," or harm-inducing.[6]

The chapter begins with a brief theoretical overview of the concept of protection. This analysis is then applied to three case studies along a single trafficking flow, that of commodities, including drugs and human beings that have and are, transiting West Africa through Guinea-Bissau, the Sahel through Mali, and North Africa through Libya. These three case studies are useful as they each provide a different example of a protection economy, but drawing on the same set of conceptual principles. To broaden this discussion, several other cases from a wider repertoire are also used, but the focus is unashamedly African; the continent is now at the forefront of a series of violent engagements underpinned by illicit markets.[7] These allow us to identify the beginnings of a typology of protection. A follow-on section examines the use of resources by these actors, drawing conclusions in relation to the financing of conflict. The chapter concludes by drawing some implications for policy.

What Is a "Protection Economy?"

Understanding how systems of power and control function can generally be traced to the resource flows that underpin them. The sets of individual transactions that define these systems are generally hidden from view and difficult for the casual observer to identify. That is particularly the case in illicit activities, where there is a strong motivation to retain a high degree of secrecy around how relationships and their structuring transactions operate.[8] In this sense, protection economies that develop around illicit flows are hidden, yet they are crucial to understanding both underlying systems of violence and related distributions of payment.

In states with functioning systems of rule of law and high levels of capacity, the state itself provides that protection in exchange for tax. Institutions of the government are designed to guarantee protection when economic transactions occur. Thus, when firms contract with each other or with individuals, there is recourse to the courts or justice system should any of the parties fail to deliver their side of the bargain. Thus, if a firm does not provide the goods required, or they are of poor quality, or if an individual does not pay in exchange for the goods, either party may approach the courts for compensation for their loss.

Charles Tilly's seminal study of European state formation noted that this process and the development of organized crime had much in common.[9] In each case, payment is extracted for protection. How that payment is distributed and absorbed into the system determines systems of power and control. In Tilly's analysis, this constituted the first steps in the formation of Western European states; the use of these resources became more equitable over time, ensuring popular acquiescence to the emerging order.

We generally define the development of "organized crime" in contrast to the state—after all, the latter is meant to counter the former. So, what happens when the state is not present; or the justice services it delivers are of poor quality; or the commodity being traded is illegal, and state intervention is either unlikely or would endanger both parties and scupper the deal? In those circumstances, parties may seek a guarantee for their investment in the commercial arrangement: in criminal economies or where the state is not present that security may take the form of a "fee" given in exchange for protection.

The process is, however, more complex than this. In cases where the state is not in a position to intervene, powerful forces that have the monopoly on violence—or at least have the capacity to engage in violence without significant opposition—are in a position to extract rents from the movement of commodities. These are mainly illegal commodities often because legal forms of economic activity have declined or have much lower profit margins. It has been shown in the Sahel, for example, that the provision of protection around more profitable illegal markets extends to legal ones in the absence of state presence and/or enforcement.[10]

It is worth noting here that this emerging arrangement around protection closely resembles that of businesses that are extorted for money; here, the entity doing the extortion may well be the same as that which disrupts the trade if payment is not made.[11] Such arrangements have been termed "the extraction of rents" in other contexts; in reality though, they can be better understood as a simple transaction for security—even if security and insecurity stem from the same source. As in the case of extortion, the extorter often makes a judgment about how much can be paid, so that the business can continue to function—even if not optimally—but will not collapse altogether, thus eliminating the source of payments.[12]

In places where resource flows transit areas of poor or weak governance, or where corruption is already endemic, structured protection economies develop. This is based on the characteristics of the trade of high value illicit goods (for example, illegal narcotics), in which there is a simultaneous need to transport the good to a variety of locations globally, through a series of local networks. Protection economies are, therefore, often relationships that are generated by the link between global markets and local control. The position can also be reversed, as in the control of mining sites or oil refineries, where the source of the commodity is under control and available for rent extraction, with the requirement to move goods to market—or often in this case—to launder them into offshore licit markets. In both cases, local control is a feature of extracting payment from wider global economic flows. In a different context, such an arrangement has been termed "glocal," and the term retains some useful explanatory power to describe these relationships.[13]

Critically, since "payment for protection" implies at least two actors who transact with each other, protection theory recognizes the difference between two important role players within in the system.[14] The first is the criminal entrepreneur, generally the individual or group engaged in the economic transactions to move or extract the goods. The second is the entity that provides protection, which has been referred to in the African context as "an entrepreneur of violence."[15] They sell protection as a commodity in its own right. While it may often be hard to distinguish between the two functions when looking at conflict dynamics from afar, it is almost always the case that the entrepreneurial trade function is separate from the entrepreneurial violent one.

Rethinking the Role of the State

A key point here, and one already highlighted at the outset, is that we should remove our prejudice about who can and cannot be a "protector." Protection may take multiple forms, constitute complex networks, and most importantly, almost always involves state actors in some shape or form. Some recent literature has suggested that a distinction should be made between "state-sponsored protection rackets" and "private protection rackets."[16] In reality, however, the provision of protection almost always takes place along a spectrum, with state institutions performing a range of functions, dependent on the degree of capacity they may have.

Table 11.1. Typologies of Protection Economies

	State Control	Devolution	Mixed Control	Indirect Links	Mixed Local Control	Local Control
Roles	Full involvement of selected state actors. Direct protection or withdrawal of state actors at crucial times.	State as "gatekeeper" for regional and local control. Selects partners.	Agreements between state and local actors on the ground where both have a presence.	Weak state positions itself as interlocutor and receives payment.	Local state security actors and local strong men based on agreement.	Full local provision of protection. No state involvement.
Provider of Protection	State security forces.	State forces and local armed groups linked to "system of devolution."	Local militias and state actors, often in agreement as to "spheres of control."	Armed group with "quiet channel" to the state in exchange for payment.	Security forces without "central permission." Local armed actors.	Local armed groups.
Resource Flow	To senior levels of the state and/or security establishment. Small payouts at lower levels.	Senior state officials take a cut or an agreed tribute paid to state to maintain the system.	Local players with tribute paid at central state level.	Armed groups, but also to state actors who facilitate outside access.	Local security actors, state and nonstate.	Local players and leadership.
Implications	"Criminal state." Limited violence is easily quashed by dominant state.	Potential to resource local strongmen or promote conflict with those excluded from the system.	Local state presence in parallel with armed groups.	State further compromised in ability to respond. Acquiring resources strengthens armed group.	Increasing delinkage from the centre. Poor local service delivery. Conflict if alliances/agreements weaken.	Acquiring resources reinforces local territorial control.

Recent research suggests that the role of state actors as vectors in the criminal economy has been greatly underestimated; where present, they act almost without fail to provide some form of protection. Absent the rule of law and effective forms of democratic oversight, state agencies that attack extortion networks often take them over.[17] And, the prevailing wisdom that state involvement in the illicit economy as "protector" is always the result of, or results in, state weakness requires a more nuanced or contextualized view. In some cases, the state can use the protection function, at least in the short to medium term, to extend its reach.[18] In the longer term though, state power may be significantly compromised when that system breaks down.

The spectrum of protection networks and the potential role of the state are illustrated in the table below.[19] Six typologies of protection economies can be identified, each defined by the degree of state strength or weakness. A measure of state strength in each case is a feature more of the state's ability to control the criminal economy than to deliver services across the territory of the state. As Meehan has dryly noted, "to derive a more accurate analysis of the relationship between drugs (or any illicit commodity) and state-building it is imperative to engage with how states actually function, rather than how they ought to."[20] Thus, in the case of Guinea-Bissau, which could be argued to be a weak state based on the measure of its presence across the territory and its ability to deliver services, state institutions—despite their weakness, but given their comparative strength compared to other forces—were able to exclusively control drug trafficking.[21]

On the left of the spectrum are cases where there is full "state control" over the protection economy. In such cases, the senior level of the state or the security forces provide total protection for illicit trade, with direct payment being negotiated between senior state functionaries and those engaged in the illicit market. As mentioned, a good recent example of such a case is Guinea-Bissau, where the military (and connected political) hierarchy were the gatekeepers for the transit trade and used the available state security resources to maintain this position.

Further along the spectrum toward the right is the case where the central state acts as a "gatekeeper" for regional and local interests, despite the fact that it has little effective presence in the region where trafficking transits. In such cases, senior levels of the state may allow some armed actors to receive payment for protecting illicit trade, while denying others the privilege of doing so. Achieving such favor may be the result of political reasons such as, for example, the requirement to bolster one regional or local actor over others. In these cases, however, it must be the case that the overarching power lies with the highest level of the state to select or support local actors who control the trade. That presupposes some capacity at the central level for outside intervention, should "the wrong" groups "tax" or receive payment from trafficking. A good example of this typology, explained briefly in the next section, is Northern Mali before the coup in 2012.

In the middle of the spectrum are cases where there is "mixed control" between central state actors who have some presence (e.g., regional offices, a military base, control of posted elements of a centralized police) in the region and local actors who occupy the ground. Such an agreement may include an understanding around "spheres of control" and the payment of tribute to other regional or central government actors. The difference between this typology and the previous one within this analytical framework is the degree of presence of the state itself in the affected area. In the former case, the state only has a weak presence; in the latter, the state has a more permanent presence, although one that relies on agreements with local actors for the state to function at all.

In systems of "mixed local control," local state security actors and local strong men agree on dividing up the spoils into different zones of influence. The difference with the models above is that local security forces do not seek or receive permission from the central state, and set out to make money on their own, often by forging agreements with local partners. In short, the state retains some influence, but largely through the actions of local commanders or lower-level officials. A good example of such a case is Northern Niger, where agreements between state actors and local power brokers have historically provided a system of protection for trafficking and migrant smuggling.

Types further to the right of the spectrum are defined by significant levels of state weakness, as illustrated by poor territorial control and limited ability to intervene from the center. In cases where there are "indirect links," state interlocutors may position themselves in relation to local power brokers to negotiate access or safe passage for the transit of goods. Unsurprisingly, such arrangements often stem from intelligence services with connections to the highest levels of the state, but which can operate in secrecy to "close the deal." Perhaps the best example is the linkage between hostage negotiations and the state in Mali, where senior officials working with "negotiators" in the field received a portion of ransoms paid. That ensured a symbiosis between state officials dealing with foreign governments and the kidnappers themselves, providing the incentive for government officials to protect kidnapping groups from external intervention, while also ensuring a parallel reliance by foreigners on state interlocutors precisely because of the protective relationship. In short, a perfect symbiosis between state and nonstate actors.[22]

Finally, in the absence of any state institutions, or where state institutions themselves are very weak, protection is provided by local militia or strongmen in the absence of any linkage with the state. That has been the case in the context of southern Libya after the fall of Qaddafi and during the ongoing period of conflict.

Protection in Practice: Violence Control and Violence Promotion

Having provided an overview of different systems of protection, in each case identifying the role of the state in any arrangement, we must now consider how each of the models begets violence, if any.

As several studies of state dominance of illicit markets have suggested, total control of an illicit economy reduces violence considerably.[23] It is in the interest of state actors to dominate the illicit market completely to secure profits; violence attracts external attention and may invite law enforcement attention or challenges from competing criminal interests. In most cases of high-level state control of illicit markets, violence is more likely to occur within the elite, rather than as a feature of a wider conflict in society. It should also be highlighted that protection markets may not generate huge resources, at least in comparison to the overall value of the illicit goods being shifted. Nevertheless, in the content of relative resource scarcity those may be considerable.

This type is well-illustrated by Guinea-Bissau, where senior officials in the military and political elite provide protection for drug traffickers through a system of agreed-upon payments per consignment. Widespread violence that has been present in other drug markets never occurred in Guinea-Bissau. That is because there was negligible local drug use, and so there was no requirement to fight over local markets for sale. Nevertheless, there was considerable violence targeted at individuals at high levels of the state, who acted to provide protection to the drug economy.[24]

The lesson of Guinea-Bissau is that, while high-level state provision of protection may ensure that violence is not widespread, the struggle for the elite to secure control of the protection economy causes significant political instability. Guinea-Bissau has been plagued by years of political fragility, with no prime minister ever finishing his term, but low levels of violence amongst the citizenry.[25] Arguably, violence would have been much greater had a single regional or local group assumed control over drug trafficking. Despite the weaknesses of the state, which has limited reach into the interior, the nature of the country's politics and the contacts that were first made between drug trafficking entrepreneurs and military officials mitigated against this.

In cases where the state acts as a gatekeeper for regional or local control, the potential for violence is greater. That is largely because such processes often favor some local actors over others, with the aim of reinforcing the central state's interests. Local "clients" are selected because they have linkages to the central state or may have been imposed from the center. Conflict is the result of two eventualities: the first is where different groups vie for control over the trafficking economy, targeting the state-selected partner; and the second is when the state may attempt to diversify the profits from the drugs economy to placate local or regional political interests.

In Mali before the 2012 coup, the central state acted through intermediaries within the presidency to allocate who in the contested northern territory could accumulate resources from the protection of cocaine trafficking through the country. That system of selection and the flows of protection profits upwards to the center achieved a degree of stability, and indeed, was designed precisely to achieve that.[26] Remarkably, interviews with traffickers in the region suggest, some of the initiative for local devolution came from the traffickers themselves, whom are said to have persuaded the government that decentralization would ensure a greater division of the spoils.[27] While that accommodation did, in fact, achieve stability for a period, it also sowed the long-term popular dissatisfaction and political instability in the state as a whole.

On Libya's western border, in contrast, significant levels of violence resulted when the old system of spoils division broke down, and local communities struggled among themselves to gain control of trafficking. Before the revolution, two communities in the towns of Jamil and Rigdaleen had been provided the exclusive mandate by the Qaddafi state to engage in trafficking. When that state imploded, the two communities fought for the right to continue with illicit trafficking and smuggling, as surrounding communities attempted to muscle in.[28]

These two cases suggest that devolution of trafficking from center to periphery is almost always a recipe for conflict and violence in the longer term. On one hand, as demonstrated, it achieves stability for a period; on the other hand, when the system begins to break down, it portends high levels of violence and may, combined with wider process of political mobilization, lead to significant levels of instability, both before (as in Mali) and after (as in Libya) shifts in power. It should be noted, too, that such forms of devolved control of trafficking take place in those jurisdictions where the state has what might be termed medium levels of reach; the central state retaining the ability to intervene with violence to manage the system of devolved control of trafficking, but not having either the political or military strength to stamp its unquestionable authority in the peripheral regions where trafficking takes place.

There are echoes of this system in cases termed "mixed control" in the table provided earlier, in which local actors and local security interests divide up turf, usually in peripheral regions. In such cases, local security interests act outside of central government authority and negotiate arrangements with local players on the ground. The implications for governance, and by implication, the potential for violence, may be severe. Different groups may compete for control of the illicit economy and conflicts between state and nonstate forces may have wider political and security implications, clothed as they may be in alternative ideological baggage.

In addition, particularly in cases of security force involvement, tensions within the security forces between those involved and accumulating illegal profit, and those seeking to act in the interests of the rule of law, may cause wider conflicts. In northern Niger, security force involvement in migrant smuggling—essentially "taxing" the flow of people, and so a form of protection in exchange for payment—caused significant internal tensions when, after numerous cases of dead migrants who were found in the desert were exposed, there was considerable external pressure for action.[29]

Ultimately, cases to the right of the spectrum, where local forms of control predominate and the writ of the central state is weak or nonexistent, have the greatest propensity for wide-scale violence. In contexts where there may be few other resources to sustain armed groups and where the provision of protection to illicit trafficking is the largest money-spinner available, access to and control over routes is a critical source of accumulating resources to ensure local control.

In Libya, sustained periods of conflict between local strongmen over access to illicit (and licit) resource streams have dominated the posttransitional period. Where no group is strong enough to control the illicit market, conflict may be sustained over a considerable period of time, mitigating against external efforts to achieve peace.[30] Where parties engaged in protection of illicit resources accumulate significant resources they may have much greater interest in maintaining the status quo, rather than participating in a negotiated settlement where power and authority will need to be surrendered to a newly constituted central authority.

In the wake of the fragmentation of the central state in Libya, numerous local militias competed for control of the trafficking economy. Importantly, those that assumed greater control were those, such as the Zintani in the west, who had higher levels of internal cohesion, who were better structured and led, and thus able to absorb external resources; in turn, resources acquired were used for the purchasing of arms to further strengthen their hold on the trafficking economy. The point here is that growing power is not automatically linked to the acquisition of trafficking resources. In some cases, the injection of external resources may lead to internal fragmentation or greater conflict *within* the group over access to resources.[31]

In summary, different constellations of protection result in different levels and types of violence. In some cases, highly organized and agreed-upon systems of protection ensure a negotiated criminal peace. But in all cases, violence manifests itself at the elite level, or more widely, when such agreements break down. The timing and type of violence, therefore, often relates to the length of time the criminal market has existed. Early periods of market consolidation may be associated with high levels of violence, as different groups offer and enforce protection economies. In cases where that control consolidates, violence may be more controlled over longer periods of time.[32]

Given that states, or at least state security institutions, are key players within protection economies, periods of political, social, or economic change, which strengthen or weaken states, generally have a profound impact on the nature of the protection that may be provided. As in the case of Qaddafi's Libya, the implosion of state security structures and systems provided opportunities for multiple players and widespread conflict as protection economies coalesced. In other cases, the weakening or strengthening of some state institutions provides new opportunities. In Georgia, for example, Alexander Kupatadze recounts how a police campaign to remove organized crime extortion of businesses (a classic protection economy) was successful in doing so, only to provide the opportunity for the police themselves to take over the business of protection payments.[33]

It is worth pausing here to consider the implications of protection economies in democratic environments. It has been suggested that democracy provides greater opportunity for the development of systems of protection, as there may be multiple points of entry and a variety of different ways for actors to seek influence from within the system.[34] At the same time, however, democratic arrangements may allow greater transparency and oversight, undercutting systems of protection, although this process may in itself lead to instability as one form of protection takes over from another, as political regimes change. While more research is required, it can be concluded then that, depending on the circumstances, democratic politics in conflict-ridden or transitional states may make protection economies more common, but less durable.

Illicit Financial Flows and the Linkage to Violence

As the section above indicates, where and how resources from illicit markets flow into the protection economy have important implications. Resource flows from protection, as we have seen, act under a number of conditions to strengthen some groups over others. This suggests that understanding and accounting for volumes of protection money is a critical component of this analysis.

The available evidence suggests that enormous amounts of money are accumulated in the illicit trafficking economy in a diversity of markets. Money from illicit narcotics remains the most significant, but multiple illegal markets now generate considerable volumes of funds, with some—albeit contested—estimates placing transnational criminal activities at an estimated 8 to 15 percent of global gross domestic product with a value of some $3 trillion.[35] Even if the sums were half that amount, it is a considerable amount of money. An important question that must be asked, therefore, is how much of this money is accumulated by those engaged in trafficking and then by those engaged in the protection economy? And, what are the implications for violence?

Accumulating good data to answer this question is challenging for obvious reasons. While estimates can be made of the volumes of funds generated by illegal economies, it is much harder to isolate the proportion that is used to sustain protection economies. It can also be assumed that such protection may vary according to the nature of the market and the strength of the protectors. Nevertheless, some more detailed work around identifying the flows of funds into the multifaceted protection economy in West Africa, suggests a perhaps surprising answer: protection is cheap, at least measured against the overall sums of money being accumulated by criminal entrepreneurs.

In Guinea-Bissau, we have relatively certain knowledge around at least one proposed protection payment. One of the key "protectors" stated that his fee for facilitating transfers of drugs would be $1 million per ton routed through the country. If all the cocaine bound for Europe in one year, an estimated 40 tonnes, passed through Guinea-Bissau (itself unlikely), then the "protector" in question would earn some $40 million from a commodity with a wholesale, as opposed to a street price, of some $2.1 billion. If the estimated wholesale price of cocaine in Europe is $52.7 million per ton, the protection fee for that amount of cocaine would be worth approximately 1.8 percent of the wholesale price.[36]

That is a steal in any market, and although it is possible that several protection payments would be required, protection seems relatively cheap. That is contrary to the general impression that vast fortunes are being made all round from illicit trafficking: they are, but not necessarily by those who sell themselves as "violent entrepreneurs" or protectors. At those levels of payments, a significant increase in price would be required before it became unaffordable for traffickers. Can the conclusions from this example be generalized? To some extent, it is possible that they can be. There are two reasons for this.

The first is that in many zones where illegal economies are present, or where flows transit, there are relatively numerous actors able to provide protection. Through the simple operation of the market, numerous alternative protectors push prices down. That may promote conflict, but keep the cost of protection down. In Libya, for example, local traffickers and smugglers suggested that in some places they had the choice of several "protectors," depending on the route selected.[37] Similarly, if Guinea-Bissau's military strongmen asked for too much, then it was possible to move the trade to Guinea or Liberia.

The second relates to comparative levels of wealth and economic activity. As in the case of Guinea-Bissau, money garnered from the protection economy may be by far the largest sources of resources in *comparative* terms. Local actors may also ask too little, being unaware of the total volumes of profits being generated in the market.[38] Better information about the overall profits being made might, therefore, push up the prices for protection. But overall, this is a seller's market: in West Africa traffickers can shift to other countries if the supply of protection ends in any one place and so, to some degree, the asking price for protection is bounded by competition in the market.

These conditions raise the question: what conditions must be in place to raise the price of protection, ensuring that, in proportion to the overall cost of movement, greater resources flow into the protection economy? Following the most basic principles of economies, monopoly control over the protection market is likely to provide an opportunity for generating greater resources. Such monopoly of control, however, may depend on another critical factor: that the flow itself must be vulnerable. As suggested above, for extortion, the vulnerability of the commodity being transported is related to the amount of the money that can be extorted without undermining the trade itself. In cases where extortion is practiced, therefore, the payment is calculated on ensuring that there is a balance found between the maximum profit available against the chances of destroying the business itself.[39]

An excellent example of a vulnerable flow is that of migrants. A recent confidential assessment of Islamic State (ISIL) activity in Syria notes how the organization now makes money by acting as the "primary" (and thus, monopoly) controller of migrants in its areas, taxing the movement of people by seeking to control the "choke points" through which they must move. The report also notes that ISIL forces, given the profitability of the migrant supply chain for the organization, seek to treat refugee flows with a view to their potential income, driving the movement of people to flow through points where they can be "taxed." The overall value of money raised from the enforcement of the protection economy around Syrian migrant flows leaving ISIL-controlled zones is estimated to be in the order of $180 to $360 million.[40]

Migrants' vulnerability is due to the fact that they are individually exposed to danger during movement and have no collective bargaining power. This is in contrast to, for example, cocaine trafficking where criminal entrepreneurs may have more bargaining power with those in a position to provide protection. Their greater influence is a feature of both the more limited number of traffickers and the fact that they control the source of the profits—in this case, cocaine.

Where they control the commodity and have a choice of protector, the relationship between criminal entrepreneur and criminal protector shifts in favor of the entrepreneur. Indeed, evidence suggests that criminal entrepreneurs may outlast criminal protectors; in the Sahel, for example, traffickers claim to have renegotiated the transfer of illegal commodities with different protectors when the latter had been displaced.[41] In Guinea-Bissau, the same set of entrepreneurs appeared to be in a position to negotiate with new sets of protectors who had displaced the old.[42] Criminal protection, then, may not only not pay well, but it may also be an unstable business to be in altogether.

This chapter has argued that a distinction should be drawn between criminal entrepreneurial functions and activities associated with protection. Protection theory and a range of case studies of organized crime suggest that these two relatively distinct functions can be found in almost all cases where illicit goods are being moved. Indeed, it has been suggested that while illicit flows are transnational, their control by definition must be local, given the requirement for local knowledge, recruitment, and the specific connections and networks that are required to exert effective governance in particular areas.[43]

It is possible that in particular circumstances the functions of entrepreneurial trade and protection may be combined. But this occurs much less often than is assumed, and where it does, may ensure greater vulnerability for the combined grouping. Such a combination of functions is only likely in places where state control has dramatically weakened, with no state functionaries to "buy off." But in such cases, as the example of Libya shows, lower-level armed groups, even when gaining in strength, lack the contacts and networks to engage in entrepreneurial trade. Indeed, in the case of Guinea-Bissau, when senior military officers tried to negotiate deals themselves for the trafficking of cocaine they, given their lack of experience, became vulnerable to external law enforcement interventions.[44]

Arguments around protection economies have sometimes been countered by suggestions that sophisticated trafficking groups simply create the armed functions themselves. In fact, that is much more difficult than it seems. While criminal entrepreneurs may retain the capacity for violence for their own immediate protection and if required to attack competing trafficking groups, there remains a broader requirement for a protection economy. Movement of goods over long distances requires a more elaborate system, where state officials are paid off, armed groups must be financed for transiting zones they control, and high-level politicians must be "bought." Colombian traffickers arriving on West African shores for the first time, for example, and with few local contacts, had little choice but to create a protection economy around their activities. Thus, the division between criminal entrepreneurs and criminal protectors is almost always present based on experience, local contacts, and capacity for violence.

If, as this review suggests, the functions of trafficking and protection are almost always separate, with protectors receiving less compensation than is often assumed, and if protectors are more vulnerable than the traffickers that pay them, important factors for policy responses are illuminated. The concluding section briefly reviews the available options.

Policy Considerations

Understanding protection economies requires a recognition of the basic economic nature of the transaction: the protection of goods has become a commodity in and of itself, subject to market forces. If the prices are a comparatively small part of the overall volume of funds generated by the trafficking economy, then protection comes cheap. If illicit economies are to be effectively tackled, then a much greater focus is required on raising the price of protection and reducing the profits from illegal activities. Effective trafficking cannot function without protection economies, so targeting these suggests an important set of possible policy approaches.

Ordinarily, achieving this objective has been the function of law enforcement. By targeting, for example, corrupt officials who provide protection, governments by implication force up the costs of protection by increasing the risks to those involved. These strategies have had a measure of success in developed states where the quality of institutions and the overall operation of the rule of law limited the number of protectors that take the risk of engaging in the practice. Even in cases where whole components of law enforcement have been corrupted in this way, the broader system has the capacity for self-correction.

But in situations, primarily in the developing world, where security or state institutions, armed groups outside the state, or senior politicians including heads of government, act in protective functions, domestic law enforcement agencies are severely constrained from acting. The options for foreign law enforcement are also limited, given that it would require external intervention which, at best, may be widely condemned, and at worst, run considerable operational risks. A good example of a success in this regard was the U.S. Drug Enforcement Administration's arrest of senior Guinea-Bissau military officials, who were lured off the West African coast. This is not a sustainable strategy and would require many such interventions to have a measure of success. Rather, in cases where protection economies have become well-developed, much greater debate is required to discern which alternative strategies are viable. These include the following:

- *Shifting the incentives for the provision of protection.* It has been suggested that the funds from protection are, in comparative terms, small. That may provide opportunities for paying protectors to change their allegiance, to shift their role from protecting traffickers (in their private interest) to protecting the interests of the state (in the public interest). Taking this route has indeed been suggested by those engaged in protection themselves. Tebu tribesmen who play a key role in the protection economy in Southern Libya, for example, have informed the author that they would be willing to change allegiance if they received the resources from the central state that they believe are theirs.[45] "Paying off" systems of protection, while it sounds radical, is not in itself new: European governments in the past paid off the Qaddafi regime to prevent migrant flows from the North African coast. It would not be inconceivable that such payments could be shifted to nonstate groups (such as the Tebu) now deeply embedded in the migrant smuggling economy.

- *Targeting the recruitment of potential "protectors."* While protection economies can be relatively diverse and use an array of tools, by far the most common of these is the requirement to recruit those with the capacity for violence. In most places, that includes young men or networks of ex-military or sportsmen who can, with relative ease, be recruited to perform protection functions.[46] In postconflict societies in particular and those where state institutions are weak, providing alternatives for such a constituency must be an overriding priority. Again, this is not a new policy intervention; disarmament, demobilization, and reintegration (DDR) programs, for example, have just such an aim and have improved in their efficacy, based on experience of numerous attempts. But the focus of such programs is demobilized military personnel only, and while these individuals are also a key constituency for "protection" provision, programs that were tailored with the express objective of undermining recruits into the protection economy have yet to be tried.

- *Political engagement and inclusion.* As the typologies have suggested, those who engage in providing protection in the case of illicit flows are often on the periphery, either geographically or politically. The example of the Tebu in Libya is a clear recent case. Undercutting the protection economy provided by these groups requires a political strategy of inclusion. That is often overlooked when the automatic response to illegality is seen as a law enforcement response only. For example, the current European rhetoric against the smuggling of migrants seeks to identify "organized crime" and "smugglers" for law enforcement activity, ignoring the fact that, in fact, these may constitute whole communities that are better described as protecting the flow of people, than as "organized criminal networks."

Understanding the linkage between violence and trafficking economies is now a priority for many international policymakers, and has been extensively debated at the

United Nations. A growing number of Security Council Resolutions have identified the linkage and have argued for more focused action to respond to the problem.[47] However, the analytical tools used in these approaches have largely misunderstood the linkage between violence and the trafficking or illicit economy; instead, the focus remains largely on law enforcement responses.

I argue that the use of the concept of "protection economies" may provide a better framework to understand the linkages between trafficking, violence, and the financial transactions that connect the two. These arrangements can be isolated into a set of types, each of which has different implications for the forms, sources, and nature of violence, and the extent of the financial payments that may result. They also suggest that undercutting the linkage between protection economies and illicit flows and markets may be possible through an integrated strategy that includes a combination of changed incentives, the alternative recruitment of potential protectors, and the effective political engagement of the same. Countering illicit trafficking and its linkage to financial flows and conflict has not previously been viewed through this framework, and until development, security, and political policy take these factors into account current policy approaches will fail.

Notes

[1] To take one example, that of the illicit trade in rhino horn in Southern Africa: this has increased by a factor of 17,000 percent over a decade-long period.

[2] In West Africa, for example, a program by the EU to focus on the cocaine trade has had mixed results (European Commission, 2013).

[3] Scott Straus, "Wars do end! Changing patterns of political violence in Sub-Saharan Africa," *African Affairs* 111, no. 443 (2012): 179-201.

[4] Frederic Lane, "Economic consequences of organized violence," *Journal of Economic History* 18, no. 4 (1958): 401-417; Diego Gambetta, *The Sicilian Mafia: The Business of Private Protection* (Cambridge, MA: Harvard University Press, 1993); Anja Shortland and Federico Varese, "The Protector's Choice: An Application of Protection Theory to Somali Piracy," *British Journal of Criminology* 54, no. 5 (2014): 741-764.

[5] Stephan Ellis and Mark Shaw, "Does Organized Crime Exist in Africa?" *African Affairs* 114, no. 457 (2015).

[6] Nils Gilman, Jesse Goldhammer, and Steve Weber, *Deviant Globalization: Black Market Economy in the 21st Century* (New York, NY: Continuum, 2011).

[7] See, for example: International Crisis Group (ICG), *The Central African Crisis: From Predation to Stabilisation,* Africa Report No. 219 (Brussels: ICG, June 2014).

[8] See the discussion in: Federico Varese, *The Russian Mafia: Private Protection in a New Market Economy* (Oxford: Oxford University Press, 2001).

[9] Charles Tilly, "War Making and State Making as Organized Crime," in *Bringing the State Back In,* ed. Peter Evans, Dietrich Rueschemeyer, and Theda Skocpol (Cambridge: Cambridge University Press, 1985).

[10] Mark Shaw and Tuesday Reitano, *The Political Economy of Trade and Trafficking in the Sahara: Instability and Opportunities,* Sahara Knowledge Exchange (Washington, DC: Fragility Conflict and Violence Group, World Bank, December 2014).

[11] Kai Konrad and Stergios Skaperdas, "Extortion," *Economica* 65, no. 260 (1998): 461-477.

[12] This is remarkably consistent across different extortion markets, as illustrated by the author's interviews with those involved in extortion economies in Johannesburg and Cape Town.

[13] Dick Hobbs, "Going Down the Glocal: The Local Context of Organized Crime," *Howard Journal of Criminal Justice* 37, no. 4 (1998).

[14] Anja Shortland, and Federico Varese, *The business of pirate protection*, Economics of Security Working Paper 75 (Berlin: German Institute for Economic Research, 2012).

[15] William Reno, "Understanding criminality in West African conflicts," in *Peace Operations and Organized Crime: Enemies or allies?* Ed. John Cockayne and Adam Lupel (London: Routledge, 2011); Vadim Volkov, *Violent Entrepreneurs: The Use of Force in the Making of Russian Capitalism* (Ithaca, NY: Cornell University Press, 2002).

[16] Richard Snyder and Angelica Duran-Martinez, "Does illegality breed violence? Drug trafficking and state-sponsored protection rackets," *Crime, law and social change* 52, no. 3 (2009): 253-273.

[17] Alexander Kupatadze, *Organized Crime, Political Transitions and State Formation in Post-Soviet Eurasia* (New York, NY: Palgrave Macmillan, 2012).

[18] Patrick Meehan, "Drugs, insurgency and state-building in Burma: Why the drugs trade is central to Burma's changing political order," *Journal of Southeast Asian Studies* 42, no. 3 (2011): 376-404.

[19] Drawn from: Tuesday Reitano and Mark Shaw, *Fixing a fractured state? Breaking the cycles of crime, conflict and corruption in Mali and Sahel* (Geneva: Global Initiative against Transnational Organized Crime, 2015).

[20] Meehan, "Drugs, insurgency and state-building in Burma," 379.

[21] Mark Shaw, "Drug Trafficking in Guinea-Bissau, 1998-2014: The Evolution of an Elite Protection Network," *Journal of Modern African Studies* 53, no. 3 (2015).

[22] Tuesday Reitano and Mark Shaw *Fixing a fractured state? Breaking the cycles of crime, conflict and corruption in Mali and Sahel0* (Geneva: Global Initiative against Transnational Organized Crime, 2015).

[23] H. Brownstein, S. Cummins, and B Spunt, "A conceptual framework for operationalizing the violence between violence and drug market stability" *Contemporary Drug Problems*, 27 (2000).

[24] Mark Shaw, 'Drug Trafficking in Guinea-Bissau, 1998-2014: The Evolution of an Elite Protection Network' *Journal of Modern African Studies*. 53, No 3 (2015).

[25] Davin O'Regan and Peter Thompson, *Advancing Stability and Reconciliation in Guinea-Bissau: Lessons from Africa's First Narco-State* (Washington, DC: Africa Center for Strategic Studies, 2013).

[26] Morton Bøås, "Castles in the sand: informal networks and power brokers in the northern Mali periphery," in *African Conflicts and Informal Power: Big Men and Networks,* ed. Matt Utas (London: Zed, 2012).

[27] Interviews by author, Bamako, July 2014.

[28] Mark Shaw and Fiona Mangan, "Enforcing 'our law' when the state breaks down: The case of protection economies in Libya and their political consequences," *Hague Journal of the Rule of Law* 7, no. 1 (2015): 99-110.

[29] Personal communication, senior military officer from Niger, Niamey, June 2013.

[30] Mark Shaw and Fiona Mangan, *Illicit Trafficking and Libya's Transition: Profits and Losses* (Washington, DC: United States Institute of Peace, 2013).

[31] Paul Staniland, *Networks of rebellion: Explaining Insurgent Cohesion and Collapse* (Ithaca, NY: Cornell University Press, 2014).

[32] Peter Andreas and Joel Wallman, "Illicit markets and violence: what is the relationship?" *Crime, Law and Social Change* 52, no. 3 (2009): 225-229; Stephen Schneider, "Violence, organized crime, and illicit drug markets: a Canadian case study," *Sociologia, Problemas e Práticas* 71 (2013): 125-143.

[33] Alexander Kupatadze, *Organized Crime, Political Transitions and State Formation in Post-Soviet Eurasia* (New York, NY: Palgrave Macmillan, 2012), 116-139.

[34] Richard Snyder and Angelica Duran-Martinez, "Does illegality breed violence? Drug trafficking and state-sponsored protection rackets," *Crime, law and social change* 52, no. 3 (2009): 253-273.

[35] World Economic Forum (WEF), "Out of the Shadows: Why Illicit Trade and Organized Crime matter to us all," Global Agenda Council on Illicit Trade & Organized Crime 2012-2014 (Geneva: WEF, 2013).

[36] Senior DEA officials, interviews by author, November 2013; Shaw, "Drug trafficking in Guinea-Bissau."

[37] Traffickers in different parts of Libya, interviews by author, April 2013.

[38] Military officers, interviews by author, Guinea-Bissau, October 2012.

[39] Stergios Skaperdas, "The political economy of organized crime: providing protection when the state does not," *Economics of Governance* 2, no. 3 (2001): 173-202.

[40] "Islamic State manipulates refugee flows to generate income – fall of Aleppo and Syria could triple refugee flows and finance to IS by 2016," INREP briefing document (Lillehammer: Rhipto-Norwegian Centre for Global Analysis, September 2015).

[41] Interviews, Bamako, July 2014.

[42] Interviews, Bissau, 2012 and 2013.

[43] Federico Varese, *Mafias on the move: How organized crime conquers new territories* (Princeton, NJ: Princeton University Press, 2011); Jan Van Dijk and Toine Spapens, "Transnational Organized Crime Networks Across the World," in *Transnational Organized Crime: An Overview from Six Continents,* ed. Jay Albanese and Philip Reichel (London: Sage, 2014).

[44] Mark Shaw, "Drug trafficking in Guinea-Bissau."

[45] This point was raised in several discussions in Libya, and in more recent telephone interviews, October 2014.

[46] Volkov, *Violent Entrepreneurs.*

[47] For an overview of recent debates, see: John De Boer and Louise Bosetti, "The Crime-Conflict 'Nexus:' State of the Evidence," Occasional Paper 4 (Tokyo: United Nations University Centre for Policy Research, 2014).

12

The Neglected Mega-Problem: Illicit Trade in "Normally Licit" Goods

Karl Lallerstedt

We are all familiar with the "War on Drugs" initiated by Richard Nixon in 1971, Pablo Escobar's infamous role in the cocaine trade, and the "Pizza Connection" trial that laid bare Cosa Nostra's historic role in the global heroin trade.[1] This recognition is well-deserved; even today, many of the most powerful organized crime groups, as well as terrorist groups, generate significant profits from the narcotics trade. As a result, counternarcotics efforts have enjoyed tremendous political and law enforcement prioritization, with estimates that over $100 billion are invested annually in combating it globally.[2]

The United Nations Office on Drugs and Crime (UNODC) estimates that the global narcotics market turns over $320 billion each year.[3] At the same time, the International Chamber of Commerce (ICC) estimates that the illicit trade in counterfeit and pirated goods will treble from around $500 billion in 2008 to $1.3 trillion in 2015.[4] Yet, despite this magnitude, have we heard of a corresponding "War on Intellectual Property Theft?"

Although it is the largest category, counterfeiting is not the only form of illicit trade in goods that displaces normally licit goods. The illicit markets for contraband excise goods, such as tobacco, alcohol, and petroleum products, are also extensive global problems. Furthermore, the illicit trade in substandard or unauthorized/unapproved (noncounterfeit) products affects such a broad range of goods, from agrochemicals to medicines, that efforts to quantify the problem as a whole have not been attempted.

This chapter addresses this illicit trade that displaces normally licit goods, its scale and impact, as well as the challenges in addressing it. For practical reasons, the focus will be on counterfeit goods and certain excise goods.

The Magnitude of Leading Forms of Illicit Trade Displacing Legal Goods

Counterfeit and Intellectual Property Infringing Goods

The most frequently stated estimate of the value of counterfeit goods is the estimate from the Organisation for Economic Co-operation and Development (OECD) of $250 billion in 2007.[5] The figure is derived from extrapolations based on customs seizures and trade data, resulting in an estimate that counterfeit and pirated goods constituted 1.95 percent of world trade. Although seizures do confirm illicit trade, extrapolating its scale based on seizure data is problematic as the true interception rate is impossible to know. Furthermore, the effectiveness of customs can vary between jurisdictions and sectors. Despite these limitations, the ICC decided to expand on the OECD study, as that did not include domestically produced and consumed products, or nontangible pirated digital products. The ICC used the OECD study as a foundation for its estimates, and added further estimates for domestically produced (nontraded) counterfeit goods, projecting that by 2015, the global turnover of counterfeit and pirated goods would be around $1.3 trillion.[6]

The OECD published new counterfeit estimates, based on more recent seizure and trade data, in 2016.[7] The previous OECD data had suggested a significant growth rate, with the international trade in counterfeit goods more than doubling in value in the five-year period from 2002 to 2007.[8] The new OECD publication supports that the prevalence of counterfeit goods and pirated goods as a percentage of world trade has continued to increase, constituting 2.5 percent of world trade in 2013, and valued up to $461 billion that year.[9]

In 2014, a clear majority of interviewed international corporations on the NASDAQ OMX 30 Stockholm Index believed that counterfeiting and intellectual property theft had increased in the past five years, and the overwhelming majority asserted that it would continue to increase in the next five years. The primary reasons cited for this increase were the growing importance of emerging markets (where the extent of counterfeiting and intellectual property theft is greater) and the role of the internet as a facilitator of counterfeit trade.[10]

This growth trend is also supported by European interception data. The European Union (EU) has reported sharp increases in detentions of intellectual property infringing products. In 2013, national customs authorities had opened almost 87,000 cases, resulting in the detention of millions of articles. Consider that in 2002, only 7,553 such cases were opened. This marks more than a tenfold increase. Moreover, the single largest source of counterfeit products is China.[11]

Contraband Trade in Excise Goods: Tobacco, Alcohol, and Petroleum Products

Certain counterfeit goods, such as cigarettes and alcohol, can also fall into another category of illicit trade, which is the contraband trade in excise goods.[12] Contraband cigarettes have long served as a significant source of income for organized crime groups. The global illicit tobacco trade has been estimated to exceed 650 billion cigarettes a year, representing 11.6 percent of total consumption. The burden of this trade, representing over a million illicit cigarettes sold every minute, is not merely placed on tobacco corporations; tobacco smuggling resulted in tax losses exceeding $40 billion annually in 2007.[13] As tobacco taxes have been on an upward trajectory since that study, due to both public health concerns as well as fiscal interests, it is logical to assume that annual tax losses are even greater today. In the EU, where data is available, illicit tobacco consumption as a percentage of total consumption increased every year but one between 2006 and 2013. On average, the total proportion of consumption that was illicit increased by over 0.3 percent every year, from 8.3 percent in 2006 to 10.5 percent in 2013.[14] In Greece and Spain, the increase was particularly extreme. In the first year of the 2008 financial crisis, illicit consumption in both countries was just over two percent. When the economic fallout hit, and pressured governments to raise revenues through tax increases, it made the less costly illicit cigarettes even more attractive to smokers. By 2013, illicit consumption in Spain had risen to 9 percent, and in harder-hit Greece, it reached almost 18 percent.[15]

Alongside tobacco, the other major category of consumer goods subject to "sin taxes" is alcohol. The World Health Organization (WHO) estimates that approximately a quarter of total global alcohol consumption is unrecorded (either homemade and/or illegal).[16] Global estimates on tax losses due to illicit alcohol do not exist, but as an example, a Euromonitor study assessing only 6 Latin American countries estimated illicit consumption at 27 percent, generating $2.4 billion in illegal revenues, and government fiscal losses in excess of $700 million.[17]

Untaxed or illicitly sourced, petroleum products are also of economic significance. Oil theft from pipelines is a major issue generating losses of billions of dollars in certain countries, such as Nigeria, Russia, and Mexico.[18] But the sale of improperly taxed and/or contraband petroleum affects many more countries and regions; two examples are Bulgaria and Northern Ireland. According to a 2010-11 Serious Organized Crime Threat Assessment report focusing on Bulgaria, financed by the European Commission, international oil companies estimated that between 20 and 40 percent of the fuel sold in the country was illegal.[19] In 2014, the executive director of the Bulgarian Petroleum and Gas Association stated that 10 to 15 percent of the fuel market was illicit.[20] In Northern Ireland, it is estimated that 13 percent of diesel tax revenues are lost due to illicit trade.[21] Illicit oil can also play a major role in conflict zones, such as Serbia's reliance on contraband oil during the conflict in the Balkans, or Islamic State of Iraq and the Levant's (ISIL) role as a petroleum product supplier in the ongoing conflict in Syria.[22] The scale of the illicit trade in petroleum products is such that it can noticeably hurt government revenues, and at the same time, provide a strategic income stream for nonstate actors. In Mexico, for example, the state petroleum company, Pemex, estimated that it lost over $1 billion due to oil theft in 2014.[23] Organizations like Los Zetas and the Gulf Cartel are known to be involved in some of this theft.[24] In 2015, it was estimated that ISIL generated $1.5 million a day on the illicit oil trade from Syria.[25]

Illicit Trade in Other Substandard and Unapproved Goods

The 2013 horse meat scandal, in which several leading European food manufacturers' beef products were found to contain horse meat, brought public awareness to the fact that that were severe shortcomings in food supply chain control mechanisms.[26] A Dutch meat wholesaler was later convicted of having sold over 300 tons of horse meat, labelled as beef, to over 500 companies.[27] A few years earlier, in 2008, the more serious milk scandal had brought concerns over food fraud to the fore in China. Chinese officials estimated that as many as 6 babies died, and nearly 300,000 infants were sickened by dairy products contaminated with melamine, which had been added to watered-down milk in order to

Table 12.1 Intellectual Property Infringing Detentions Reported by National Customs Organizations to the World Customs Organization[31]

COMMODITY	Quantity (pieces)	Value (USD)	Quantity (pieces)	Value (USD)
	2012		2013	
Pharmaceutical products	4,140,318	14,405,404	2,325,247,466	19,388,693
Electronic appliances	3,423,896	170,355,748	470,821,728	74,551,410
Other	17,722,180	121,548,149	95,242,873	56,865,587
Foodstuff	1,316,034	6,924,2475	50,338,796	5,715,238
Clothing	12,090,266	111,511,445	32,877,929	115,400,776
Transportation and spare parts	4,120,790	27,887,278	17,957,325	13,945,339
Cigarettes	47,322	214,095	17,636,183	21,819
Accessories	6,712,922	426,677,500	10,079,064	229,781,652
Games and toys	3,505,486	22,649,732	5,750,377	28,882,956
Toiletries/cosmetics	3,268,255	60,318,717	5,663,822	65,590,057
Mobile phone and accessories	1,736,595	21,293,090	3,783,787	24,427,504
Footwear	2,352,318	69,881,973	3,139,816	112,530,010
Computers and accessories	848,578	11,593,152	2,227,604	22,461,541
Textiles other than clothing (towels, bed sheets etc,)	4,106,207	8,653,485	1,716,023	7,916,642
Watches	602,911	391,236,132	1,510,171	229,694,582
Phonographic products	3,929,684	6,483,568	736,330	6,112,419
Soft drinks	0	0	11,580	22,168
Alcoholic beverages	24,041	118,340	9,864	179,522
TOTAL	69,947,803	1,471,652,055	3,044,750,738	1,013,487,915

Source: WCO

fool protein content tests.[28] In Italy, mafia organizations were estimated to generate illicit revenues exceeding €15 billion in 2014 in the food and agriculture sector alone. Counterfeit and substandard foods production and sales are amongst their many illicit activities in the sector, which can also interconnect with other food-related crimes from which the mafia also profits, such as: control of distribution and transport of food products, forcing the sale of certain products in retail outlets, price fixing, money laundering through food-related businesses (particularly restaurants, of which mafia organizations control at least 5,000 in Italy), and illegal slaughtering.[29] As Italy is a major exporter of food products, substandard foods also affect consumers beyond the country's borders.

An example of the connection between organized crime and major food manufacturers is provided by Giuseppe Mandara, also known as Italy's "Mozzarella King" and "the Armani of Mozzarella." Mandara, who headed Italy's largest mozzarella manufacturer, was arrested in 2012 accused of having sold contaminated and falsely labelled cheeses, as well as having received money from a Camorra clan.[30] He was acquitted due to lack of evidence only to be arrested again in 2014 accused of involvement in Camorra extortion of landowners and money laundering.[31]

Substandard and noncompliant items range from illicit foodstuffs (such as Chinese gutter oil, falsified olive oil, and nonorganic foods passed off as organic), to nonfood items like noncompliant agrochemicals, vehicle components, and electronics.[32] Although there are huge numbers of products affected over a wide range of geographies, there is no collected holistic meta-analysis of the extent of the problem and to what extent organized crime is involved in the different form of illicit trade in substandard and unapproved products.

Impacts

Even though the overall prevalence of illicit goods displacing legal goods and its overall scale are unknown, we do have enough insight to determine that it is a problem of enormous magnitude. The consequences of this illicit trade are significant, and affect a number of areas.

Economic

The most obvious economic impact is the direct effect of illicit products displacing legitimate products on the market. As illicit products are generally (although not always) provided at lower prices than their legitimate equivalents, they squeeze legitimate products out of the market, both reducing revenues for law-abiding companies, and if the product is improperly taxed or untaxed, also reducing government revenues.

Reduced business revenues and tax revenues in turn have a negative multiplier effect on the economy. Both government and private sector investments and expenditures are reduced, having a negative impact on job creation and economic growth. Furthermore, there appears to be a correlation between a state's protection of intellectual property rights and foreign direct investment, suggesting that counterfeiting and piracy have a detrimental effect on foreign direct investment.[33] This appears to be supported by the view of multinational corporations, whose representatives have stated that their companies are less willing to invest in markets that lack effective intellectual property rights enforcement.[34] Jobs producing counterfeit products are also likely to be low-wage jobs, possibly without employment rights and characterized by unsavory working conditions.[35]

The economic importance of enforcing intellectual property rights is reinforced by Benoît Battistelli, president of the European Patent Office, who, in 2014, stated:

> For innovating companies the protection of their intellectual property (IP) has become extremely important: one in three jobs in the EU today is created in industrial sectors with an above average use of IP rights. These sectors account for almost 40 percent of the GDP and 90 percent of exports of the EU. They are a pillar of the competitiveness of

the European economy at global level. Similarly, continued violation of these rights puts a serious threat to Europe's capacity to innovate and compete, and to lastingly secure economic growth and employment for its citizens. It is necessary, therefore, to improve and strengthen the use of IP rights not only in Europe, but also internationally.[36]

On the other side of the Atlantic, the Commission on the Theft of American Intellectual Property writes the following about economic costs incurred by IP theft in the United States:

> The annual losses [due to intellectual property theft] are likely to be comparable to the current annual level of U.S. exports to Asia—over $300 billion. The exact figure is unknowable, but private and governmental studies tend to understate the impacts due to inadequacies in data or scope. The members of the Commission agree with the assessment by the Commander of the United States Cyber Command and Director of the National Security Agency, General Keith Alexander, that the ongoing theft of IP is "the greatest transfer of wealth in history."

The Commission's report further stated that enhanced intellectual property protection globally would add millions of jobs, boost research and development, facilitate investment, and increase the growth of the U.S. economy.[37]

The strong interest in enhancing intellectual property rights enforcement is reflected in the Trans-Pacific Partnership (TPP) trade agreement, which, at the time of writing, still has not been signed and ratified by the potential members. TPP raises the bar significantly beyond the intellectual property rights protection provided by the World Trade Organization (WTO) Agreement on Trade-Related Aspects of Intellectual Property Rights (TRIPS).[38]

The Philippines offers an illustrative example of tax losses from illicit trade. A Global Financial Integrity study estimated that 25 percent of all import value in the country over the last decade was undeclared. As more than one-fifth of Philippine tax revenues is generated by international trade, the fiscal impact is major.[39] A government report in Ghana estimated that the state loses 36 billion Ghanaian Cedi (approximately $9.5 billion) annually in revenues due to the fraudulent activities of importers.[40] To put this in perspective, Ghana's gross domestic product (GDP) in 2014 was 113 billion Cedi.[41] Even if the estimate exaggerates the scale of the problem, it is indicative of a problem of immense proportions. Global Financial Integrity has also identified trade misinvoicing as a major issue for a number of countries, and can be assumed to be a major source of revenue loss in several more. The higher the incidence of illicit trade, the greater the income loss for the state, undermining its capacity to deal with the multiple challenges it faces, including illicit trade itself.

Income for Criminal Actors

While illicit trade deprives the state and the legitimate economy of revenues, criminals stand to benefit. Europol and the EU's Office for Harmonization of the Internal Market states in the "2015 Situation Report on Counterfeiting in the European Union" that:

counterfeiting is now regarded by criminals as having lower risks and providing higher returns than drug trafficking. It has emerged as an ever-increasing and profitable transnational business in which organized crime networks manufacture and distribute counterfeit products widely, taking advantage of advances in technology and the rise of e-shopping and e-commerce.[42]

That same year, a study by the organized crime research institute Transcrime estimated that organized crime in the European Union generated greater revenues from counterfeiting than all forms of narcotics put together.[43] It is worth bearing in mind that counterfeiting is a more extensive problem in many other parts of the world, as compared to the European Union.

Europol estimates that there are approximately 3,600 international organized crime groups in the European Union, and over 1,000 of these are so-called "poly-crime groups," deriving their profits from multiple criminal activities.[44] This highlights the point that revenues generated from counterfeiting and other contraband trade in consumer goods empowers organizations involved in other crimes, which may be of greater public concern. Europol has also claimed that "it is difficult to identify any organized crime group involved exclusively in the trafficking or production of counterfeit goods, as they are inevitably linked to other crime areas."[45] Europol has identified several organized crime groups, with different national origins and ethnic compositions, that are involved in counterfeiting activities in the European Union, and Chinese crime syndicates are among the more prominent.[46]

The Chinese diaspora, spread across Europe, is leveraged by Chinese organized crime groups that utilize legitimate businesses involved in the import and sale of textiles, kitchenware, appliances, and other products. In order to facilitate money laundering, they have developed collusive relationships with money transfer agencies, enabling them to send large amounts of money to China. Similarly, relationships with corrupt shipping agents enable the transportation of illicit goods. Beyond regular consumer goods, Chinese organized crime syndicates are also involved in the production and distribution of counterfeit pesticides, medicines, and tobacco across the European continent. In the case of pesticides and tobacco, they also control manufacturing operations inside the European Union. In Italy, Chinese organized crime groups enjoy close collaborative relationships with the Italian Camorra, where it is also believed that Chinese groups supply human trafficking victims to work in counterfeit textile sweatshops.[47]

The synergistic effects between the illicit trade in counterfeit excise goods with other crimes is also illustrated by the Balkan route, a key channel for Afghan heroin reaching Central and West European markets. The UNODC estimates that 60 to 65 tons of heroin flow into Southeast Europe annually, and cannabis cultivation in the region is also of growing importance, with cannabis seizures reaching 48 tons in 2012.[48] The Balkans are also an important source of small arms and explosives for the European black market.[49] Albanian mafia clans play a key role in the smuggling of both heroin and cannabis, as well as the illicit arms trade, and play a role in several other criminal activities. According to

Europol, they are involved in the manufacture and trade of counterfeit cigarettes, smuggling the cigarettes along the same routes they use for marijuana and heroin. In the process, they use false documents for the cargo and bribe law enforcement officers to turn a blind eye to transports.[50] The Kosovar Albanian Naser Kelmendi, designated a "foreign narcotics kingpin" by the U.S. president in 2012 and arrested in 2013, has described himself as a trader, and has allegedly profiteered on numerous criminal activities, including contraband trade in both tobacco and oil.[51]

The aforementioned involvement of Chinese crime syndicates in European counterfeit trade points to the Camorra as a stakeholder in the counterfeit business. The Camorra, along with the Sicilian Cosa Nostra and 'Ndrangheta, are the most established and powerful Italian mafia-type organizations. Like the Cosa Nostra and 'Ndrangheta families, Camorra clans control or profiteer on the multitude of criminal activities taking place in the territories under their control, implying that profits from any one activity empower groups active in multiple other crime areas. Although the Sicilian Cosa Nostra and 'Ndrangheta are also known to profit on the illicit counterfeit trade, the Camorra has gained more attention for their involvement in counterfeiting, including large-scale operations targeting international markets outside Italy with different kinds of counterfeit products.[52]

In the United States, Federal Bureau of Investigation (FBI) Operation *Smoking Dragon* illustrates how the same networks used to smuggle contraband consumer goods can be leveraged for narcotics and weapons. The FBI operation resulted in the dismantling of a smuggling ring that initially smuggled contraband tobacco, then progressed to smuggling narcotics, false medicines, and counterfeit currency into the United States. Before being taken down, the criminals offered to smuggle weapons from China, including QW-2 surface-to-air missiles.[53] Terrorists, rather than organized crime groups, would appear to have a stronger interest in shooting down aircraft in the United States. Fortunately, the buyer was neither a criminal nor a terrorist, but the FBI. An undercover agent described one of the targets in the case as "the most dangerous man in America." In the agent's view, "Whatever we wanted, whatever we brought up, he was capable of getting." He stated plainly that "this guy could put you together with anybody to make any deal. He said he could get us any weapons—anything but nuclear weapons—and I think he could have. Whatever China had, this guy was capable of getting."[54]

Undermining Border Security

In addition to organized crime groups being economically empowered by illicit trade, the facilitation of illicit trade undermines border security in general. The routes established to move one type of contraband goods across borders can be leveraged for other uses. An illustrative example is the 700-meter-long tunnel, equipped with an electric railway between Ukraine and Slovakia, discovered in 2012. Police found contraband cigarettes in the tunnel, but according to the Slovak Interior Minister, the small train in the tunnel was also "capable of transporting various kinds of goods and we suspect also people."[55]

Another example is the autonomous aircraft detained by the Russian Federal Security Service (FSB) in 2014, carrying a 10-kilogram payload of cigarettes, used for smuggling between the Russian and Lithuanian border.[56]

Beyond techniques for physically bypassing border controls, illicit trade networks corrupt border guards to facilitate passage of smuggled cigarettes and other consumer goods. The more extensive the smuggling of "low priority" items (such as cigarettes, alcohol, counterfeits, etc.), the more extensive smuggling routes are established that can be leveraged for other uses, perhaps unbeknownst to some of the facilitators involved. Such corrupt border guards could also be used to facilitate the passage of other materials that are of greater concern.[57]

The illicit tobacco consumption in the European Union, for example, was estimated at 58.6 billion cigarettes in 2013, representing approximately a volume of 5,860 20-foot containers.[58] The breadth of the illicit cigarette market, with the majority of products originating outside the EU, illustrates not only how porous the borders are, but also that an established smuggling infrastructure is in place, which can likely be leveraged for other purposes. The large-scale illicit supply of alcohol and counterfeit goods creates similar vulnerabilities.

Terrorism

Just as organized criminals find illicit trade in counterfeit and contraband excise goods an attractive high profit and low risk revenue generator, so do terrorist groups. Irish Republican Army (IRA) factions, the Kurdistan Workers Party (PKK), Hezbollah, Hamas, the Revolutionary Armed Forces of Colombia (FARC), ISIL, various al-Qaeda affiliated groups, and others, have all profited from illicit trade in goods that displace normally legal goods. These terrorist organizations have profited from a diverse range of activities. Some examples are: the smuggling of sugar in East Africa, charcoal from the Horn of Africa to the Gulf, contraband across Turkish borders, interstate cigarette smuggling in North America, illicit tunnel trade between the Gaza Strip and Egypt, illicit trade in excise goods in Ireland and Northern Ireland, and the international trade in counterfeit medicines.[59]

The connection between the narcotics trade and the funding of terrorist organizations has long been established. Yet, if organized crime is increasingly moving into the expanding counterfeit market, due to high profits and lower risks, it is a logical assumption that it is of growing importance to terrorists, too.

Dr. Louise Shelley testified before the Congressional Task Force to Investigate Terrorism Financing in the U.S. in 2015. In her view, the concept of narco-terrorism has meant that there has been focus on the large-scale financial generators, such as the drug trade, while other income streams have been neglected. She states that "…increasingly smaller-scale illicit trade in commodities such as counterfeit goods, fuel, cigarettes, food, medicine, textiles, and clothing are used by terrorists to fund themselves in the United States, Europe, North Africa, and the Middle East." In her view, "the limited penalties attached to trade in consumer goods such as counterfeit pharmaceuticals, food, alcohol, cell phones, cigarettes have made these important growth areas for terrorist revenues." Shelley also points out that when terrorists function as criminal entrepreneurs they require a number of services, including those provided by corrupt officials, as well as witting and unwitting facilitators in the corporate world. Furthermore, "they also require professional services from the criminal world as they retain the services of human smugglers and specialists in nontraceable communications, forgers, and money launderers. Without hiring this expertise, they cannot make their business function."[60]

Bigger black markets sustain larger numbers of "criminal service providers," who are as essential to the regular criminal as to the terrorist who wants to profiteer on illicit trade. Furthermore, terrorists engaging in illicit trade will likely establish useful criminal connections that can be leveraged for terrorist purposes (without the "criminal service providers" needing to be aware of what they are truly facilitating). In the same way that illicit arms dealers may be useful to criminals and terrorists alike, so are dishonest shippers, suppliers of fraudulent documentation, and corrupt customs officials.

Other Costs of Illicit Trade

The ill effects of illicit trade do not stop at the economic losses and its undermining of national security. A significant proportion of the global timber and fish supplies are illicitly sourced, which contributes to the depletion of ecosystems, while generating several billions of dollars in illicit revenue annually.[61] Other forms of illicit trade with negative environmental impact include chemicals used in products, or production processes, which may not comply with environmental standards (e.g., such as consumer goods containing CFCs/HCFCs, or other compounds detrimental to the environment).[62] Illicit mining is another area of concern, where mining operations frequently do not comply with environmental standards.[63]

As it is logical for profit-maximizing counterfeit producers to cut corners on quality and safety standards, their products may pose dangers to their consumers, who frequently are unknowing victims. Electrical goods, for example, may pose fire hazards; faulty vehicle components may affect road safety; and unregulated food products expose their consumers to danger.

The clearest public health danger posed by illicit goods, however, is that of substandard medicines. These counterfeit medicines may contain incorrect amounts of active ingredients, no active ingredients, or the wrong ingredients. Worryingly, there are no exact estimates of global prevalence. Although the problem also exists in developed countries, it is particularly prevalent in the developing world, and affects most categories of medicines.[64] In a 2012 meta-study published in the Lancet, 35 percent of sampled antimalarials in Southeast Asia and sub-Saharan Africa failed chemical analysis, while 36 percent of the examined medications in Southeast Asia were falsified, as were 20 percent of the African samples.[65] With hundreds of thousands of deaths every year due to malaria, substandard medicines have a major impact on preventing effective treatment, and have further been identified as a factor contributing to increased drug resistance.[66] The development community has devoted significant attention to improving the means of delivering malaria drugs to affected communities. It is clear, however, that medical supply chains remain vulnerable to the abuse of counterfeit and substandard medicine dealers, to the detriment of the local populations and global public health efforts.

As with other forms of illicit trade, such as the narcotics trade, corruption can also play a role in facilitating flows of illicit products that displace legitimate products.[67] Corruptive payments can be made to facilitate passage of the products, prevent apprehension by law enforcement, avoid detection by public inspectors, and prevent convictions in the courts. Higher-level "protection" can also be bought from politicians and other socially prominent actors.[68] The larger the illicit flows and the more entrenched the operators, the more logical it is to assume that entrenched corruption is associated with the flows.

The prevalence of illicit cigarettes and contraband products is generally higher in developing countries, which are generally subject to higher levels of corruption.[69] The causal link could go both ways. Corrupt environments (many of which in the developing world also suffer from weak state capacity) would appear to be conducive for illicit trade, but extensive illicit trade may also be a contributing factor (amongst other factors) towards corruption.

Illicit trade can play a nefarious role in the corruption of public officials. Yet the state itself, or the leading individuals in it, can also play a central role in controlling illicit flows. Anecdotally, senior corporate executives have complained about the relative of a head of state controlling illicit flows of counterfeit goods, and a customs official explained that his government had employed a strategy of favoring certain organized crime groups engaged in illicit trade, in order to negatively impact the income stream for a major terrorist organization in the country.[70] There are several public sources alleging the corrupt involvement of states or heads of state (or their family members) in illicit trade. Contraband tobacco alone provides some striking examples. The president of Paraguay owns the company producing the greatest number of smuggled cigarettes in Latin America.[71] Montenegrin Prime Minister (and former President) Milo Djukanovic was accused by Italian prosecutors of having run a cigarette smuggling operation worth more than $1 billion, but charges were eventually dropped in 2009, due to his immunity from prosecution as head of state.[72] Since 2010, Belarus has remained the largest single identifiable source of contraband cigarettes that are smuggled into the European Union.[73] It is worth noting that Belarus only has two state-regulated tobacco manufacturers, and the largest of the two is state-owned.[74]

State involvement in illicit trade can also be "noncorrupt," in the sense that it may be actual state policy, as opposed to primarily generating revenue for corrupt individuals within it. One such example is North Korea, a state-controlled economy, where the production of counterfeit goods in the country and the involvement of diplomats in illicit trade can be assumed to be pursued under government orders in order to generate foreign currency revenue.[75] Production of counterfeit goods, and the diplomatic corps involvement in contraband trade, is by no means unique to North Korea, but presumptions relating to the state sanctioning the activity may be less clear-cut in other cases.[76]

Regimes such as apartheid South Africa, Saddam's Iraq, and more recently Iran, have all utilized illicit trade as a sanctions-busting tool to undermine the will of the international community. The Iranian President Hassan Rouhani even turns this into a virtue, having stated on national television that "of course we bypass sanctions. We are proud that we bypass sanctions because the sanctions are illegal."[77]

Challenges

Data

A major challenge, due to the multifaceted nature of the problem, is a lack of data to measure the scale of the various black markets, and how they impact individual nation-states. Without this elementary foundation, it is very difficult to justify investing the prerequisite resources—and political will—to combat illicit trade. There are some rough global estimates of relevance, such as the OECD and ICC estimates relating to counterfeiting, but global estimates are not directly relevant from the state's perspective.[78]

At the national level, there are useful estimates in some countries relating to particular categories of goods, such as the KPMG estimates for illicit tobacco consumption in Europe, which are updated on an annual basis and based on a systematic methodology.[79] However, such illicit tobacco estimates do not exist in all countries and this estimate is only for one sector of illicit trade. Publicly available nongovernment estimates regarding illicit alcohol are less common, and like in the tobacco sector they are, to some degree, driven by industry. However, by virtue of being excise goods, tobacco and alcohol (and, to some extent, petrochemical fuels) do enjoy some degree of government interest, from a revenue perspective. The picture, when it comes to the much bigger category of counterfeit and substandard goods, is much more problematic. In terms of public studies, there may be individual studies in individual geographies focusing on a particular category of goods. Most of these are not regularly recurring (although some efforts are now being made for some categories of pharmaceutical drugs, in some countries).[80] This lack of data means that it is very difficult to assess the scale of the problem from the perspective of individual states and to, thus, develop country-specific assessments, as well as strategies to counter these flows.

Furthermore, counterfeits and intellectual property infringing goods do have negative ramifications beyond a nation's borders. It is not only the states subjected to illicit trade that suffer. The owners of the intellectual property, frequently based in different jurisdictions, can suffer reduced sales revenues for their legitimate products and erosion of their "brand equity" as low quality products flood the market. This means that, in addition to estimating the loss due to illicit trade within a national territory, it is also important to assess how illicit trade impacts a country's business activities globally. The Japan Patent Office (JPO) conducts an annual survey of how leading companies are affected by counterfeiting, which provides an indication of the proportion of companies affected. It is quite a basic survey, but the JPO has conducted it every year for about two decades. Its finding, that over 20 percent of responding companies state that they are affected by counterfeiting, has potentially been a causal factor in the Japanese government's decision to prioritize combating counterfeiting.[81]

From the perspective of capturing the extraterritorial impact of counterfeiting, the intellectual property-owning companies themselves are best placed to assess the scale and impact of the problems they face. However, they face a dilemma. Publicly flagging counterfeiting problems risks undermining not only consumer confidence in their products, but also shareholder and investor sentiment. Consequently, they are unlikely to disclose their problems with complete transparency, and furthermore, may even be reluctant to commission effective studies as this could lead to unpleasant disclosures.[82] This implies that although companies themselves are likely key sources of data, these data sources are unlikely to be fully utilized without governments playing a role in facilitating the collection.[83]

Professor Tom Berglund, head of the Centre for Corporate Governance at the Hanken School of Economics (Svenska Handelshögskolan) in Helsinki and chairman of the Nordic Corporate Governance Network, states:

> Counterfeiting and intellectual property related crime is a significant challenge for many companies. But to demand that companies should account more transparently about this is problematic. It goes against the companies' commercial interests to openly account for issues such as the prevalence of cheaper counterfeits which are difficult to differentiate from the original. It can therefore be very difficult to find a trustworthy basis upon which to evaluate the problem's breadth, costs and risks. That makes it unreasonable to expect that companies shall become better at accounting for how they are affected by illicit trade. And this creates a larger societal responsibility, which ultimately means that the state must assume responsibility for deriving the foundation upon which it can evaluate how the national interests are affected by IP crimes.[84]

Fundamentally, without data it is not possible to make a diagnosis about the extent, effects, and best means of preventing counterfeiting. Better data could also highlight the mutual interests of developing countries and developed countries in reducing the prevalence of counterfeit goods. Weaker states with a high prevalence of counterfeit goods may have limited capacity to enhance their enforcement, whereas richer exporting states of intellectual property may be in a position to assist. Visualizing present losses could further incentivize developed countries to enhance their assistance efforts in this space.

The Global Trade System is Better at Facilitating Flows than Regulating Them

Global merchandise trade has expanded rapidly in the past decades, constituting an increasing proportion of global GDP, rising from 32 percent in 1990 to 49 percent in 2013. This has been a tremendous success story in terms of boosting economic development. There is, however, a darker underside to this story. While the facilitation of trade has generated positive results, the need to regulate illicit flows has been neglected.

The WTO, the multilateral body responsible for regulating global trade, not only lacks effective tools to prevent illicit trade, but some countermeasures to prevent illicit trade can actually be deemed to contravene WTO obligations. For example, it was ruled that Colombia had acted in contravention of WTO obligations when it attempted to reduce smuggling and under-invoicing of textiles and footwear from Panama and China by restricting imports to certain ports, in order to increase the efficiency of customs controls. Similarly, efforts to reduce tobacco smuggling in the Dominican Republic, by requiring that tax stamps be attached to imported tobacco within the country's territory under the supervision of that country's tax officials, were found to be in contravention of WTO obligations.[85]

In terms of substantive legal rules, the clearest linkage between the WTO and illicit trade is the Agreement on TRIPS, which sets out global minimum standards on the protection and enforcement of intellectual property rights. TRIPS sets out general standards, to be implemented according to the framework determined by each member, and it recognizes the existence of different standards in the enforcement of intellectual property rights among countries. Consequently, the protections required under TRIPS are more of a "floor," rather than a "ceiling," requiring only minimum standards. As the WTO has 161 members, these standards vary significantly.[86]

In a WTO case brought by the United States against China, the WTO panel stated that "China has a level of protection higher than the minimum standard required by the TRIPS Agreement."[87] Considering that China has been identified as the number one source of global counterfeits, it is clear that the TRIPS Agreement provides very limited protection. And under the WTO trade framework, there is very little else providing potential protection against illicit trade. Consequently, if the legal framework of the global trading system is to be efficient in terms of reducing illicit trade, it would need a major overhaul—a difficult and lengthy process if it were to be attempted. The seed of the TRIPS Agreement was planted in 1978, by the "anti-counterfeiting code." Yet, it took another 16 years for the TRIPS Agreement to come to fruition.[88]

Among those calling for reform of the current trade system are critics, such as Professor Amir Attaran, who holds the Canada Research Chair in Law, Population Health and Global Development Policy at the University of Ottawa, and has focused specifically on medicines. He describes the present situation as globalization having gone only halfway. In his view, there have been significant steps to facilitate global trade flows, but the mechanisms to regulate these flows are not in place. Historically, the biggest trade flows were between advanced economies, and a liberalization of trade between states—like Canada and Germany, which have strong regulatory systems in place—does not present the same challenges that increased flows between states with weak regulatory systems pose. As South-South trade has grown, countries like China and India have become major global pharmaceutical suppliers to developing countries. One problem associated with these growing exports is substandard and falsified medicines, something the importing states are not well-placed to handle. The result is a high prevalence of substandard medication, with a high public price in terms of public health in the developing world. In short, Attaran's proposed solution is to put the onus of quality control on the exporting state through a new treaty framework, whereby states that cannot not live up to the required quality standards for their pharmaceutical exports could be barred from exporting pharmaceuticals. This would place a strong incentive on pharmaceutical exporting countries to ensure that only compliant products were exported.[89]

The practicality of such a solution needs to be investigated, bearing in mind the WTO obligations on states regarding obstacles to trade. A joint study by the WTO and WHO, *WTO Agreements and Public Health*, points out that the General Agreement on Tariffs and Trade (GATT) guarantees member states the right to take measures to restrict imports and exports of products when those measures are necessary to protect the health of humans. Furthermore, TRIPS "does not contain an exception for health purposes per se, but it does allow measures necessary to protect public health and nutrition, provided they are consistent with other TRIPS provisions." In the assessment of the WTO and WHO, "WTO jurisprudence has clearly established that WTO Members have the right to determine the level of health protection they deem appropriate...."[90]

For a WTO member state to make use of health exceptions, the health measures may not be more trade-restrictive than necessary.

> Determining whether a measure is "necessary" involves a process of weighing and balancing a series of factors which include the importance of the interests protected by the measure, its efficacy in pursuing the policies, and its impact on imports or exports. The more vital or important the policies, the easier it would be to accept as "necessary" a measure designed for that purpose. Human health has been recognized by the WTO as being "important in the highest degree."[91]

The study's conclusion states that, "To monitor and evaluate the health impacts of existing WTO agreements and assess the potential health effects of proposed WTO rules and disciplines, there is a need for research and analysis. A current obstacle to analysis is the absence of systematic data collection...."[92] This reiterates the previously identified challenge regarding illicit trade, namely, the need for better data. It will be hard to justify measures that may be deemed detrimental to trade, unless there is reliable data demonstrating the deleterious effects on public health.

Whether a treaty to reduce false and substandard pharmaceuticals, as advocated by Professor Attaran, would be deemed compliant under existing WTO obligations would need thorough analysis. If such a proposal is deemed to be practical for medicines, there may even be other categories of illicit and substandard goods, such as foodstuffs and certain consumer goods that have human health implications, which could justify a similar approach.

The list of challenges to address illicit trade is long: addressing weak enforcement capacity and corruption in developing countries, the need to mobilize a more robust and coordinated global approach dealing with the problem, addressing specific issues such as the roles of intermediaries (e.g., payment and transportation providers) and suppliers, regulation of free trade zones, ensuring effective implementation of technical track-and-trace and product authentication solutions, addressing fundamental economic drivers of illicit trade (such as prohibition, taxation, subsidies, and sanctions), raising consumer and societal awareness of fake and substandard products, and more. Many of these challenges, however, will remain hard to address without policy-relevant data, which are a prerequisite to mobilize governments and the international community.[93] Likewise, mobilization behind a more effective global legal framework for preventing illicit trade will not be possible without better data.

A paradigm shift in how to address certain forms of illicit trade is desirable. The present approach, largely dependent upon preventing illicit trade by trying to catch individual perpetrators, is ineffective. Instead, the states responsible for the origins of the illicit flows could be held accountable. A new legal framework could motivate states, through self-interest, by making their continued access to free trade conditional upon compliance requirements for their exports.

The Future: A Growing Problem

Although the illicit trade in goods displacing normally licit goods is already a major problem that merits international attention, there are reasons to believe that it will continue to grow in magnitude.

Intellectual Property and the Evolution of the Digital Economy

As mentioned in the section describing the magnitude of counterfeiting earlier in this chapter, historical estimates from the OECD indicate a growing scale of trade in intellectual property infringing counterfeit and pirated goods.[94] Leading multinationals have identified a growth trend during the past five years, which they see as driven predominately by the role of the internet as a facilitator of illicit trade, and the growing significance of emerging markets, where counterfeiting is much more prevalent than in developed economies. These corporations also expect that the growth trend in counterfeiting and intellectual property theft will continue over the coming five years.[95]

In addition to the expected continued growth of counterfeiting, there are ongoing transformational changes relating to the growing importance of digital content, which will further increase opportunities for intellectual property theft-driven illicit trade. Digital piracy is already a major issue, affecting music, film and television, software, e-books, and other sectors of the global economy. According to a 2013 report by NetNames, intellectual property infringing material stood for almost a quarter of the internet bandwidth in North America, Europe, and Asia, and the piracy-focused websites facilitating this traffic are run for profit.[96]

The IT research firm Gartner predicts that by 2018, 3D printing will result in losses of at least $100 billion per year through global intellectual property infringement.[97] Leading manufacturing firms have already identified the threat, that in the future, a critical commodity being sold will be digital blueprints, where the customer is also the manufacturer. This has raised concerns that even manufacturing industries may suffer like other industries that have "turned digital," such as the music industry.[98] Since the emergence of the early peer-to-peer sharing services in 1999, U.S. music industry revenues have practically halved, declining from \$14.6 billion to \$7.9 billion in 2013.[99]

The failure, so far, in regulating and policing the "digital ecosystem" may very well have consequences in the future, for industries previously unaffected by digital piracy. It is perhaps understandable that pirated music, film, and computer games did not raise critical alarm bells in governments. But, perhaps these should have been seen as canaries in a coal mine. As more and more industries become increasingly dependent on the digital ecosystem, the economic and social costs of complacency may become unacceptable.

Only targeting the actors who make pirated content available, or those who buy it, is a very limited approach. A broader approach, which also includes the actors who facilitate illicit activities, is needed. Success requires fostering a digital ecosystem where internet service providers, search engines, payment solution providers, and online advertisers all behave responsibly, rather than aid and abet intellectual property theft. There are reasons for optimism. In certain countries, internet service providers have started blocking access to pirate websites, advertisers have taken measures to avoid advertising on such sites, and payment providers have implemented policies to block payments to such sites.[100] But with estimates that a quarter of internet bandwidth usage relates to intellectual property infringing content, much remains to be done, and the growth of dark net transactions and cryptocurrencies poses some tough challenges ahead.[101]

Excise Goods

If the illicit trade in counterfeit and other intellectual property infringing products is expected to continue on an upward trajectory, there are few reasons to expect a different development for illicitly traded excise goods. "Sin taxes," such as those on tobacco, are a politically attractive source of raising revenues, as they can be justified as "good" on the grounds of their expected positive impact on public health. Consequently, sin taxes frequently enjoy a long-term growth rate over and above inflation. In terms of economic logic, this also implies that the profit potential for criminals supplying illicit cigarettes is on a long-term upward trajectory, coupled with a growing incentive for the consumers to switch to a cheaper illicit alternative.

Although there are reasons to believe that both the illicit trade in both intellectual property infringing goods and excise goods will grow, there are also developments that may mitigate these expected trends. A more effective global coordinated response may not yet be around the corner, but the continued development and broader application of product authentication (as well as track-and-trace) technologies is empowering customers, public servants, and businesses to more effectively identify illicit products; it is one such area that could have significant impact, even if other enforcement efforts remain insufficient. The solutions provided by companies such as Sproxil and MPedigree enable customers to verify whether a consumer product is genuine or not, using a mobile phone and a scratch code on the product itself. Such solutions may not address the underlying causes of illicit trade, but they at least provide a tool to prevent individuals from becoming unwitting victims of counterfeits. They can also provide real-time data and analytics on the illicit market. Furthermore, if authentication solutions help an increasing number of customers discover that they are purchasing fakes, they could in turn pressure retailers to reconsider using the suppliers providing fakes.

Nonetheless, bearing in mind the likely growing scale of illicit trade in goods displacing normally legal goods, the price of underprioritizing this already enormous problem will likely continue to grow. Unless these trends can be reversed, an already large and neglected income stream for the criminal underworld will grow even larger, empowering nonstate actors, while further undermining state capacity through its detrimental impact on economic development.

It is characteristic of human behavior to prioritize, and even exaggerate, threats with a high "fear factor," such as terrorism, mass shootings, or narcotics. Highly visible human suffering engages our emotions and pressures leaders to act. The illicit trade in normally legal goods does not directly generate fear-inducing, high-profile incidents. It is largely an invisible flow, remaining below the radar, yet incurring enormous economic, human, and security costs.

Consequently, success in tackling illicit trade requires leadership that takes us beyond the politically intuitive threats. A holistic approach towards tackling the underlying enablers, or accelerators, of organized crime, corruption, and terrorism must address all its major components. Thus, we can no longer afford to neglect the mega-problem of illicit trade in "normally licit" goods.

Notes

[1] The longest criminal jury trial in U.S. history, which exposed Cosa Nostra's major transatlantic role in the drug trade.

[2] "50 Years of the War on Drugs, Wasting Billions and Undermining Economies," *Transform Drug Policy Foundation*, Count the Costs economics briefing, available at <http://www.countthecosts.org/sites/default/files/Economics-briefing.pdf>.

[3] United Nations Office on Drugs and Crime, *World Drug Report 2011* (New York, NY: United Nations Publications, 2011).

[4] Frontier Economics, "Estimating the global economic and social impacts of counterfeiting and piracy," commissioned by *Business Action to Stop Counterfeiting and Piracy* (London: Frontier Economics Ltd., February 2011).

[5] "Magnitude of Counterfeiting and Piracy of Tangible Products: An Update," *OECD*, November 2009, available at <http://www.oecd.org/sti/ind/44088872.pdf>.

[6] Specifically, an estimate of US$1,160 – 1,530 billion.

[7] OECD, *Trade in Counterfeit and Pirated Goods: Mapping the Economic Impact* (Paris, OECD Publishing, 2016).

[8] "Magnitude of Counterfeiting and Piracy of Tangible Products," *OECD*.

[9] OECD, *Trade in Counterfeit and Pirated Goods: Mapping the Economic Impact* (Paris, OECD Publishing, 2016).

[10] Karl Lallerstedt and Patrick Krassén, *How Leading Companies Are Affected by Counterfeiting and IP Infringement: A Study of the NASDAQ OMX 30 Stockholm Index* (Stockholm: Swedish Confederation of Enterprise and Black Market Watch, 2015), available at <http://www.svensktnaringsliv.se/migration_catalog/Rapporter_och_opinionsmaterial/Rapporter/omx30_english_webbpdf_617515.html/BINARY/OMX30_English_webb.pdf>.

[11] Europol and the Office for Harmonization in the Internal Market, '2015 Situation Report on Counterfeiting in the European Union,' April 2015, available at <https://www.europol.europa.eu/content/2015-situation-report-counterfeiting-european-union>.

[12] The majority of illicit alcohol and tobacco are not counterfeit products; but, rather, products that either evade taxation altogether or evade taxes due in the country of consumption.

[13] Luk Joossens and Martin Raw, "Strategic directions and emerging issues in tobacco control: From cigarette smuggling to illicit tobacco trade," *Tobacco Control* 21, no. 2 (2012): 230-234.

[14] "A study of the illicit cigarette market in the European Union: 2013 Results," *KPMG Project SUN*, 2014.

[15] Ibid.

[16] According to the World Health Organization, "unrecorded alcohol refers to alcohol that is not taxed in the country where it is consumed because it is usually produced, distributed and sold outside the formal channels under government control." See "Global Status Report on Alcohol and Health 2014," *World Health Organization* (Geneva: WHO, 2014).

[17] The six countries were Colombia, Ecuador, Peru, Honduras, El Salvador, and Panama. Euromonitor Research, "Latin American Illegal Alcohol Market Valued at US$2.4 Billion," *Euromonitor International*, May 12, 2014, available at <http://blog.euromonitor.com/2014/05/latin-american-illegal-alcohol-market-valued-at-us24-billion.html>.

[18] A 2013 report by Chatham House states that the best estimate is that an average of 100,000 barrels of oil a day vanished from onshore, swamp, and shallow-water areas in Nigeria in the first quarter of 2013. The report estimated that only Russia has a higher level of oil theft, estimated at 150,000 BPD, primarily in the Caucasus; Pemex, the Mexican state oil company, reported that it lost US$1.14 billion to oil theft in 2014; See Christina Katsouris and Aaron Sayne, *Nigeria's Criminal Crude: International Options to Combat the Export of Stolen Oil* (London: Chatham House, September 2013); "Mexico Pemex to stop gasoline via pipeline to curb theft," *Reuters*, February 17, 2015, available at <http://www.reuters.com/article/mexico-pemex-theft-idUSL1N-0VS00X20150218>; Nayeli González, "Crece 58 percent el robo de comustibles," *Dinero en Imagen*, April 27, 2015, available at <http://www.dineroenimagen.com/2015-04-27/54592>.

[19] Center for the Study of Democracy, "Public Discussion Serious and Organized Crime Threat Assessment," 2012, available at <http://www.csd.bg/artShow.php?id=15995>.

[20] Clive Leviev-Sawyer, "Seasonal fuel price hikes in Bulgaria 'could be affected by Ukraine' – report," *Independent Balkan News Agency*, May 5, 2014, available at <http://www.balkaneu.com/seasonal-fuel-price-hikes-bulgaria-could-affected-ukraine-report/>.

[21] "Report from Committee A (Sovereign Matters) on Cross-border Police Cooperation and Illicit Trade," *British-Irish Parliamentary Assembly*, 2015.

[22] Marko Hajdinjak, *Smuggling in Southeast Europe: The Yugoslav Wars and the Development of Regional Criminal Networks in the Balkans* (Sofia: Center for the Study of Democracy, 2002). Investigative journalists writing for the *Financial Times* in October 2015 estimated that ISIL earned US$1.5 million daily from the oil trade. See Erika Solomon, Robin Kwong, and Steven Bernard, "Inside Isis Inc: The journey of a barrel of oil," *Financial Times*, December 11, 2015, available at <http://ig.ft.com/sites/2015/isis-oil/>.

[23] "Mexico Pemex to stop gasoline via pipeline to curb theft," *Reuters*.

[24] Kathryn Haahr, *Addressing the Concerns of the Oil Industry: Security Challenges in Northeastern Mexico and Government Responses,* Wilson Center Case Study (Washington, DC: Wilson Center Mexico Institute, January 2015).

[25] Investigative journalists writing for the *Financial Times* in October 2015 estimated that the ISIL earned 1.5 million USD daily on the oil trade. Erika Solomon, Robin Kwong and Steven Bernard, "Inside Isis Inc: The journey of a barrel of oil," *Financial Times*, December 11, 2015, <http://ig.ft.com/sites/2015/isis-oil/>.

[26] European Commission Press Release Database, "Horsemeat: one year after – Actions announced and delivered!" *European Commission*, February 14, 2014, available at <http://europa.eu/rapid/press-release_MEMO-14-113_en.htm>.

[27] Dan Mitchell, "Justice Handed Down in Horsemeat Scandal," *Time,* April 7, 2015, available at <http://time.com/3774528/justice-handed-down-in-horsemeat-scandal/>.

[28] Andrew Jacobs, "Chinese Release Increased Numbers in Tainted Milk Scandal," *New York Times*, December 2, 2008, available at <http://www.nytimes.com/2008/12/03/world/asia/03milk.html>.

[29] "Agromafia. Third report on food crime in Italy 2015 (Agromafie. 3° Rapporto sui crimini agroalimentari in Italia 2015)," *Institute of Political, Economic and Social Studies (EURISPES),* available at <http://www.eurispes.eu/content/agromafie-rapporto-crimini-agroalimentari-eurispes>.

[30] Nick Squires, "Italy's 'Mozzarella King' arrested over 'contaminated cheese,'" *The Telegraph*, July 17, 2012, available at <http://www.telegraph.co.uk/news/worldnews/europe/italy/9406507/Italys-Mozzarella-King-arrested-over-contaminated-cheese.html>.

[31] "Italy's 'mozzarella king' arrested over mafia links," *The Local,* May 14, 2014, available at <http://www.thelocal.it/20140514/italys-mozzarella-king-arrested-over-mafia-links>.

[32] The term "gutter oil" actually covers a range of oils not fit for human consumption, which enter the food supply. But the term likely originates from the practice of recycling oil from the gutter. A *Radio Free Asia* film clip from China showing sewer refuse being used as an ingredient in gutter oil gained significant attention. See "'Gutter Oil' Video Goes Viral," *Radio Free Asia*, November 4, 2013, available at <http://www.rfa.org/

about/releases/gutter-oil-11042013112018.html>. Mislabeling of lower quality olive oil as "extra virgin olive oil" is common, but there have also been several cases of mixing olive oil with other vegetable oils, or adulteration of other cheaper oil types in order to pass them off as olive oil. The highest profile oil adulteration case occurred in Spain in the early 1980s, having caused almost 700 deaths and leaving 25,000 victims with lifelong ailments. Victims consumed what they thought was olive oil, which was in fact rapeseed oil adulterated with aniline (a coal tar extract). See Alan Riding, "Trial in Spain on Toxic Cooking Oil Ends in Uproar," *The New York Times*, May 21, 1989, available at <http://www.nytimes.com/1989/05/21/world/trial-in-spain-on-toxic-cooking-oil-ends-in-uproar.html>. With prices for organic products commanding a premium over nonorganic products, there is an incentive to cheat and present nonorganic produce as organic. Organic soy beans, for example, command a price that is over 300 percent higher than their nonorganic counterparts. See Edward C. Jaenicke and Iryna Demko, "Preliminary Analysis of USDA's Organic Trade Data: 2011 to 2014," *Organic Trade Association Policy Conference Infographic*, April 2015, available at <http://ota.com/sites/default/files/indexed_files/PolicyConference2015_Infographic_8.5x11_1a_0.pdf>.

[33] Walter G. Park and Douglas C. Lippoldt, *Technology Transfer and the Economic Implications of the Strengthening of Intellectual Property Rights in Developing Countries*, OECD Trade Policy Working Paper No. 62 (2008).

[34] Representatives of companies listed on the NASDAQ OMX 30 Stockholm stock index interviewed by this author in 2014 explicitly stated that extensive counterfeiting and IP theft in particular markets acted as a deterrent to investment. Krassén and Lallerstedt, *How leading companies are affected by counterfeiting and IP infringement.*

[35] "The Illicit Trafficking of Counterfeit Goods and Transnational Organized Crime," *United Nations Office on Drugs and Crime*, Vienna, <https://www.unodc.org/documents/counterfeit/FocusSheet/Counterfeit_focussheet_EN_HIRES.pdf>.

[36] Karl Lallerstedt and Mikael Wigell, *Illicit Trade Flows: How to deal with the neglected economic and security threat*, Finnish Institute of International Affairs Briefing Paper No. 151 (Helsinki: Finnish Institute of International Affairs: March 12, 2008).

[37] The Commission on the Theft of American Intellectual Property, "The IP Commission Report," *The National Bureau of Asian Research*, 2013.

[38] *U.S. Trade Representative*, The Trans-Pacific Partnership, Chapter 18, "Intellectual Property," available at https://medium.com/the-transpacific-partnership/intellectual-property-3479efdc7adf#.ptv7ataxp.

[39] Dev Kar and Brian LeBlanc, *Illicit Financial Flows to and from the Philippines: A Study in Dynamic Simulation, 1960-2011* (Washington, DC: Global Financial Integrity, 2014).

[40] A copy of the government report was provided to the author by a high-level Ghanaian public official in 2015. The report had previously been leaked to the media, and the estimate was published in news sources.

[41] "Revised 2014 Annual Gross Domestic Product," (Ghana Statistical Service: Accra, 2015).

[42] Europol and the Office for Harmonization in the Internal Market, "2015 Situation Report on Counterfeiting in the European Union."

[43] Ernesto U. Savona and Michele Riccardi, "From Illegal Markets to Legitimate Businesses: The Portfolio of Organised Crime in Europe," final report of *Project OCP Organised Crime Portfolio, Joint Research Centre on Transnational Crime*, 2015.

[44] "EU Serious and Organised Crime Threat Assessment (SOCTA) 2013," *Europol*, March 19, 2013, available at <https://www.europol.europa.eu/content/eu-serious-and-organised-crime-threat-assessment-socta>.

[45] Europol and the Office for Harmonization in the Internal Market, "2015 Situation Report on Counterfeiting in the European Union."

[46] Ibid.

[47] Ibid.

[48] United Nations Office on Drugs and Crime, *The Illicit Drug Trade through South-Eastern Europe* (Vienna: United Nations, March 2014).

[49] Frontex, "Western Balkans Annual Risk Analysis 2015," May 2015, available at <http://frontex.europa.eu/assets/Publications/Risk_Analysis/WB_ARA_2015.pdf>.

[50] Europol and the Office for Harmonization in the Internal Market, "2015 Situation Report on Counterfeiting in the European Union."

[51] "Naser Kelmendi," *Reporting Project*, available at <https://www.reportingproject.net/peopleofinterest/profil.php?profil=18>;

"Naser Kelmendi Criminal Organization," *U.S. Department of the Treasury*, March 2015, available at <https://www.treasury.gov/resource-center/sanctions/Programs/Documents/20150324_kelmendi.pdf>.

[52] Eurojust: The Hague, "Camorra-type clan behind massive counterfeiting of power tools and clothing," press release, May 21, 2010, available at <http://www.eurojust.europa.eu/press/PressReleases/Pages/2010/2010-05-21.aspx>; Gregory F. Treverton, Carl Matthies, Karla J. Cunningham, Jeremiah Goulka, Greg Ridgeway, and Anny Wong, "Film Piracy, Organized Crime, and Terrorism," *RAND Safety and Justice*

Program and the Global Risk and Security Center, (Washington, DC: RAND Corporation, 2009); Europol and the Office for Harmonization in the Internal Market, "2015 Situation Report on Counterfeiting in the European Union."

[53] "Operation Smoking Dragon," *Federal Bureau of Investigation*, 2011, available at <http://www.fbi.gov/news/stories/2011/july/dragon_070511>.

[54] Ibid.

[55] Martin Santa, "Slovaks find railway smuggling tunnel to Ukraine," *Reuters*, July 19, 2012, available at <http://www.reuters.com/article/us-slovakia-ukraine-tunnel-idUSBRE86I0ZO20120719>.

[56] Sean Gallagher, "Update: Russians capture cigarette-smuggling drone," *ars technica*, May 16, 2014, available ar <http://arstechnica.com/tech-policy/2014/05/russians-capture-cigarette-smuggling-drone/>.

[57] Between 2007 and 2010, 15 EU member states identified corrupt involvement of their border guards in the trafficking of cigarettes, and 13 states identified corrupt involvement in the smuggling of other consumer goods; see "Study on anti-corruption measures in EU border control," *Center for the Study of Democracy*, 2012, available at <http://frontex.europa.eu/assets/Publications/Research/Study_on_anticorruption_measures_in_EU_border_control.pdf>.

[58] "A study of the illicit cigarette market in the European Union," *KPMG Project SUN*.

[59] Mark Leon Goldberg, "How the Charcoal Trade Fuels Terrorism," *UN Dispatch*, February 22, 2012, available at <http://www.undispatch.com/how-the-charcoal-trade-fuels-terrorism/>; Amanda Ferguson, "Real IRA 'is ninth richest terror group in the word,'" *Belfast Telegraph*, November 17, 2014 available at <http://www.belfasttelegraph.co.uk/news/northern-ireland/real-ira-is-ninth-richest-terror-group-in-the-world-30748913.html>; "Report from Committee A (Sovereign Matters)," *British-Irish Parliamentary Assembly*; "How Can Counterfeit Drugs Be Useful to Terrorists?" *The Partnership for Safe Medicines*, available at <http://www.safemedicines.org/counterfeit-drugs-and-terrorism.html>; "Operation Smokescreen: The Crime and Investigation," *Bureau of Alcohol, Tobacco, Firearms and Explosives*, available at <https://www.atf.gov/our-history/operation-smokescreen>; Paul Alster, "The Truth about Hamas' Smuggling Tunnels," *The Investigative Project on Terrorism*, February 15, 2013, available at <http://www.investigativeproject.org/3915/the-truth-about-hamas-smuggling-tunnels>; "Cigarette smuggling biggest source of income for PKK," *Daily Sabah*, September 26, 2015, available at <http://www.dailysabah.com/investigations/2015/09/26/cigarette-smuggling-biggest-source-of-income-for-pkk>; Benoit Faucon and Ayla Albayrak, "Islamic State Funds Push Into Syria and Iraq With Labyrinthine Oil-Smuggling Operation," *The Wall Street Journal*, September 16, 2014 available at <http://www.wsj.com/articles/islamic-state-funds-push-into-syria-and-iraq-with-labyrinthine-oil-smuggling-operation-1410826325>; Drazen Jorgic, "Kenya wages war on smugglers who fund Somali militants," *Reuters*, June 21, 2015, available at <http://www.reuters.com/article/us-kenya-security-somalia-insight-idUSKBN0P105320150621>.

[60] Statement of Dr. Louise Shelley, "Could America Do More? An Examination of U.S. Efforts to Stop the Financing of Terror," testimony before the Congressional Task Force to Investigate Terrorism Financing in the U.S., U.S. House of Representatives, September 2015.

[61] "Momentum gathers for international agreement to combat rogue fishing," *Food and Agricultural Organization of the United Nations*, July 30, 2015, available at <http://www.fao.org/news/story/en/item/318052/icode/>; "Scale of illegal logging," *Illegal Logging Portal*, available at <http://www.illegal-logging.info/topics/scale-illegal-logging>; "Momentum gathers for international agreement to combat rogue fishing," *Food and Agricultural Organization of the United Nations*, July 30, 2015, available at <http://www.fao.org/news/story/en/item/318052/icode/>.

[62] "Illegal Trade in Environmentally Sensitive Goods," *OECD Trade Policy Studies*, 2012, available at <http://www.oecd.org/tad/envtrade/illegaltradeinenvironmentallysensitivegoods.htm>.

[63] "Risk Assessment of Illegal Trade in HCFCs," *United Nations Environment Programme*, 2011.

[64] "Medicines: spurious/falsely-labelled/falsified/counterfeit (SFFC) medicines," World Health Fact Sheet No. 275, *World Health Organization*, 2012.

[65] Gaurvika M.L. Nayyar, Joel G. Breman, Paul N. Newton, and James Herrington, "Poor-quality antimalarial drugs in southeast Asia and sub-Saharan Africa," *The Lancet* 12, no. 6 (2012): 488-496.

[66] "Counterfeit and Substandard Antimalarial Drugs," *Centers for Disease Control and Prevention*, September 22, 2015, available at <http://www.cdc.gov/malaria/malaria_worldwide/reduction/counterfeit.html>.

[67] An illustrative example is provided in the records on the seized computer of the arrested narco-trafficker, Juan Carlos Ramírez Abadía, detailing corrupt payments made between 2003 and 2006. There were three admirals, seven army and police colonels, and two naval captains on his payroll. See "Illicit Networks and Politics in Latin America," *Netherlands Institute for Multiparty Democracy and Netherlands Institute of International Relations*, 2014.

[68] As an example, representatives of political parties, religious institutions, and tribal chiefs solicit payments from prominent illicit traders. From: public officials in Ghana, interviews by author for a forthcoming OECD publication, 2015.

[69] A 2012 paper by Joossens indicates that the prevalence of illicit tobacco is higher in low-income countries. Joossens and Raw, "Strategic Directions and Emerging Issues in Tobacco Control." In a 2015 paper,

the author interviewed corporations stated that counterfeiting and the illicit trade in intellectual infringing goods was much more extensive in non-OECD states than in OECD states. Lallerstedt and Krassén, *How leading companies are affected by counterfeiting and IP infringement*; "Corruption Perceptions Index 2014," *Transparency International*, 2014, available at <https://www.transparency.org/cpi2014/results>.

[70] Discussions with corporate representatives and a senior customs representative.

[71] James Bargent, "Colombia Criminals Use Paraguay Contraband Cigarettes to Launder Drug Money," *InSight Crime*, March 24, 2014, available at <http://www.insightcrime.org/news-briefs/colombia-criminals-use-paraguay-contraband-cigarettes-to-launder-drug-money>.

[72] Liz MacKean and Meirion Jones, "Documents tarnish Montenegro's EU bid," *BBC News*, May 29, 2012, available at <http://www.bbc.com/news/world-europe-18237811>.

[73] "A study of the illicit cigarette market in the European Union," *KPMG Project SUN*.

[74] "Tobacco in Belarus," *Euromonitor International*, 2015, available at <http://www.euromonitor.com/tobacco-in-belarus/report>.

[75] Paul Rexton Kan, Bruce E. Bechtol, Jr., and Robert M. Collins, *Criminal Sovereignty: Understanding North Korea's Illicit International Activities* (Carlisle, PA: Strategic Studies Institute, 2010); North Korean diplomats at the Stockholm Embassy in Sweden were discovered to be involved in contraband tobacco and alcohol sales as early as 1976, and in a more recent case, two North Korean diplomats were caught smuggling cigarettes into Sweden in 2009. See Lovisa Lamm, "Smuggling ring at North Korean Embassy (Smuggelhärvan på nordkoreanska ambassaden, P3 Dokumentär)," *Sveriges Radio*, April 11, 2010; Micke Ölander, "Diplomats caught for smuggling (Diplomater fast för smuggling)," *Expressen*, November 20, 2009.

[76] In 2014, for example, seven Gambian diplomats were convicted for their involvement in illicit tobacco sales, defrauding the HM Treasury of £4.7 million. See Kunal Dutta, "Gambian diplomats found guilty of using London embassy for tobacco fraud," *Independent*, December 8, 2014, available at <http://www.independent.co.uk/news/uk/crime/gambian-diplomats-found-guilty-of-using-london-embassy-for-tobacco-fraud-9911414.html>.

[77] Colum Lynch, "U.S. Accuses Iran of Secretly Breaching U.N. Nuclear Sanctions," *Foreign Policy*, December 8, 2014, available at <http://foreignpolicy.com/2014/12/08/us-accuses-iran-of-secretly-breaking-un-nuclear-sanctions-exclusive/>.

[78] "Magnitude of Counterfeiting and Piracy of Tangible Products," *OECD*; Frontier Economics, "Estimating the global economic and social impacts of counterfeiting and piracy."

[79] "A study of the illicit cigarette market in the European Union," *KPMG Project SUN*.

[80] There are some systematic efforts to analyze the quality of pharmaceutical drugs, such as the results stored in the U.S. Pharmacopeial Convention's Medicines Quality Database, which contains test results for select categories of drugs for select countries. But with more extensive testing, and ensuring that more results are made public, could enhance the value of the present efforts. "Medicines Quality Database," *The United States Pharmacopeial Convention*, 2015.

[81] Japan Patent Office, "FY2014 Survey Report on Losses Caused by Counterfeiting has been Compiled," *Ministry of Economy, Trade and Industry*, March, 11, 2015. The prioritization of intellectual property rights enforcement by the Japanese government is reflected by its Intellectual Property Strategy Headquarters (IPSH), whose meetings are directed by the prime minister and include cabinet ministers as well as external experts. The IPSH is responsible for setting intellectual property strategy, as well as ensuring its coordination and follow-up. Japan is also a generous donor to both the World Customs Organisation (whose secretary-general happens to be Japanese) and the World Intellectual Property Organisation, and it has provided bilateral support to help other states enhance their intellectual property rights enforcement, and Japan was also a strong proponent of the Anti-Counterfeit Trade Agreement (ACTA). See Japan Patent Office, "Formulation and Implementation of National IP Strategy in Japan – Intellectual Property Strategic Program 2011," *Ministry of Economy, Trade and Industry*, February 3, 2012.

[82] This is an assumption based on the author's personal experiences, having worked in the private sector with anti-illicit trade activities, and what individuals working at multinational corporations have stated in informal discussions with the author.

[83] This implies a particular challenge in countries with weak state capacities, which the international community could play a role in addressing.

[84] Krassén and Lallerstedt, *How leading companies are affected by counterfeiting and IP infringement*.

[85] Karl Lallerstedt, Elva (Jing) Zhang, and Joost Pauwelyn, "The world trade system must arm itself to fight illicit trade," *The Global Initiative Against Transnational Organized Crime*, available at <http://www.globalinitiative.net/the-world-trade-system-must-arm-itself-to-fight-illicit-trade/#_ftnref1>.

[86] Ibid.

[87] "China – Measures Affecting the Protection and Enforcement of Intellectual Property Rights (WT/DS362/1)," panel report, *First Submission of the United States of America*, January 30, 2008, available at <http://www.ifta-online.org/sites/default/files/58.pdf>.

[88] Lallerstedt, Zhang, and Pauwelyn, "The world trade system must arm itself to fight illicit trade."

[89] Amir Attaran, Donna Barry, Shamnad Basheer, Roger Bate, David Benton, James Chauvin, Laurie

Garrett, Ilona Kickbusch, Jillian Clare Kohler, Karnal Midha, Paul N. Newton, Sania Nishtar, Paul Orhii, and Martin McKee, "How to achieve international action on falsified and substandard medicines," *British Medical Journal* 345 (November 2012); Professor Attaran, interview by the author, 2015.

[90] "WTO Agreements and Public Health: A joint study by the WHO and the WTO Secretariat," *World Trade Organization/World Health Organization*, 2002, available at <https://www.wto.org/english/res_e/booksp_e/who_wto_e.pdf>.

[91] Ibid.

[92] Ibid.

[93] The OECD Task Force on Charting Illicit Trade, established in 2013, is a manifestation of this recognition.

[94] "Magnitude of Counterfeiting and Piracy of Tangible Products," *OECD*, (2009).

[95] Krassén and Lallerstedt, *How leading companies are affected by counterfeiting and IP infringement.*

[96] "Nearly 25 percent of global Internet traffic involves the illegal distribution of copyrighted work," *Net-Names*, February 2011, available at <http://www.netnames.com/news/nearly-25-global-internet-traffic-involves-illegal-distribution-copyrighted-work>.

[97] "Gartner Says Uses of 3D Printing Will Ignite Major Debate on Ethics and Regulation," *Gartner Press Release*, January 29, 2014, available at <http://www.gartner.com/newsroom/id/2658315>.

[98] Krassén and Lallerstedt, *How leading companies are affected by counterfeiting and IP infringement.*

[99] "Scope of the problem," *Recording Industry Association of America*, available at <http://riaa.com/physicalpiracy.php?content_selector=piracy-online-scope-of-the-problem>.

[100] The UK is one example, but internet service providers act similarly in a number of other countries. "ISPs asked to block more pirate sites," *BBC News*, October 20, 2015, available at <http://www.bbc.com/news/technology-34581801>.

In a country like Sweden, where internet service providers do not block pirate websites, the Association of Swedish Advertisers (Sveriges Annonsörer) has produced a so-called "blacklist" of websites through which it recommends its members to avoid advertising. See "The Black list – movie sites you don't want to advertise on (Svarta listan – filmsajterna du inte vill synas på som annonsör)!" *Association of Swedish Advertisers*, May 16, 2015. Online payment solutions providers, such as PayPal, have implemented policies to prevent payments to entities infringing intellectual property rights, and have introduced a reporting mechanism for rights owners. See "Infringement Report Policy," *PayPal*, August 28, 2012.

[101] "Darknet Marketplace Watch – Monitoring Sales of Illegal Drugs on the Darknet (Q1)," *Digital Citizens Alliance*, April 24, 2015, available at <http://www.digitalcitizensalliance.org/cac/alliance/content.aspx?page=Darknet>; "Money is no object: Understanding the evolving cryptocurrency market," *PwC*, 2015.

13
Cybercrime: The Evolution of Traditional Crime

Raj Samani

"Cybercrime is a growth industry. The returns are great, and the risks are low. We estimate that the likely annual cost to the global economy from cybercrime is more than $400 billion."[1]

While the actual figure may be debated, there is no question that cybercrime is a growth industry. It seems the only debate we have regarding the growth of cybercrime is whether it is an evolution of traditional crime or a revolution. Making this distinction is based on one's perspective; having worked within this industry for some time now, I would argue that the former conceptualization of cybercrime—as evolutionary—is most accurate. However, when compared with crimes that have been wreaking havoc for centuries (e.g., smuggling, theft), their transition to or dependency on the cyber domain started overnight, and might seem to justify characterizing this transition as revolutionary. Regardless of this distinction, criminals that have adopted a digital arm are netting enormous sums of money. Take, for example, the recent case entitled Operation *Carbanak*, in which financial institutions were reported to have lost $1 billion through an "Advanced Persistent Threat" (APT).[2] Though the name "Advanced" infers sophistication, the reality is that these banks were initially infected by nothing more than an employee receiving an email with a link to a malicious site. In fact, what is even more worrying is that this simple approach of infection is behind most of the breaches that we hear about on a daily basis.

The opportunities cybercrime present have caught the attention of traditional criminals; now "traditional Mafia groups are increasingly outsourcing their specialist operations to highly skilled freelance cybercriminals who promote their services on hidden websites."[3] These new partnerships have altered the modus operandi of traditional criminals. According to Rob Wainwright, director of Europol, "We're seeing a move away from the traditional model of organised Mafia-like structures that are very hierarchical.... They make use of this industry of criminals with a particular service or product to feed to organized crime groups. It's the same as the legitimate commercial world: outsourcing is what happens."[4] The outsourcing of cyber-related operations means that the barriers for becoming a cybercriminal are probably at the lowest ever. The reduction in these technical barriers is important, because when combined with the huge revenues, as well as the ability to perpetrate a crime in another country without ever leaving your desk, the perfect storm for society is created. In this case, the introduction of such factors means that every business, and every user is susceptible to digital crime. This may seem like overstating the issue, after all not every person is online; however, recent cyber-attacks have even disrupted the availability of electricity in thousands of homes in Eastern Europe.

This chapter will survey a limited sample of the novel cyber techniques that have made the cyber domain easily accessible to those with criminal intent, and a danger zone for legitimate actors, both official and private. However, this is only part of the story, because any such cyber-attack, and certainly those that involve the theft of data, will result in criminals being in possession of huge amounts of sensitive information. One question that is often asked is, "What to do they do with this stolen data?" Answering this question is particularly difficult as we are not privy to deals criminals may have with one another; however, in many cases this data is stolen without a prearranged buyer. In these circumstances, we can clearly see this data being sold in online marketplaces. What is even more remarkable is that much of this data is sold on the Surface Web—the same internet that you and I use every day—using the same popular search engines. It is important to make this distinction, because this chapter focuses on the accessibility of tools that facilitate cybercrime; therefore, easily getting to these tools becomes important. The hosting of these tools will vary; some of which are available on the Surface Web, others are within Dark Web, and some of which are in the Deep Web. In many cases, the latter two are used interchangeably but strictly speaking this is inaccurate. To be clear, the "Surface Web" refers to the portion of the internet that can be accessed via the standard browser, and found with a normal search engine. The "Dark Web" refers to those sites that can still be accessed via a normal browser, but are not searchable via search engines. Finally, the "Deep Web" refers to a portion of the internet that is hidden on purpose and cannot be accessed via a standard web browser. Accessing this portion of the internet requires a specialized web browser, which can be easily downloaded onto standard operating systems. Once downloaded, the user can then use this browser to access sites on this hidden portion of the internet.

Cybercrime-as-a-Service

Describing the emergence of a marketplace for those wishing to carry out cybercrime in one chapter is no simple task. There are a lot of sellers selling all manner of products to a range of buyers. The following, therefore, only provides a sample of the types of products available, but does so within four categories:

1. *Research-as-a-Service.* Unlike other categories, buying in research does not have to originate from illegal sources; there is room for a gray market. Take, for example, the *Carbanak* case study referred to above; criminals clearly had email addresses of their victims. Buying such lists is a simple task that can be done by commercial companies selling what they call "marketing databases."

2. *Crimeware-as-a-Service.* Once criminals have identified their target, in many cases they will attempt to install malicious software on the victim's computer. This could be with the intent to steal using malware, which incorporates the

identification and development of the exploits used for the intended operation—and may also include development of ancillary material to support the attack.[5] This category also includes the availability of hardware that may be used for financial fraud (e.g., card skimming) or equipment used to hack into physical platforms.

3. *Cybercrime Infrastructure-as-a-Service.* Once the toolset has been developed, cybercriminals are faced with the challenge of delivering their exploits to their intended victims. An example is rental of a network of computers to carry out a denial-of-service (DoS) attack. Other examples are included within the text below.

4. *Hacking-as-a-Service.* Acquiring the individual components of an attack remains an option; alternatively, there are services that allow for outsourcing of the attack entirely. This path requires minimal technical expertise, although it is likely to cost more than acquiring individual components. This category also supports the availability of information to be used for identity theft, for example, requesting information such as bank credentials, credit card data, and login details to particular websites.

While many services within the cybercrime ecosystem will fall within the preceding categories, there are many more that will not neatly fit within these descriptions. There is no intention to attempt to define every possible service, particularly because the environment is fluid and new products and services emerge on a constant basis. The intention of this chapter is to demonstrate the ease with which anybody has the ability to become a cybercriminal through illustration of the ecosystem and of the "as-a-Service" (aaS) nature of cybercrime today. Furthermore, it will focus on the sheer scale and volume of stolen data becoming available for sale as an indication as to the problem cybercrime has become for us as a society.

Cybercrime Exposed

In 2013, I co-authored a report entitled, "Cybercrime Exposed" that presented this burgeoning economy and provided examples where cybercrime is replacing or evolving from traditional crime in multiple ways. Even in the short time since that paper was published, the cybercrime market has grown tremendously. This chapter might be considered as an update to the earlier report, addressing the subsequent reports that considered other factors such as payment and cashing out.[6]

The central question that both "Cybercrime Exposed" and this chapter seek to address is: Are we witnessing a fundamental transformation of traditional crime, and if so, what are the characteristics of the criminal activities replacing it? The Federal Bureau of Investigation (FBI) reported a decline in physical crimes, such as bank robberies, in 2012, as opposed to cybercrime, which has increased significantly.[7] In the report, it was reported that there were 5,000 bank robberies in 2011, but that number dropped to 3,870, marking a

23 percent decrease. Compare this to cybercrime: a recent Pricewaterhouse Coopers (PwC) study of cybercrime in the United States stated that "the U.S. Secret Service has reported a marked increase in the quality, quantity, and complexity of cybercrimes targeting both private industry and critical infrastructure."[8] In terms of monetary value, the global cost of cybercrime is estimated to be $445 billion per year.[9] Clearly, the risk associated with physical crimes, such as bank robberies, contributes to such a shift; cybercriminals enjoy the luxury of carrying out their crimes from a physical location of their choosing.

The Center for Strategic and International Studies (CSIS) report's "Net Losses: Calculating the Cost of Cybercrime" suggested that "the combination of high value, low risk, and low 'work factor' (the amount of effort it takes to break into a network) makes cybercrime a winning proposition."[10] The latter part of this statement refers to the ease with which cybercrime tools are made available. Moreover, the cybercrime market now affords potential criminals with a multitude of services. The result is that deep technical expertise is not a prerequisite to engage in cybercrime. In "Cybercrime Exposed," the head of the European Cybercrime Centre notes that "today's cybercriminals do not necessarily require considerable technical expertise to get the job done, nor, in certain cases, do they even need to own a computer. All they need is a credit card."[11]

The services-based nature of cybercrime allows greater efficiency and flexibility when conducting business. The ability to provide technology solutions as a service to businesses has allowed organizations to focus on their core competencies. An unintended consequence of this evolution has been the rise of the aaS model and a marketplace offering multiple variants of hosted services. Although this approach may seem innovative, the aaS model itself is nothing new. The underground economy established by cybercriminals has been using a services-based model for some time. Although the term "Crimeware-as-a-Service" may be relatively new, the services-based nature of cybercrime has been in effect considerably longer than its descriptive acronym. Moreover, the services-based approach extends well beyond hiring individuals to undertake specific tasks (e.g., coding an exploit). With a broad variety of products and services available either to buy or rent, such an economy has evolved with more products coming online. Having addressed the overall services orientation of the cybercrime industry, it is now fitting to delve into the specific services available.

Research-as-a-Service

The available services within this category include the identification of a previously unknown vulnerability within a targeted system, otherwise known as "zero-day vulnerability." Despite the threat of legal action by affected software vendors in certain countries, the sale of vulnerabilities has recently become a growth area for researchers and brokers alike. Today, security researchers are presented with a number of options when they identify previously unknown zero-day vulnerabilities. Each option is accompanied by differing outcomes in publicity and monetary compensation. Today's marketplace provides those looking to acquire zero-day vulnerabilities with many options. At first glance, this may appear to be detrimental to underground marketplaces. For example, one particular vendor defines its eligibility requirements as being limited to only public sector organizations, in particular, law enforcement. Furthermore, restrictions are placed on the geographic location of the agency; its customers can only be in predefined countries. However, though many organizations selling zero-day vulnerabilities actually limit their sale to specific buyers, the underground market continues to thrive.

It is worth noting that the acquisition of zero-day vulnerabilities is a somewhat gray area, since sellers will not disclose their customer lists; however, recent breaches of companies that sell such services revealed that their customers included many government agencies.[12] Important among these actors are "exploit brokers." Although the acquisition of vulnerabilities can be conducted through a commercial entity, there is an opportunity to connect with a brokering service, which can be defined as a single individual who acts as a middleman to facilitate the sale to a third-party. By acting as middleman for the sale of zero-day vulnerabilities, exploit brokers are able to charge a commission by facilitating the sale of said vulnerabilities. For example, in one scenario the sale of an Apple iOS exploit for $250,000 allowed the broker to pocket 15 percent in commission.[13] Customers of such vulnerabilities are not necessarily committing a crime in acquiring such products, and as the example above has demonstrated, their acquisition is limited to certain organizations.

Spam Services

In the majority of successful cyber-attacks, the initial infection vector will use a spear-phishing message to someone within the target organization. What this usually means is that somebody within the targeted organization will receive an email invoking some form of action (e.g., clicking onto a link), which in turn results in malware being installed onto their computer. This was demonstrated in Operation *Carbanak*; in this case "all observed cases used spear phishing emails with Microsoft Word 97 – 2003 (.doc) files attached or CPL files."[14] An example of the phishing email that was sent to employees within the targeted organization is illustrated in Figure 13.1.

Figure 13.1. *Carbanak* Phishing Email (with Translation)

ORIGINAL:

```
Хорошего дня!
Посылаю вам наши контактные данные
Сумма депозита 32 000 000 миллионов рублей 00 копеек за период 366
дней,% год — КОНЕЦ вкладчиками термин
С уважением, Сергей Кузнецов;
+ 7 (953) 3413178
f205f @ mail.ru
```

TRANSLATED:

```
Good Day!
I send you our contact details
    The amount of deposit 32 million rubles and 00
kopecks, for a period of 366
days, % year---end contributors term
    Sincerely, Sergey Kuznetsov;
+ 7 (953) 3413178
f205f @ mail.ru
```

Source: Kaspersky

Whether the target for the attack is a consumer or an employee within a large enterprise, the modus operandi for many cybercriminals is to leverage some form of social engineering to coerce the user into an action facilitating malware infection.

However, with the prevalence of social engineering in many publicly disclosed cyber-attacks, there exists either an inherent weakness in the acumen of victims to distinguish malicious communications, or cybercriminals are using more complex methods to bypass the "human firewall." The answer, of course, likely lies somewhere in between these two statements, but regardless of the root cause, it does demonstrate that the first line of defense is evidently failing. More importantly, the default position—to simply blame users as the cause for breaches—is not entirely fair. Indeed, while there will be examples where clearly unsafe practices are being employed, as this chapter will demonstrate, the techniques used by attackers are intended to bypass the consciousness of their targets and attempt to manipulate victims through leveraging subconscious levers of influence.

For the purpose of this chapter, I define "social engineering" as a deliberate application of deceitful techniques designed to manipulate someone into divulging information or performing actions that may result in the release of that information.[15] During a social engineering interaction, the target is not consciously aware their actions are wrong or ill-advised. The social engineer exploits the natural instincts, not the criminality of the target. This is done using a variety of methods to trick the user into divulging useful information or performing an action, such as clicking a link. Social engineering exploits subconscious behavior of the target, unlike conscious techniques such as bribery, threat of violence, and so on, which do not fall within the scope of social engineering.

Any successful spam campaign relies on a number of factors, many of which are discussed below. One fundamental element belongs in the Research-as-a-Service category: the identification of targets. Having to manually gather together an email list can be a time-consuming exercise—fortunately, the would-be spammer has the luxury of simply purchasing a list of email addresses. Aside from the customization of the message in a particular language, the unsolicited email may require more granularity. For example, there is also the opportunity to acquire email addresses from a specific geography, with the level of granularity even to state or city level.

Such examples are merely the tip of the iceberg. Indeed, since the publication of "Cybercrime Exposed," we have identified an enormous marketplace offering email addresses at considerably low prices within commercial environments. In fact, on commercial auction sites, 50,000 email addresses that are available on a per-town basis sell for as little as $10. These are advertised as marketing databases; however, they give no indication of whether consent has been confirmed (as per Data Protection requirements). Attempting to understand the scale and volume of sales for unauthorized personal information is difficult; however, one indication was revealed in the "What Price Privacy" report by the United Kingdom Information Commissioner which states, "We know of one private investigation firm receiving some £50,000 a month from just one finance company for tracing new addresses at £35 a time, and £55 for a new employer and new address. The same firm was also undertaking checks for other companies, which gives some idea of the scale of operations."[16]

While granularity based on geographic location is common, the would-be spammer has the opportunity to be even more specific about the potential targets. For example, a campaign may target specific users of a service, a particular bank, or an internet service provider. Or a campaign may target specific professions, or even a specific gender. In such instances, the underground marketplace supports the acquisition of such lists, as depicted in Figure 13.2. As this illustrates, it is possible to identify a specific profession as well as a certain geography, such as U.S. doctors. Another consideration should be the manner in which the service is offered. The presentation of the service is similar to those offered by legitimate companies selling legal products; some even offer commercial payment mechanisms.

Figure 13.2. Selling Email Addresses by Profession

Source: McAfee Labs

Identifying a list of targets is the first step, developing a convincing message and the requisite malware will follow. However, as we will see in the following sections, these are all accessible and available for sale.

Crimeware-as-a-Service

In the past, the distinction between the accessibility of crimeware and Research-as-a-Service options was that the former was only accessible in underground forums while the latter was accessible on the Surface Web. Focusing our attention on the availability of tools necessary to infect victims, we find a multitude of products available for either sale or rent. Below are many of the Crimeware-as-a-Service tools available today.

Professional Services

Simply purchasing details of where vulnerabilities may exist within a specific application or computer operating system is only the first step. There will then exist a requirement to develop specific code to exploit this vulnerability, which can be outsourced. An example of this was seen as early as 2005, with the Zotob worm. In this example, a programmer was paid to develop the malware; indeed, after the arrest of two individuals, according to the assistant director of the FBI Cyber Division, "the Moroccan was responsible for writing the code, he had a financial relationship with the Turkish man."[17] Other professional services include translations. In the Research-as-a-Service category, we saw how it was possible to acquire email addresses for a specific country; if the attacker is a native speaker, then crafting an email to entice victims is relatively simple. However, the cybercrime market has evolved such that not knowing a language is no longer a hindrance in targeting particular populations. Services provide translations to support non-native speakers in their efforts to communicate with potential victims. What is even more remarkable is that much like the modern social media tools we use today, there is an indicator as to the reputation of the individual behind the profile offering such services.

Malware Services

Developing a convincing phishing email is the first step, but as we have seen in the *Carbanak* example, developing malware that can infect the computer of an employee within the targeted organization is also critical. For the development of such nefarious code, there is a burgeoning marketplace available that negates the need for technical knowledge on the part of the cybercriminal. This takes the form of numerous malware variants available for sale, or even available for rental. Purchasers can acquire developed code to conduct their attacks. For example, attackers who want to acquire information can buy a Trojan Horse, a malicious program concealed within a legitimate file. Other examples include:

- *Rootkit Services.* Surreptitious code that conceals itself within the compromised system and performs actions as programmed.
- *Ransomware Services.* Software that restricts the user from conducting further activity until a specific action is taken (such as providing credit card details). An example of this ransomware aaS phenomenon under the name of "Tox" was uncovered in recent research conducted by McAfee Labs.[18] This program provides visitors a free service to develop a ransomware campaign and target any number of users. Once the visitor has registered, the only required action is to enter the amount of money that will be charged to unfortunate victims (the ransom amount), a particular cause, and of course, the obligatory captcha (the graphic that is used in the registration process, usually asking the person to type in certain characters). Once this information is provided, the newly registered user has to enter a Bitcoin address, and all profit (less a 20-percent service fee) is transferred to them. Tox is not an isolated case; many of the most prevalent ransomware campaigns are based on this aaS model which may, in part, explain the reason for such a growth in this digital threat—ease of use, and of course, profits. Consider that "between February and April 2015, the perpetrators extorted

$76,522 from 163 victims," using the TelsaCrypt variant of ransomware.[19]

- *Exploits.* There are many options to purchase exploits that take advantage of vulnerabilities. Their prices vary based upon the target system and whether the vulnerability has been previously identified. There is also the opportunity to rent as opposed to buying. The CritX tool kit, for example, charges by the day, and recently advertised for $150 per day. While acquiring individual exploits is an option, a more commonly adopted method has been the use of exploit kits (EKs). These kits are used "in a process known as a drive-by download, which invisibly directs a user's browser to a malicious website that hosts an exploit kit. The exploit kit then proceeds to exploit security holes, known as vulnerabilities, in order to infect the user with malware. The entire process can occur completely invisibly, requiring no user action."[20] To get a sense of the growth of exploit kits, Figure 13.3 illustrates the number of detections on a weekly basis by security firm Sophos, of the most popular kit (as of the time of writing) known as Angler.

Figure 13.3. Growth of Angler EK

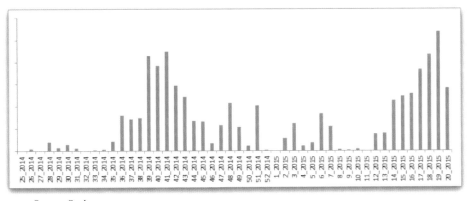

Source: Sophos

Such investment on the part of the cybercriminal will motivate the acquisition of services to evade possible detection (note that Tox has these included by default). To avoid possible detection, there exists many services that can assist in obfuscation; this includes the cryptor packs that can encrypt particular files, as well as counter-antivirus (AV) services. The latter are services that allow would-be attackers to run their exploits against the most popular AV products to ensure that they are not identified as malware. While the manual process of installing every possible AV program and running the exploit can be costly, it will certainly be incredibly time-consuming. Counter-AV services are an extremely cost-efficient method in ensuring that the attack is not easily detected, and one such service allows the cybercriminal to run newly acquired exploits against the most popular AV software packages for less than $0.15 per check.

Cybercrime Infrastructure-as-a-Service

Identifying a list of suitable targets, and developing the necessary tools to facilitate an infection may be perceived as all of the necessary steps required to conduct a successful cyber-attack. However, the question of delivering the attack remains. For a would-be cybercriminal, the option to leverage the aaS model remains available. As expected, the marketplace offers a number of infrastructure services to support a cybercrime operation. These range from the availability of services to conduct DoS attacks (attacks that aim to impact the availability of victim services by overwhelming them with excessive volume), to hosting malicious content.

Botnets

A robot network, or botnet, is a network of infected computers under the remote control of an online criminal. The botnet can be used for a number of services, such as sending spam, launching DoS, and distributing malware. As one would expect, the cybercrime marketplace offers the would-be cybercriminal a multitude of options in terms of leveraging the botnet for the facilitation of an attack. Indeed, the rental options of a botnet show a relatively cost-effective mechanism. These prices will of course vary, but according to Havocscope:

> The price to rent 1,000 infected computers in the United States costs $180. If the hosts are located in the United Kingdom, the price is $240. France and Russia both costs [sic] $200, Canada costs $270, and 1,000 infected computers located around the world costs $35. There is a daily limit of 20,000 hosts.[21]

What we tend to find is that the variants, or rather, configurable options, will impact the price paid. As the above quote demonstrates, geographic location makes a difference; however, alternate options include regular updates (so that the code will regularly change and reduce the likelihood of detection and clean-up amongst the infected computers) and the modules that are provided. These modules refer to the function of the botnet; for example, if you want your botnet to steal passwords from infected computers, then this module may be added.

Hosting Services

A "bulletproof" hosting provider is a company that knowingly provides web or domain hosting (or other related services) to cybercriminals, intending to ignore complaints by turning a blind eye to the malevolent use of their services. Such services provide a multitude of options for the buyer. In one particular example, an individual known as "Matad0r" provided three levels of service, ranging from $50 per month to as much as $400 per month. The variable pricing is based on the specification of the system provided; a more powerful system with more options corresponds to a higher price. This demonstrates that, much like the commercial environment, a myriad of hosting services is available—the only constraint is the amount of money one is willing to pay, and, in some cases, the ethics of the hosting provider. The term "ethics" refers to what the hosting provider is willing to host on the systems they offer to market. Some sellers will be specific about what they are not willing to host, for example, information related to terrorist activities or related to child pornography. There are, of course, sellers of such services that have no concern about hosting such data.

The importance of hosting services should not be underestimated. In the ongoing game of cat and mouse, those behind cyber-attacks will go to extraordinary lengths to ensure that the systems hosting their malicious content are not blocked by security controls. Indeed, this was realized in 2013, in an attack that was described as the biggest cyber-attack of its kind in history.[22] The attack impacted the nonprofit organization Spamhaus, whose team of volunteers maintains blocklists of systems it believes are used for malicious purposes. After blocking servers owned by a Dutch provider known as "Cyberbunker," Spamhaus experienced a significant Distributed Denial-of-Service (DDoS) attack intended to prevent legitimate traffic. Such was the ferocity of the attack, that the chief executive for Spamhaus commented, "they haven't been able to knock us down. Our engineers are doing an immense job in keeping it up. This sort of attack would take down pretty much anything else."

Delivering Unsolicited Mail

Numerous services are available for the would-be spammer. These include services that support the sending of unsolicited mail, alternatively known as a "mail relay." Certain services are capable of managing particularly large volumes; research identified one service capable of sending 30 million emails in a one-month period. What was remarkable was that the service offered a live chat option with a customer service agent, as well as payment options many of us are accustomed to in the legitimate world.

Of course, simply having an infrastructure is not enough to support an unsolicited email campaign. There is also a need for the email addresses themselves, as well as a back end set of systems to continue the deception. The latter could be hosted through bulletproof hosting services, and the former was addressed under the Research-as-a-Service category.

Hacking-as-a-Service

If the budget allows, a budding cybercriminal can skip the process of conducting research, building appropriate tools, and developing an infrastructure, to launch a cyber-attack by choosing a service that will outsource the entire process.

Password Cracking Services

There is a multitude of services available within the Hacking-as-a-Service category. The following examples illustrate how little technical knowledge is required for buyers to try their hand at cybercrime. This includes the availability of services that allow the prospective buyer to retrieve an email password of their intended victim. To illustrate the point made earlier about cybercriminals not requiring any technical expertise, the would-be buyer would only need the email address and name of the target. After that, all that remains is to pay for the service.

Denial-of-Service (DoS)

The press has been awash with stories of hacktivists bringing down large companies with sophisticated hacking techniques. The reality is very different. Although many attacks may be sophisticated, many of them are simply DoS or DDoS attacks. These DoS services aim to send a huge volume of traffic to the victim to overwhelm and disrupt normal business operations. Building a cyber-army capable of generating enough traffic does, at the very least, require an investment in time that the would-be cybercriminal may not have. Fortunately for them (and unfortunately for the rest of us), the aaS cybercrime market is there to help. The price list for a "Cheap Professional DDoS Service" will vary; however, the level of technical knowledge required to recruit such services is very low. It only requires the buyer to inform the service of which site they wish to launch a DDoS attack against, decide how much they are willing to pay, and then initiate the service. What is remarkable however is the cost; one such example is a DDoS attack lasting an hour that only costs $2.

One of the biggest challenges law enforcement faces in combating cybercrime is its global nature, and the above example highlights this. Take for example an attack by one business on its neighbor, where a traditional crime would be carried out in the same physical jurisdiction (e.g., smashing the windows of a neighboring business). In a digital DoS attack, a third-party service may be used. Such a service could be hosted by a provider outside of the geographic jurisdiction of the victim. Introduction of this added level of complexity makes any investigation considerably more difficult.

Hidden Data Economy

The previous examples are of technical services. Yet, also available is an economy in which multiple forms of data are made accessible for sale. Published in late 2015, the new white paper by McAfee (co-authored by myself) uncovers an industry in which for a small fee almost every conceivable stolen record is available.[23]

This underground marketplace has evolved to include almost every conceivable cybercrime product for sale or rent. We correctly predicted that the rise of this aaS model would act as a key driver in the growth of cybercrime. The "McAfee Labs Threats Report: May 2015" provides evidence of this with the rise of the ransomware CTB-Locker.[24] The authors of CTB-Locker established an affiliate program as part of their business strategy: affiliates use their botnets to send spam to potential victims, and for every successful infection in which the victim pays the ransom, the affiliate gets a percentage of the money.

The growth of the aaS economy across all components of an attack (research, cybercrime tools, and infrastructure) continues to grow and none more so than Hacking-as-a-Service, and in particular the component in which stolen data is made available. It is important to highlight why apathy among victims of a data breach, and ultimately those data subjects whose information is being sold, may be costly. A sad side effect of reading about data breaches is the concept of "data breach fatigue," which is another way of saying apathy. The recent article "I Feel Nothing: The Home Depot Hack and Data Breach

Fatigue" provides a wonderful example of such apathy. The author writes, "Because banks are responsible for making us whole if our credit cards are misused, and we are simply issued new cards (an annoying hassle, but not life altering), I join you in reacting to news of these hacks with a shrug."[25] "We are in the trough of disillusionment," says Gartner analyst Avivah Litan.[26] Although such a view may be understandable due to the steady stream of breach notifications and stories detailing the theft of millions of records, it is important to recognize that this is data about us. Our information is being openly sold, and the individual repercussions may not be felt for some time.

Financial Data

Selling stolen financial data is a relatively broad topic, with a multitude of data types for sale and marketplaces that vary between the visible web via a standard browser and the Dark Web through other access methods.

Data breaches involving the theft of financial data, particularly payment card information, continue to dominate headlines. Particularly impacting retailers, the theft of such information invariably results in this data appearing on the visible web. Payment card information made available in those marketplaces will vary in price based on a multitude of options. A snapshot of these options is shown in the following table.

Table 13.1. Estimated Per-Card Prices for Stolen Payment Card Data (Visa, MasterCard, Amex, Discover)

Payment Card Number with CVV2	United States	United Kingdom	Canada	Australia	European Union
Random	$5–$8	$20–$25	$20–$25	$21–$25	$25–$30
With Bank ID Number	$15	$25	$25	$25	$30
With Date of Birth	$15	$30	$30	$30	$35
With Fullzinfo	$30	$35	$40	$40	$45

The preceding categories relate to the information available along with the payment card number:

- "CVV" is the industry acronym for "card verification value." CVV1 is a unique three-digit value encoded on the magnetic stripe of the card, and CVV2 is the three-digit value printed on the back of the card.
- "Random" means that the number is automatically generated (via software).
- "Fullzinfo" means the seller supplies all of the details about the card and its owner, such as full name, billing address, payment card number, expiration date, PIN number, social security number, mother's maiden name, date of birth, and CVV2.

Occasionally, additional information is available for sale. Payment card data that includes "with COB" refers to those cards with associated login and password information. Using these credentials, the buyer can change the shipping or billing address or add a new address. Some sellers fail to deliver the data after purchase. After all, whom will the buyer complain to in the event that the stolen information is not delivered?

Payment Card Data with Additional Information

Buyers have many options, including the geographic source of the card and the card's available balance. Both of these options impact the price of a card, as we see in the following table:

Table 13.2. Dump Track Prices per Card

Dump Track with High Balance	Price
Track 1&2: PinATM United States	$110
Track 1&2: PinATM United Kingdom	$160
Track 1&2: PinATM Canada	$180
Track 1&2: PinATM Australia	$170
Track 1&2: PinATM European Union	$190

The term "dump" refers to information electronically copied from the magnetic strip on the back of credit and debit cards. There are two tracks of data (Track 1 and Track 2) on each card's magnetic stripe. Track 1 is alphanumeric and contains the customer's name and account number. Track 2 is numeric and contains the account number, expiration date, the CVV1 code, and discretionary institution data. List prices are variable, based on supply, balance, and validity.

The sale of payment card data is common, and is well documented in a recent series of McAfee blogs.[27] However, such payment cards are not the usual type of financial data targeted and subsequently sold on the open market. Much like cards, PayPal accounts are also sold on the open market, with their prices determined by additional factors. Such factors are, however, considerably more limited than those of payment cards, with the balance the only defining factor influencing prices, as we see in the following table:

Table 13.3. PayPal Accounts for Sale here

PayPal Account Balance	Estimated Price per Account
$400–$1,000	$20–$50
$1,000–$2,500	$50–$120
$2,500–$5,000	$120–$200
$5,000–$8,000	$200–$300

The prices in this table are estimates, though we have seen many examples of services for sale that fall outside of these price ranges. Everything is available. This includes bank-to-bank transfers offered for sale, and the availability of banking login credentials.

There will always be suspicions about the validity of the products for sale, as many individuals have paid for stolen financial data, only to not receive what they expected. One seller refers to this dishonor among thieves within their opening pitch:

> ARE YOU FED UP OF BEING SCAMMED, AND RIPPED?
> ARE YOU TIRED OF SCAMMERS WASTING YOUR TIME,
> ONLY TO STEAL YOUR HARD-EARNED MONEY?

This particular seller, though not offering free credit cards that a buyer could use as a test, does offer a replacement policy for any cards that do not provide the advertised balance. Other methods of ensuring a seller's honesty include the use of social validation, with positive feedback from other buyers. Forums are full of helpful advice from buyers that have successfully negotiated purchases as well as which sellers to avoid.

> Hey man, don't know if you know this, but ████████ pulled a exit scam on evo?
> as far as i know, he pulled an exit scam, then he came back saying his friends had screwed him over, asked people to pay like 4BTC to join his official priviate reselling club. he then just disspeared again.
> in fact theres a guy called Underwebfullz (or somthing like that) whos doing the same thing on alpahbay, so people think its him 😳

Sellers who employ sophisticated sales and marketing efforts are leveraging YouTube to advertise their wares to potential customers. The videos often attempt to provide some degree of visual confirmation for prospective buyers that they can be trusted, although such approaches can backfire through comments associated with the videos.

Login Access

Other types of data for sale include access into systems within organizations' trusted networks. The types of entry vary, from very simple direct access (e.g., login credentials) to those that require a degree of technical competence to carry out (e.g., vulnerabilities). One seller, for example, was selling access to bank and airline systems located in Europe, Asia, and the United States for a fee.

As with the sale of financial data, sellers strive to offer a degree of proof to prospective buyers that their offers are valid. Recent research by cybercrime expert Idan Aharoni suggests that the types of systems criminals sell access to now include critical infrastructure systems. In his article "SCADA Systems Offered for Sale in the Underground Economy," Aharoni included one example in which a seller provided a screenshot that appears to be a French hydroelectric generator as evidence that the seller had access to it, as depicted in Figure 13.4.[28]

Figure 13.4. For Sale: Access to a Critical Infrastructure System

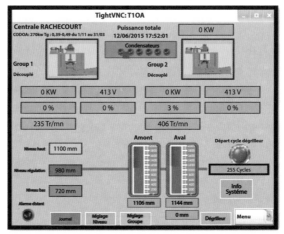

As with previous examples, a buyer can question whether the access offered is indeed valid. It would not be particularly difficult to produce a screenshot and imply this represents access; yet, this message does represent a very worrying trend, as Aharoni points out.

Access to Online Services

Many people subscribe to digital services, including music, videos, loyalty programs, and others. Because such accounts are relatively inexpensive, one might assume that information from them would not offer a sufficient return. Nonetheless, such accounts are widely available across multiple marketplaces which would suggest a demand amongst prospective customers.

When a stolen online account becomes compromised, the legitimate owner can be impacted in a variety of ways. The account can be held or closed due to malicious activity by the buyer—sometimes causing weeks of support calls. A victim could also suffer financial losses from the purchase of items with stored credit card information, or lose access to free perks such as loyalty points collected during the lifetime of the account. Worse, there are circumstances in which the impact is quite disturbing. The use of a stolen Uber account can bypass many safety checks to protect patrons of the service. An owner of the Uber account can suffer reputation, information, and monetary damage, and a customer of the service could face great personal risk. It is unclear exactly how many valid accounts have been sold, but prices as low as $5 offer anyone the opportunity to masquerade as a driver.

The sad reality is that access to just about every conceivable online service is available. We found one Hulu account selling for $0.55. With single accounts to digital services selling for less than a dollar, criminals must move a lot of Netflix or Hulu accounts to make their efforts worthwhile. Many other streaming entertainment media services are commonly sold. Both HBO NOW and HBO GO accounts can be found for less than $10, as well as the TV-streaming service Xfinity. Clearly, video streaming services are in high demand. Even live sports streaming services, such as MLB.tv, can be found for $15. We also found other online accounts being sold, including lifetime subscriptions to premium pornography accounts, as well as free referral links to the Dark Web market, Agora.

Even free online accounts attract criminals. For example, a hotel loyalty account with 100,000 points can sell for $20. Customers legitimately open these accounts at no cost, and yet there is a market for them, resulting in the loss of accumulated perks that sometimes take years to accrue. One motivation for purchasing stolen online account access is to hide the buyer's reputation, either due to bad business practices or outright fraud. A buyer wishing to acquire a new eBay business identity can pay plenty, but an established account with good history can be valuable. For less stringent needs, eBay accounts are available in packs of 100 for a range of account types.

Figure 13.5. eBay Accounts for Sale

eBay No Limits Aged Unlimited Allowances 0 to 200 feedbacks Account/Business For Sale

£1,400.00

ADD TO CART

eBay accounts with no limits and aged with unlimited selling allowances 0 to 200 feedbacks ebay business account for sale. This account is from 8 to 10 years old.

Add To Wishlist

Add To Compare

Identities

Beyond the sale of these accounts, identity theft is burgeoning aspect of cybercrime. The sale of a victim's identity is the most frightening category because it is so personal. Intel Security recently collaborated with law enforcement agencies in Europe to take down the Beebone botnet.[29] This botnet was able to download malware—including ZBot banking password stealers, Necurs and ZeroAccess rootkits, Cutwail spambots, fake AV, and ransomware—onto the systems of unsuspecting users. We are dismayed at the lack of remedial action taken by users, and in particular those based outside of the United States and Europe. A vast section of society fails to appropriately protect their data—often with significant ramifications. In one case, a particular seller was found offering the complete digital identity of an individual based in the UK; this allowed the prospective buyer to take control of this individual's digital life—social media, email, and more.

Closely related to the marketplace for stolen identities is the marketplace for stolen medical information. Such data is not as easy to buy as payment card data, but sellers of medical information are online. Security journalist Brian Krebs discussed this in his article, "A Day in the Life of a Stolen Healthcare Record," in which a "fraudster leaked a large text file [that] contained the name, address, social security number, and other sensitive information on dozens of physicians across the country."[30]

Figure 13.6. Medical Data for Sale

Conclusion

It would appear that almost not one day goes by without another story of another organization that has fallen victim to another data breach. The only question that is often asked, aside of course from speculation regarding attribution, is the number of records that have been compromised. Indeed, as this chapter has demonstrated, it is not only about payment cards, what with marketplaces selling all forms of data and with ready and willing buyers.

This chapter has intentionally avoided the discussion of attribution for the simple reason that determining the true source of attack within the digital realm is unsuitably technical for this book. As we have seen earlier, there are multiple products, tools, and services available for sale. These are sold to all manner of buyers, from those within the criminal realm—and as we have seen from exposures from zero-day sellers—to those in government institutions. What this demonstrates is that the rather dominant view of hacktivists, criminals, and nation-states is simplistic and naïve. Individuals with specific technical expertise are available for hire. They are available for hire to anybody with the necessary funds, and through criminal investigations we have seen this outsourcing model used more frequently for everything from the attack itself to even laundering the funds gained from the attack.

This is not a new form of crime; rather, it is an evolution of traditional crime. It seems somewhat trite to end this chapter with this sentence, partly because it is not entirely true. Wars are waged online by nation-states at significantly lower cost than traditional weapons, and more importantly, with plausible deniability. No longer is there a need to send physical assets into hostile territory when it is possible to make a digital attack appear to come from another place in seconds, disrupting an entire nation. Make no mistake—cybercrime is no longer about the lone teenager looking to show off his technical prowess (although this still exists); it is a big business and it is here to stay.

More ominously, all that is available in the Cybercrime-as-a-Service marketplace is available to any malicious actor regardless of their motivations. As we learn more regarding the 2015 Ukrainian power outage, we see that malicious actors are beginning to exploit the connected nature of critical infrastructure. Equally, money gained through cyber-related crimes can be used to advance the goals of such malicious actors against their targeted states. It should be a priority of every state to at least keep up with, if not outpace, the evolving wave of cybercrime.

Notes

[1] Center for Strategic and International Studies, *Net Losses: Estimating the Global Cost of Cybercrime* (Santa Clara, CA: McAfee, June 2014).

[2] Limor Kessem, "Carbanak: How Would You Have Stopped a $1 Billion APT Attack?" *Security Intelligence,* February 23, 2015, available at <https://securityintelligence.com/carbanak-how-would-you-have-stopped-a-1-billion-apt-attack/>.

[3] Paul Peachey, "Mafia cybercrime booming and with it a whole service industry, says study," *Independent,* September 29, 2014, available at <http://www.independent.co.uk/news/uk/crime/mafia-cybercrime-booming-and-with-it-a-whole-service-industry-says-study-9763447.html>.

[4] Ibid.

[5] An exploit is a software that takes advantage of a vulnerability or flaw.

[6] Raj Samani and Francois Paget, *Cybercrime Exposed: Cybercrime-as-a Service*, McAfee White Paper 2013 (Santa Clara, CA: McAfee, 2013), available at <http://www.mcafee.com/us/resources/white-papers/wp-cybercrime-exposed.pdf>.

[7] "Robberies decrease as cyber crime increases, FBI says," *WMBF News,* February 5, 2013, available at <http://www.wmbfnews.com/story/20972727/robberies-decrease-as-cyber-crime-increases-fbi-says>.

[8] PricewaterhouseCoopers LLP, *US cybercrime: Rising risks, reduced readiness Key findings from the 2014 US State of Cybercrime Survey* (New York, NY: PwC LLP, 2014), available at <http://www.pwc.com/us/en/increasing-it-effectiveness/publications/assets/2014-us-state-of-cybercrime.pdf>.

[9] Chris Strohm, "Cybercrime Remains Growth Industry With $445 Billion Lost," *Bloomberg Business,* June 9, 2014, available at <http://www.bloomberg.com/news/articles/2014-06-09/cybercrime-remains-growth-industry-with-445-billion-lost>.

[10] Center for Strategic and International Studies, *Net Losses.*

[11] Samani and Paget, *Cybercrime Exposed.*

[12] Joshua Kopstein, "Here Are All the Sketchy Government Agencies Buying Hacking Team's Spy Tech," *Motherboard,* July 6, 2015, available at <http://motherboard.vice.com/read/here-are-all-the-sketchy-government-agencies-buying-hacking-teams-spy-tech>.

[13] Andy Greenberg, "Shopping For Zero-Days: A Price List For Hackers' Secret Software Exploits," *Forbes,* March 23, 2012, available at <http://www.forbes.com/sites/andygreenberg/2012/03/23/shopping-for-zero-days-an-price-list-for-hackers-secret-software-exploits/#bbc53f060335>.

[14] "The Great Bank Robbery: Carbanak cybergang steals $1bn from 100 financial institutions worldwide," *Kaspersky,* February 16, 2015, available at <http://www.kaspersky.com/about/news/virus/2015/

Carbanak-cybergang-steals-1-bn-USD-from-100-financial-institutions-worldwide>; Kaspersky, *Carbanak APT: The Great Bank Robbery* (Moscow: Kaspersky Lab HQ, February 2015), available at <https://securelist.com/files/2015/02/Carbanak_APT_eng.pdf>.

[15] Raj Samani and Charles McFarland, *Hacking the Human Operating System* (Santa Clara, CA: McAfee, 2013), available at <http://www.mcafee.com/us/resources/reports/rp-hacking-human-os-summary.pdf>.

[16] Information Commissioner's Office, *What price privacy? The unlawful trade in confidential personal information* (Norwich: The Stationery Office, 2006), available at <https://ico.org.uk/media/about-the-ico/documents/1042393/what-price-privacy.pdf>.

[17] Robert Lemos, "Zotob suspects arrested in Turkey and Morocco," *The Channel,* August 30, 2005, available at <http://www.channelregister.co.uk/2005/08/30/zotob_suspects_arrested/>.

[18] McAfee Labs, "Meet 'Tox': Ransomware for the Rest of Us," *McAfee Blogs,* May 23, 2015, available at <https://blogs.mcafee.com/mcafee-labs/meet-tox-ransomware-for-the-rest-of-us/>.

[19] John Leyden, "Hi! You've reached TeslaCrypt ransomware customer support. How may we fleece you?" *The Register,* May 20, 2015, available at <http://www.theregister.co.uk/2015/05/20/teslacrypt_ransomware_scam_dissected/>.

[20] Downloads, "A closer look at the Angler exploit kit," *Sophos,* July 21, 2015, available at <https://blogs.sophos.com/2015/07/21/a-closer-look-at-the-angler-exploit-kit/>.

[21] "How Much it Costs to Rent a Botnet to DDoS," *Havocscope,* 2014, available at <http://www.havocscope.com/how-to-ddos-by-renting-botnet/>.

[22] Dave Lee, "Global internet slows after 'biggest attack in history,'" *BBC News,* March 27, 2013, available at <http://www.bbc.com/news/technology-21954636>.

[23] Charles McFarland, Francois Paget, and Raj Samani, *The Hidden Data Economy: The Marketplace for Stolen Digital Information*, McAfee White Paper 2015 (Santa Clara, CA: McAfee, 2015), available at <http://www.mcafee.com/us/resources/reports/rp-hidden-data-economy.pdf>.

[24] McAfee, *McAfee Labs Threats Report 2015* (Santa Clara, CA: McAfee, 2015), available at <http://www.mcafee.com/us/resources/reports/rp-quarterly-threat-q1-2015.pdf>.

[25] Elise Hu, "I Feel Nothing: The Home Depot Hack And Data Breach Fatigue," *NPR,* September 3, 2014, available at <http://www.npr.org/sections/alltechconsidered/2014/09/03/345539074/i-feel-nothing-the-home-depot-hack-and-data-breach-fatigue>.

[26] Ibid.

[27] Raj Samani, "New Year's Sales; Big Discounts on Stolen Data," *McAfee Blogs,* January 29, 2014, available at <https://blogs.mcafee.com/mcafee-labs/new-years-sales-big-discounts-stolen-data/>.

[28] Idan Aharoni, "SCADA Systems Offered for Sale in the Underground Economy," *Infosec Island,* June 22, 2015, available at <http://www.infosecisland.com/blogview/24608-SCADA-Systems-Offered-for-Sale-in-the-Underground-Economy.html>.

[29] McAfee, *Catch Me If You Can: Antics of a Polymorphic Botnet* (Santa Clara, CA: McAfee, 2015), available at <http://www.mcafee.com/us/resources/reports/rp-catch-me-if-you-can.pdf>.

[30] "A Day in the Life of a Stolen Healthcare Record," *Krebson Security,* April 28, 2015, accessed at <http://krebsonsecurity.com/2015/04/a-day-in-the-life-of-a-stolen-healthcare-record/>.

IV. A Toolbox for the 21ˢᵗ Century

14

Leviathan Redux: Toward a Community of Effective States

Clare Lockhart and Michael Miklaucic

If the 20th century was consumed by the global struggle between incompatible ideologies—fascism, communism, and democratic capitalism—the 21st century will be consumed by the epic challenge of creating and sustaining viable, effective states. Viable, effective states are the only form of collective governance that has a proven ability to contain and reverse a trajectory of growing entropy driven in part by the illicit networks described throughout this book. States have successfully fought off powerful illicit adversaries in all regions of the globe, from Colombia in South America to the Philippines in Asia. Some authors have argued that the state itself is a significant contributor to growing global entropy, and that is possibly true.[1] Yet, the global state system has enabled great prosperity and security, and ready alternatives to state-based governance are few.

Over the past 25 years, the threat to international security posed by failed and failing states has been widely acknowledged. Scholar Robert Rotberg writes that such states "pose dangers not only to themselves and their neighbors but also to peoples around the world."[2] Indeed, in 2002, it was explicitly stated by then President George W. Bush in his introduction to the National Security Strategy of the United States "…that weak states, like Afghanistan, can pose as great a danger to our national interests as strong states."[3] Only recently, however, have analysts begun to examine the contributions to state weakness and failure from converging terrorist, criminal, and insurgent networks.

Previous chapters in this book have described visions of a world experiencing pandemic state failure. This chapter will attempt to persuade not only that a rule-based system of sovereign states provides a hospitable environment for human endeavor, but that it is the best remedy against the destructive effects of illicit actors and networks. The authors will argue that the unsuccessful cases of Iraq and Afghanistan should not be permitted to discredit the state-building enterprise, as they were, in fact, cases in which no coherent state-building strategy was ever developed, attempted, or sustained. The chapter will provide evidence supporting two theses; that successful state-building is possible; and that external intervention, though never without risk, if appropriately designed, can effectively support successful state-building efforts. The final section will identify and describe a range of critical operational approaches to successful state-building to be considered by either local or external agents. These include:

- establishing a monopoly of the legitimate use of force;
- establishing the rule of law and mechanisms for articulation, adjudication, and redress of grievances;
- prohibiting and punishing corruption;
- creating an inclusive national narrative promoting citizenship;
- nurturing an economic environment in which citizens can meet their economic needs;
- establishing mechanisms for managing state assets and budgets and securing state financing based on tax-paying constituents;
- creating accessible mechanisms for interaction between civil society and the state;
- investing in human capital; and
- creating a reliable and competent civil service to administer official state functions and manage state assets.

In Defense of Westphalia

For those critical of the modern state and its failures, there is much to lament; from the failure of many of the world's states to provide citizens with fundamental public goods—not least security and safety—to the collective failure of the state system to effectively address global threats, such as climate change and food insecurity. States big and small, rich and poor, advanced and underdeveloped have left much to be desired.

For the sake of objectivity, however, let us look at the other side of the ledger. The great innovation of Westphalia, the field in Germany where 100 delegations met in 1648 to negotiate the end of the Thirty Years' War that had devastated Central Europe, was not the invention of the state. Functioning bureaucratic administration had existed for eons in various forms and regions.[4] The innovation of Westphalia was a rule-based system of states, based on the concept of sovereign equality. Diverse in wealth, populations, endowment, cultures, and religions, the Westphalian states were equal in their sovereignty, each with a right to a voice, to its own religion, customs, and practices with regard to its citizens. This resulted in a framework of widely respected norms for interstate behavior—to be sure, never for long unchallenged—that permitted states to turn greater attention toward their internal development.

While the Peace of Westphalia did not end war, let alone conflict, it did establish a culture of rule-based international relations conducive to unprecedented advancements in the human condition—as measured by the most widely accepted indicators of human development. Human longevity which had lingered between 30 and 40 years for millennia shot up, from 35 in 1650 to 67 years today.[5] Income, which hovered far and wide at subsistence levels—barely enough to survive—catapulted from $615 to $10,000 per year between 1700 and 1900.[6] Finally, literacy, long the preserve of the rich or of the clergy, reached over 80 percent by the 2000s, from about 35 percent in 1650.[7] Versions of these three indicators are the most commonly used today to measure human development.[8] Can the innovation of the Westphalian state system lay claim to these unprecedented advances? Perhaps not exclusively—but the conditions and rules of interstate behavior produced an environment conducive to these impressive achievements. Building upon pre-Westphalian innovations in bureaucratic governance and the Westphalian innovation, this system expanded, until the post-World War II order extended it on a global basis with investments to attain sovereign self-governance in all countries.

The state is by no means the only, let alone the earliest, set of institutions established to order political relations among populations. From time immemorial, political activity has been governed by more archetypal structures, such as the family, tribe, and clan—forms which are arguably so profoundly embedded in our human nature, that they continually resurface.[9] These forms of political organization evolved over centuries in some regions into various iterations of autocracy such as monarchy, kingdoms, and empires. At other times, in other places, political life has been governed by religious orders, ranging from the micro-monastic orders of medieval Christianity to the great caliphates of Islam. The 20th century witnessed a radical effort to organize political life along economic class lines; but by 1991, the great communist/socialist experiment failed. And the others—the clan, tribe, feudal, monarchy, caliphate, or empire—have failed to deliver the public goods needed for people to prosper in the 21st century. Thus, paraphrasing Churchill, the state is perhaps the worst form of government—except for all the others.

We would go well beyond Churchill; there is a great deal to be proud of in the accomplishments of modern states—but as earlier chapters in this book have made clear, none of these accomplishments can be taken for granted. There is much to lose, and the prospect of losing much or even all we have achieved in recent centuries—if history is any guide—is not as remote as some might think.[10] Great advances of civilizations past have been reversed by war and by decay. The risk of losing these extraordinary and unprecedented achievements is no less than a return to the life characterized by Hobbes so famously as nasty, brutish, and short.[11] The tremendous human progress we have experienced in recent centuries is grounded in the rules-based system of sovereign equality among states. The best hope to retain and build on this progress is to preserve the system of effective sovereign states, ensure that states can perform required functions, and adapt this system to the needs and realities of the 21st century—confronting globalization as well as the empowered citizen with higher expectations. Certainly a new kind of state will be required—innovative and adaptive to meet the challenges of the emerging threat environment, rather than just a recreation of the 19th-century state.

The Sovereignty Deficit

Our contemporary state system is far from complete. The illusion of sovereignty, derided by one scholar as "organized hypocrisy," has created in many minds a misleading world map of neatly divided countries (each with clear boundaries and its own bright color). In terms of governance, however, these many countries are, in fact, significantly diverse in legitimacy and effectiveness.[12] To what extent does this world map, with which we are all familiar, represent realities across continents? South Korea differs vastly from Papua New Guinea, Botswana from South Sudan, and Switzerland from Greece. In 1994, Robert D. Kaplan wrote of the "lies of mapmakers," describing the inadequacy of "this inflexible, artificial reality."[13] Of the 193 members of the United Nations general assembly, how many can be characterized as exercising full and effective sovereignty? If sovereignty is to mean not only recognition by the family of nations, but effectively governing the population and territory within a specific set of borders, balancing the budget and controlling corruption, policing crime, providing opportunities for citizens, and guarding traffic through those borders; how many are truly sovereign?[14] *The Economist* reported in 2013 that 65 countries were at a high or very high risk of social unrest, 19 more than in 2009.[15] Depending on the indicators used for sovereignty, the count of less than sovereign states could range from 40 to 140; either way leaving a considerable number outside the domain of sovereignty, indeed leaving the world with a substantial sovereignty deficit.[16]

As others in this book have argued the limited domain of effectively governed states—the Westphalian domain—is under attack. The illicit networks described throughout this book, in collusion with allied agents in faltering—and even within robust—states, work relentlessly to subvert the integrity of the system. They prey on states, both weak and strong, like parasites, and infect undergoverned spaces, turning these into black holes that attract all manner of criminal activity.[17]

We know of only one antidote to this degenerative threat, and that is the establishment of states capable of effective governance throughout the world. But many contemporary states are not on a trajectory toward legitimate and effective governance. States such as Honduras, the Democratic Republic of the Congo, and Iraq may have the desire, but lack the capacity to withstand countervailing forces. Their consolidation as legitimate and effectively governed states will require a concerted effort by capable, well-resourced, and well-intentioned external partners working with internal leaders, institutions, and constituencies. Such a concerted effort can only succeed if leaders have the will—the commitment and courage—to succeed in this challenge.

State-building regrettably has been discredited over the past two decades.[18] Both the cost and difficulty inherent in a coherent and realistic approach to state-building have understandably soured policymakers and budget setters to the proposition of trying to stand up states. Indeed, the catastrophic state-building failures in Iraq and Afghanistan, where budgets were unprecedentedly large, have proven to many the futility of the effort and the concept. It is true that the coalitions engaged in the recent conflicts in Iraq and Afghanistan failed to establish effective, legitimate states; and even before those were disappointing failures in Somalia and Haiti. The dismissal of state-building as a vital national security tool, however, is a strategic folly of historical proportions. None of these failures was inevitable. There are numerous examples of successful state-building efforts, including Jordan, South Korea, the Republic of China (Taiwan), Singapore, and Colombia (a successful case of state-rebuilding), among others.

Iraq and Afghanistan, however, are not examples of the futility of state-building. In neither case was there any strategic analysis of the requirements of state-building. Efforts were half-hearted, misguided, and ad hoc. Some successful initiatives were tried, but then, not sustained. But these failures were neither complete nor inevitable; better strategy, execution, and judgment would have made success far more likely. Iraq and Afghanistan are both, in fact, excellent examples of the costs of the failure to develop a coherent and realistic state-building plan. Today, they both exemplify the multi-tiered costs of state failure; humanitarian tragedy, regional destabilization, and feral cultivation of toxic ideologies and organizations.

Delusions of Security: Afghanistan

Afghanistan is often dismissed as the poster child of why efforts at nation-building or state-building are futile. Yet for most of the 15 years since the 9/11 attacks and the Bonn Agreement, state-building and nation-building were explicitly rejected as goals and methods.

In the aftermath of 9/11, then President Bush and senior officers of his Administration explicitly rejected nation-building as an approach in Afghanistan. Instead, U.S. efforts had focused on a military mission to collapse the Taliban regime, by using U.S. airpower partnered with Northern Alliance militias to rout the Taliban from Kabul, and after that, from a base at Bagram outside Kabul, to defeat remaining Taliban forces in the countryside.

When a case was made to the State Department that, once the Taliban had fled, a critical issue would be the formation of a government that was acceptable to the Afghan public and could carry out minimal activities—not least holding the country and preventing a civil war between warlord factions—that team was encouraged to work from the UN to establish negotiations. But these objectives were never more than peripheral for the United States. Instead, the U.S. effort continued to focus on partnering with warlords to raise and maintain militias to accomplish near-term security objectives. Initially, the bulk of the remaining effort was given to humanitarian relief, through a network of UN agencies and nongovernmental organizations (NGOs) that operated in parallel to, and often at odds with, the Afghan state.

Preparations to establish a new government took place through the auspices of the UN. The experienced UN diplomat-statesman, Lakhdar Brahimi, who had also tried to broker a successor to Najibullah's government in Afghanistan in the early 1990s, took the helm of the team. It quickly became clear that the Taliban would fall much more swiftly than initially anticipated. Then Secretary of State Colin Powell said he had three instructions for the team: "Speed, speed, speed." But although the team mobilized quickly and raced against time to assemble a group of Afghans at the St. Petersburg castle in Bonn to conduct political negotiations to establish a new government, facts on the ground were already being created. By the time the UN team assembled in Bonn on November 20, 2001, the Northern Alliance, backed by U.S. air power, had advanced into Kabul and had seized the key ministries.

The group assembled around the table at Bonn consisted of four factions—the Northern Alliance, the Rome Group (a set of courtiers to the former king), and two other political groups operating in exile—the Cyprus Group and the Peshawar Group. Absent was representation from Afghanistan's southern regions and the Taliban. Aware of the limited representation, the group, facilitated by Brahimi, agreed on a set of governing principles and a phased process for establishing legitimate institutions of government over time. It would start with an Interim Administration, then an Emergency *Loya Jirga* would be convened. This would be followed by a Transitional Administration. Next was to be the drafting of a constitution by a Constitutional Commission that would put this constitution to a Constitutional *Loya Jirga* for ratification, followed by elections after some three years.

This political process held to its timetable and worked remarkably smoothly. It was accompanied by an effort to build—or resuscitate—the organs and institutions of state that would make the high-level political promises a reality in the towns and villages across the country. Resuscitate, because contrary to many widely held assumptions of a "blank slate" or "tabula rasa," there, in fact, remained in 2001, a remarkably resilient set of institutions and organizations of government across Afghanistan. A World Bank report found that there were "anywhere between 250,000 and 350,000" civil servants in place, in ministries and provincial and district offices across the country.[19] While there had been considerable human capital flight from the country to neighboring countries and beyond, the backbone of the civil service had persisted from the time of the monarchy through the communist period, the civil war period, and the Taliban period. As in many other contexts, the initial international assistance effort did not stop to take stock of what was already in place, but rather, assumed there were no institutions—or at least none worth considering building on or preserving—and instead built up alternative parallel delivery structures.

Despite this, a nascent initiative to build on existing structures and invest in reestablishing Afghanistan's state institutions began. It started with mapping out existing ministries and institutional and organizational arrangements, and reviving basic processes of governance and decisionmaking, ranging from paying civil servants their basic salaries on time, to establishing weekly cabinet meetings with agendas, establishing a civil service commission to sift through competing factional claims to positions in the civil service, and getting a basic budget agreed, funded, and operational. From there, it proceeded to take stock of ministries and their capacities, and to design a small number of priority initiatives and actions to build the public's confidence, demonstrate tangible progress and lay the foundations for the long term. Examples of initiatives included:

- Convening of two *Loya Jirga* on the basis of district elections together with some selected delegates, and establishment of the Constitutional Commission.
- Changing the currency, to move from three separate currencies to a single currency, with new symbols of national unity. This was accomplished by using the Afghan *hawala* dealers and their networks, at a cost far lower than an alternative proposal by the UN system using UN personnel.

- Building the budget process as an instrument of policymaking, putting in place measures to raise revenues, creating a secure payroll system for government employees, and using a simple IT system to reconcile national accounts (across the countryside).
- Creating a trust fund for development partner contributions that would be heavily audited and guarantee high levels of accountability for contributions.
- Building out nationwide programs to reach all villages, including the National Solidarity Program (NSP), which gave block grants to 34,000 villages to manage themselves.
- Establishing the National Health Program, a consortium of the United States, EU, and World Bank that agreed on common standards for basic health care and contracted out service provision to each province to those with capacity to deliver—whether through NGOs or the private sector.
- Designing the transportation network infrastructure, including starting on the ring road, rail infrastructure, and power linkages that would lay the basis for Afghanistan as a connector, rather than a black hole in connectivity in the region.

Most of these initiatives had a few interesting characteristics that stand in contrast to the mainstream way of approaching aid and development today. First, the initiatives were prepared in such a way that they could be designed, led, and managed by Afghan leaders and managers, with either no or very minimal foreign expert technical assistance. Second, they involved a large degree of citizen consultation, co-design, and participation. Third, while they were intended to enhance the legitimacy of the state, this did not mean that the state had to implement them entirely through its bureaucracy. Rather, the initiatives allowed for rules of the game to be set and policies determined with full government leadership, but funding could flow to NGOs, citizen groups, or private sector entities depending on capability and appropriateness. Fourth, the programs themselves, if implemented using domestic capacity, would cost a fraction of the price tags we are familiar with today. For example, a school built through the NSP would cost $20,000, versus up to 10 times that amount if built by a foreign contractor. This is not to say that foreign contractors are not sometimes needed—for instance, in the design and building of large infrastructure projects—but that alternative systems for small infrastructure and social programs can be built at much lower costs and with much greater domestic participation.

Many of these initiatives survive until today and are lauded as the programs that work in the country, but they—and the program of legitimizing the state-citizen compact—did not become the dominant reality in the successive years. There were two major obstacles to this. First, the privileging of support to militias and strongmen over the reform movement, missing the point that short-term security might be able to be bought through the former method, but long-term stability comes through building trust with the 95 percent of the public who are the reform constituency. When the insurgency began in 2004 and lasted through 2006, a large driver was the loss of trust of the population in fairness, rule of law, and holding back predatory interests. Second, the operations of the aid system. While individuals usually had the best of intentions, the way the system as a whole operated served to fragment and fracture the Afghan state, paradoxically in the name of capacity building.

Sadly, failure to reach an agreement that would give some more space to the reform movement and bring into balance the operations of the other two spheres meant that the reform team lost hope and steam. Some key leaders convened around the "Cairo memo" in December of 2004, which predicted that the narco-mafia state would continue to consolidate its stranglehold on the state and instability would set in.[20] At the time, minimal measures were needed to address this. The reformers left; legitimacy dwindled and the insurgency returned. As this became clear, the international community swung into action, finally—by 2006—taking heed of recommendations to bolster the legitimacy of the state. But it was too little—or perhaps too much—too late. The "Golden Hour" had given the Afghan public a tantalizing taste of what was possible.

But as the aid machine lumbered into gear for large-scale implementation, some key design lessons were lost. It veered from small footprint to massive footprint, and in the process, its tens of thousands of small and large projects overwhelmed and splintered the still fragile Afghan institutions, helping to import the culture and practice of corruption, as the availability of funding increased and the likelihood of consequences for corrupt activity diminished. The state became a parody of itself—creating the sovereignty paradox—and as its leaders increasingly claimed sovereignty, the trust of the population declined. The country was kept afloat—and its people employed—on the back of a large, and unsustainable, foreign presence.

As the international military and development presence began to draw rapidly down in 2014, the public came out to vote in huge numbers, for a reform message. This was a clarion cry from the Afghan public cutting across ethnic, conservative/modern, and pro- and anti-government lines, calling for legitimate institutions, an anticorruption agenda, and reform. It demanded a real chance for the "Afghanization" of the state-building project. The National Unity Government has taken on this challenge, with the huge impediments of a turbulent election and unconstitutional government arrangement, rapid international security drawdown, the economic shock following such a drawdown, an inherited condition of corrupt administration intertwined with criminal networks, and tough regional and security dynamics. Despite this, the public and the leaders have formed a constituency for reform and it is progressing—the results will be judged in years ahead.

Some key lessons can perhaps be drawn from this experience. First, that efforts to establish minimally viable institutions are the efforts of decades, not months nor years. Our benchmarks, yardsticks, and measurements must change accordingly. With the realism that most change happens over decades, rather than to the metronome beat of one-year log frames, comes the opportunity to sequence activities carefully over time. Perhaps a country will not reform its health system, its education system, and its military at the same time;

leaders will tackle reforms in stages. But committing to a destination of a fair, rule-based government and state that strives to meet the expectations of the citizenry is a goal worth pursuing.

Second, much of what it takes to create institutions and organizations effectively is vastly cheaper than suggested by current methods of aid engagement. The price tags of tens of billions are not a fair reflection of real costings if implementation methods were redrawn and country systems used appropriately.

Third, to do so means drawing down our addiction to donor circuses of tens of thousands of individual projects that fragment the state, and investing more in policy and program design before project mayhem. If we are talking about coordination, it is too late. The designers of the Marshall Plan (which, in its original form, we are not advocating) considered six designs and picked three. They judged that the implementation of the three they rejected would have meant that Europe would never be reconstructed. They rejected the vehicle of thousands of small projects, managed from Washington and other capital cities, dependent on grant aid and micromanagement. Instead, they realized that unless the leaders of a country (and wider constituencies within that country) agreed on their own plans, implementation would not be assured. So they created a set of mechanisms that required country ownership of a plan, which would include a plan for enhancing domestic revenue so that the country would become self-reliant.

Fourth, to take this approach will require investment in the leaders and managers of the country and real engagement with its people. Investment—through scholarship programs, training, and mentorship. It will pay dividends. Imagine if South Sudan had, upon signing the Comprehensive Peace Agreement, set up a leadership academy and trained 10,000 of the next generation in management, leadership, accountancy, agriculture, oil management, and other key skill sets. This could have required a substantial sum to get right, but would have been a fraction of the hundreds of millions spent on importing technical assistance or providing security when the new nation lapsed back into conflict, for reasons mainly of poor governance—not to mention, the loss of life and human suffering. It will also take real engagement with the country's public. Through civic participation, through two-way communications, through the construction of national narratives that allow for national identity to be built.

Fifth, there are strong grounds for hope and confidence, despite real reasons for caution and re-examination. It was possible for Afghan leaders to establish national (by which we mean countrywide rather than focused on the capital city) programs at reasonable cost that took institutional hold and delivered real things to real people. At the same time, many countries—including in the last two decades—have turned the corner from conflict, instability, and endemic corruption and made huge strides forward. There is no cookbook with a recipe for doing this, but there are certainly examples and lessons.

Sixth, the budget matters. Financial management seems technical and boring to many, but it is the glue that holds governance together, and is, in fact, intensely political. In many countries budget day is the political event of the year, as this is where real political trade-offs and fights occur. For an emerging nation, the key marker of success is increasing revenue generation, and allocating the budget to critical needs of the public, rather than relying on donor largesse.

Seventh, assuming (as many in the peacebuilding community have admitted) that the economy can wait—with politics and security as the only priorities—is a terrible mistake. The imperatives of security and safety, and the role of politics, are essential and undeniable. But on interviewing those who have brokered and implemented peace deals, and reflecting on shortcomings in ones that worked and why the others slipped back into conflict, all these diplomats realized that their biggest blind spot had been this assumption. First, young men did not have jobs, and took up arms again. Second, they promised all sorts of things in the peace agreement, that without the International Monetary Fund (IMF) and World Bank, and country's budget behind, they could not pay for. Finally, if one does not explicitly set about to create the policy and institutional conditions for legitimate market activity, what you get is not spontaneous legitimate growth, but a deepening of the criminal and war economy, which tips the country back to war.

Lastly, in many countries, there is a reform constituency and a warlord constituency, often deeply intertwined with the war economy and criminal networks. Security built on alliances with the latter may be necessary and expedient, but enduring stability will come from building trust with the public through institutions that serve and protect the public interest. This may seem harder, but there are no shortcuts. The recent fascination with "elite deals" comes at the peril of ignoring a much bigger bargain with the public interest, expressed through a process that commits to a destination of a just society. We collectively ignore the real grievances of large segments of the public, mainly stemming from predatory government behavior. In Afghanistan, a civic leader said that in his country, there are 4 percent thugs, 1 percent extremists, and 95 percent ordinary people. He said, the problem with the foreigners is that they tend to cut deals between the 4 percent and the 1 percent, over the heads of the 95 percent. These last for a while, but the population suffers so much, that they are eventually driven to extreme measures to protect life and limb. And so, insecurity grows. To reverse the trajectory, no one is asking for Switzerland overnight. Just that year on year, there be a little less predatory government, and a little more fairness and public value.

No Plan at All: Failure in Iraq

In this section we describe four critical errors over four distinct phases that led to catastrophic state failure in Iraq between 2003 and 2009. The failure of the state-building effort in Iraq predates the 2003 U.S. invasion of Iraq itself. In a presidential debate with then Vice President Al Gore in October 2000, then Governor George W. Bush stated, "I don't think our troops ought to be used for what's called nation-building.... I mean, are we going to have some kind of nation-building corps from America? Absolutely not."[21] His future National Security Advisor and Secretary of State Condoleezza Rice was equally skeptical of state- or nation-building, as indicated in her pre-election warning against permitting American armed forces to become "the world's '911.'"[22] In an interview with *The New York Times* she famously opined, "We don't need to have the 82nd Airborne escorting kids to kindergarten."[23] These strong statements reflected unequivocal views that were widely shared among then President Bush's national security team. This skepticism constituted the starting disposition of the Bush Administration entering into office in 2001, and boded poorly for any serious consideration of state- or nation-building as an element of U.S. national security policy or practice.

The al-Qaeda attack on the United States on September 11, 2001, was a traumatic shock for then President Bush and his national security team. As discussed above, the initial impetus in Afghanistan was one of revenge and retaliation against the Taliban regime for its continuing hospitality to al-Qaeda. When, in early 2003, the Bush Administration turned its sights on Iraq, some were surprised; the war in Afghanistan was ongoing and no direct link between al-Qaeda, the Taliban, and the Saddam Hussein regime had been identified.[24] Other senior Bush Administration leaders, however, had been advocating an invasion of Iraq even before taking office in 2001.[25] Among the most aggressive in advocating regime change in Iraq were then Secretary of Defense Donald Rumsfeld and his then Deputy Paul Wolfowitz, veterans of the President George H.W. Bush Administration (1989-1993), who experienced the frustration of Saddam's survival after his universally condemned invasion of Kuwait precipitating the first Gulf War.[26]

The assumption among Bush Administration leaders was that the invasion of Iraq and subsequent stand-up of a new government would be rapid and inexpensive. Initial estimates of the anticipated length of the U.S. commitment in post-Saddam Iraq were naively optimistic. Rumsfeld predicted the war would last, "Five days or five weeks or five months, but it certainly isn't going to last any longer than that."[27] Then Vice President Cheney, an avid advocate of regime change in Iraq, told *Meet the Press*, "My belief is we will, in fact, be greeted as liberators.... I think it will go relatively quickly...[in] weeks rather than months."[28]

Bush's leading national security officials were confident the costs of installing and consolidating a new regime in Iraq would be underwritten by Iraqi oil revenues. According to Wolfowitz, "There's a lot of money to pay for this. It doesn't have to be U.S. taxpayer money. And it starts with the assets of the Iraqi people.... We are dealing with a country that can really finance its own reconstruction and relatively soon."[29] Cheney was again naïve, confidently stating, "every analysis said this war itself would cost about $80 billion, recovery of Baghdad, perhaps of Iraq, about $10 billion per year. We should expect as American citizens that this would cost at least $100 billion for a two-year involvement."[30]

It was assumed that following a rapid and decisive military victory over Saddam's armed forces, the functions of transitional governance would be turned over to prominent, returning Iraqi exiles, like Ahmed Chalabi. Believing that the Iraqi population, grateful for their liberation from the tyrannical Saddam regime, would fall in line under the returning expatriates, "prewar planning for postwar governance following Saddam Hussein's fall concentrated on the role of key Iraqi exiles and Kurds in the north to assume control of the government's infrastructure and pave the way for democratic transition."[31]

In view of the Bush Administration's intuitive hostility to state-building and the expectation of a rapid victory and transition to a self-paying Iraqi transitional regime run by returning Iraqi exiles, it should come as no surprise that there was no U.S. plan for postwar Iraq. In a 2006 interview with the *Newport News Daily Press*, Brigadier General Mark Scheid, an Iraq war planner, confided that Rumsfeld said, "he would fire the next person" who talked about the need for a postwar plan. Rumsfeld's instruction to the planning team was, "everything we write in our plan has to be the idea that we are going to go in, we're going to take out the regime, and then we're going to leave…. We won't stay."[32] Ultimately, there was no cogent "Phase IV" plan for postinvasion operations including security, stability, and reconstruction. Though civilian agencies had voiced concern and developed concepts for postinvasion governance in Iraq, both independently and under the leadership of the National Security Council, no consolidated or integrated state-building plan emerged.[33]

With euphemistically speaking lukewarm support from the international community, the United States invaded Iraq in the spring of 2003, with no coherent or realistic plan for addressing postinvasion governance. This was the first critical error leading to the failure of history's costliest effort at state-building. The absence of a plan left forces on the ground without a roadmap for organizing the postconflict state, compelling them to improvise blindly based on limited understanding of Iraq, while deconflicting competing plans and activities implemented by multiple U.S. government agencies and fighting a lethal and growing insurgency.

The initial effort was headed by the ill-starred Office of Reconstruction and Humanitarian Assistance (ORHA), led by retired Lieutenant General Jay Garner. To the extent that ORHA had any concept or plan for governance it was based on Ministerial Advisory Teams consisting of U.S. officials, returned Iraqi expatriates, and "the last Iraqi standing," referring to those Iraqi ministerial officials remaining after the purging of top Saddam loyalists. These leadership teams were to sustain the civil functions of the Saddam state apparatus with the support of the presumed vast civil service bureaucracy intact. Often operating at cross-purposes with the U.S. armed forces, obstructed by bureaucratic resistance from the U.S. civilian agencies, and besieged by a deteriorating security environment and ongoing combat operations, ORHA's planning for the future Iraqi state was marginal. To be clear, there was no serious thought or plan for building the Iraqi state after the toppling of Saddam Hussein for a simple reason. Then President George W. Bush and his national security leadership envisioned a swift military demonstration of shock and awe, and an early exit from Iraq. The Iraqi state would simply restore itself.

In May 2003, ORHA was replaced by the Coalition Provisional Authority (CPA), with Ambassador L. Paul Bremer appointed as the Presidential Envoy to Iraq. Paradoxically, while the Bush Administration senior national security leadership essentially abdicated responsibility for postinvasion Iraqi governance, the CPA disbanded the most robust and capable Iraqi national institutions. According to Bremer, "At liberation, there was no political structure on which to build a new Iraqi system."[34] Bremer and the CPA must accept some of the responsibility for this situation, having on May 16, 2003 issued CPA Orders Number 1 and 2, disbanding the Ba'ath Party and all Iraqi military entities respectively.[35]

The contention that the Iraqi Army and other organizations of state had already "self-demobilized" is disingenuous; as a result of unanticipated circumstances as well as its own actions, the CPA decided to begin a state-building process in Iraq from scratch. The second critical error condemning state-building efforts in Iraq to failure was the decision to dismantle or disregard the existing Iraqi institutional inventory—the "muscle memory" of the country. It might have been understandable to believe the demented survivors of Saddam's "republic of fear," having endured decades of pathological totalitarianism would be incapable of carrying out the functions of the new Iraqi state.[36] However, unrooted in the organic plasma of Iraqi life, the returned exiles were not able to reanimate governance at any level. One inadvertent consequence of the de-Ba'athification policy was an estimated 15,000 to 30,000 unemployed civil servants, in addition to creating a shortage of qualified workers to run the government. As the occupying authority the United States and its creature the CPA left themselves few alternatives to taking the responsibilities of governance upon themselves.

In the context of a continually deteriorating security environment Bremer plowed forward, outlining a seven-step blueprint for restoring full Iraqi sovereignty, in which he argued, "the process is straightforward and realistic," and that "knowing how to turn Iraq into a sovereign state" is not a problem.[37] This blueprint was developed with minimal Iraqi participation. Having arrived in Baghdad with no plan at all for postwar governance, the United States then attempted to impose a hastily conceived plan onto a sullen Iraqi population. The exclusion of the Iraqi governing elite and reliance on expatriates, coupled with complete opacity of the transitional process to ordinary Iraqis led predictably to popular alienation from the CPA and its transitional project. In other words, planning, design, and leadership for the new Iraqi state were provided by non-Iraqis. The literatures of development, management, and other disciplines are unanimous in their acknowledgement of "buy-in" and "ownership" as central to any kind of transition or transformation effort.[38] A process based on substitution and replacement as opposed to buy-in and ownership was bound to fail, and was the third critical error in the state-building process in Iraq.

The Bush Administration's misplaced hope in the leadership capacity and legitimacy of Iraqi expatriates was ultimately replaced by faith that a group of Iraqi leaders selected through an American designed and managed process would assume the burdens of governance. But how would the right individuals be identified? Bremer argued that "Elections are the obvious solution to restoring sovereignty to the Iraqi people. But at the present elections are simply not possible. There are no election rolls, no election law, no political parties law and no electoral districts."[39] Bremer was under pressure from Washington, however, to accomplish a transition to full Iraqi sovereignty before the November 2004 U.S. presidential election, as the war was already beginning to take a heavy toll on the American public, and thus, on the White House. And Grand Ayatollah Ali Sistani, the Iraqi Shia spiritual leader, insisted that popular elections should be held at all levels of government, from the national to the municipal.[40]

By summer of 2003, the state-building effort in Iraq had become consumed by the search for legitimate Iraqi leadership to assume the responsibilities of future governance. Neither appointed proconsuls nor U.S.-selected Iraqis could meet the legitimacy test among a restive and divided population. A feverish effort ensued to find the right balance between Sunni and Shia, Kurd and Arab, and expatriates and those who remained under the Saddam regime. The ill-fated 25-member Iraqi Governing Council (IGC), announced by the CPA on July 13, 2003, suffered from a lack of credibility from the beginning, due in no small part to the fact that its members were appointed by the CPA. By June 2004, the IGC was replaced by the Iraqi Interim Government (IIG), another CPA-appointed body responsible for the taking the transfer of governing authority from the CPA. Thus ended the first of four phases of Coalition engagement in Iraq.

Phase two was characterized by a panicked, massive response that included the establishment of the Iraq Reconstruction Management Office (IRMO), the Project and Contracting Office (PCO), Provincial Reconstruction Teams (PRT), and a proliferation of projects under the Commander's Emergency Response Program (CERP). These programs and offices injected over $20 billion into a shattered country in a short period, guided still by neither strategy nor master plan. The IRMO and PCO were the reconstruction management units of the State and Defense Departments respectively, established to assume the responsibilities formerly managed by the CPA. PRTs constituted an innovative U.S. military initiative to combine both "hard" and "soft" power in extending the writ of government by empowering local governance. Begun originally in Afghanistan in 2003, 10 PRTs were established in Iraq in November 2005.[41] The CERP funds were used at the discretion of military commanders as "walking around money," for the purpose of winning support for reconstruction and stabilization from the local Iraqi population. Feverish "catch-up" project saturation overwhelmed Iraqi capacity.

Beginning in 2004, a debate raged between those insisting on immediate election of Iraqi officials, and those who argued the requisite security conditions and resources (e.g., census, voter rolls) were not yet conducive to free and fair elections. Though marred by low Sunni turnout, a national parliamentary election was finally held in January 2005. In the meantime, however, the security environment throughout the country was deteriorating dramatically. "Beginning with the rampant looting and violence throughout the country following the fall of Baghdad on April 9 (2003), coalition forces lost time and the trust of the population by failing to control the security environment."[42] The initial neglect was due to the coalition military forces assuming that policing and maintaining order were not their responsibilities. "At no point do we really see becoming a police force," Brigadier General Vincent Brooks stated at a briefing in Qatar on April 11, 2003.[43] Policing and stabilization operations have always been orphans in the U.S. foreign affairs apparatus, and absent a plan for postinvasion Iraq, little thought was given to what might happen once the regime was overthrown. In response to questions about the collapse of the rule of law Rumsfeld famously responded, "Stuff happens."[44] The stuff that happened in Iraq is that 2.5 percent of the Iraqi population died as a result of the conflict between March 2003 and June 2006—nearly 655,000 casualties.[45]

Although states historically have been formed in the crucible of conflict, this level of lethality is obviously nonconducive to the many tasks necessary to form and consolidate the institutions of an effective and legitimate state. The fourth critical error condemning the postinvasion state-building effort in Iraq to failure, was the failure to understand the critical importance of establishing a much higher degree of security from the outset. To accomplish that may have required U.S. military units taking on police or paramilitary functions (such as gendarmerie or carabinieri in Europe), but dismissing the growing insecurity as an inevitable collateral cost and as a manifestation of the stabilization process was in retrospect woefully misguided. The premature focus on establishing mechanisms for the play of competitive politics at the expense of establishing the rule of law handicapped the state-building effort from the beginning. Phase two eventually ended with a premature troop drawdown that derailed whatever progress had been achieved in the reconstruction effort.

By 2007, the security situation had deteriorated to such a degree that, against the advice of many of his confidantes and widespread popular opposition, then President Bush decided to change course in Iraq, and double down with a new strategy. The third phase of the U.S. engagement in post-Saddam Iraq was built on a new strategy embodied in a troop surge, a leadership change, and a population-centric approach to counterinsurgency. "The biggest of the big ideas that guided the strategy during the surge was explicit recognition that the most important terrain in the campaign in Iraq was the human terrain—the people...."[46] The so-called "surge" sent more than 20,000 additional U.S. troops, primarily to Baghdad and Anbar Province.[47] Military command of the Iraq War was given to General David Petraeus, while the political and diplomatic lead was given to seasoned diplomat Ryan Crocker. Despite initial skepticism, in retrospect, phase three—with the new leadership, troop surge, and counterinsurgency strategy—was successful in reducing violence in Iraq and in making progress in building the Iraqi state possible.[48]

Phase three, from 2007 to 2009, offered a brief window during which the alignment of politics and reconstruction advanced the project of building the Iraqi state. Belatedly a significant effort to help rebuild the atrophied Iraqi judicial system included "construction of judicial facilities, training of judicial security elements, and support for reestablishment of judicial systems and structures."[49] New laws on elections, provincial powers, and de-Ba'athification were passed. A measure of hope was restored, however all too abruptly snuffed out with the November 2008 passage by the Iraqi Parliament of the Status of Forces Agreement (SOFA) requiring all U.S. forces to depart Iraq by December 31, 2011.[50]

Phase four, initiated by the SOFA, consisted of the consequent troop drawdown under newly inaugurated President Barack Obama, the dismantling of the PRTs, and the end of the CERP. This was consistent with the new president's campaign commitment to "ending this war."[51] The agreement, negotiated during the second George W. Bush Administration and implemented by President Obama, aborted the progress made during phase three and left the reconstruction and state-building project undone.

Ultimately, in view of the disregard for the centrality of state-building to the mitigation of conflict and stabilization of states and regions that the campaigns in Afghanistan and Iraq demonstrated, it is unjustifiable to conclude that these represent failed efforts. State-building was, in both cases, an inconvenient afterthought, never seriously embraced by the leading powers in either theater. Profoundly misguided abdication of responsibility for planning state-building efforts—rooted in an ideological predisposition against state-building within the George W. Bush Administration—and an ongoing series of execution and judgment errors are to blame for these epic failures, not the concept of state-building.

State Building is Possible

Jordan

According to Charles Tilly, "War wove the European network of national states, and preparation for war created the internal structures of states within it." In other words, states emerge from the crucible of conflict, which hardens and aligns their geographical borders where they can be defended effectively from external adversaries, while exercising sovereignty over their subjects and interior territory.[52] Maintaining the security forces necessary to defend large territories (and to maintain internal law and order) is a substantial undertaking requiring commensurate appropriation of national financial and human resources. Large governmental organizations are established to provide for national defense and to maintain law and order—public goods which citizens are historically willing to underwrite. States are built around these institutions and processes.

The 20th century shows us there are many paths to statehood, though conflict and war—either external or internal—remain powerful shaping forces. Consider the cases of Jordan, Singapore, and Colombia—each unique in its path to consolidated statehood, and none without some degree of conflict. Each is worthy of brief examination for indications of what institutions, processes, and attributes external agents (e.g., the United States or the international community) may wish to foster.

Liberated from the Turks following their defeat in World War I, Arab leaders struggled to establish the integrated Arab homeland to which they aspired, but were ultimately frustrated by the strategic requirements of the victorious European powers. France and Britain competed for territory and control in the Levant, dividing it into mandates defined at the 1920 Conference at San Remo, Italy, based on the notorious Sykes-Picot Agreement.[53] Palestine and Iraq were accorded to Britain, and Syria and Lebanon to France. With no history as an autonomous polity, Jordan rose from the ashes of the Ottoman Empire as part of the Palestine Mandate. Governed for centuries by the Ottomans within the Damascus *vilayet*, the land that became Jordan was dismissed as "the vacant lot which the British christened the Amirate of Transjordan."[54] With no capital city nor population center, mandatory governance was managed by the British High Commissioner in Palestine, with Abdullah bin Al-Hussein, a descendant of the Prophet and key leader of the Arab Revolt, designated as the emir under British supervision. Britain provided an annual subsidy of £2 million.

Basic governance functions in Transjordan were assumed by British administrators posted to the mandate, who established an army and police force, organized finances and border demarcation, and established civilian governmental ministries.[55] The British recognized Transjordan as an independent government in 1923, but maintained control over financial, military, and foreign policy matters. In 1925, the mandatory territory of Transjordan was determined by the Permanent Court of International Justice and the International Court of Arbitration to be one of the newly created successor states to the Ottoman Empire. The state was further consolidated in 1928, by the enactment of an "Organic Law" and a "Nationality Law," "authorizing the new state in its territorial and temporal claims and in its control of the bodies over which it rules."[56] It was only in 1946 that the Hashemite Kingdom of Jordan was internationally recognized as a fully sovereign state, and only in 1955 as a full member of the UN.

Britain's first priority in the mandatory territory was to establish stability and order among the Bedouin tribes. At its inception, Jordan (then Transjordan) became a de facto British protectorate, with Britain guaranteeing both internal and external security. In 1920, the 100-strong "Mobile Force" was formed under the command of Captain F.G. Peake. By 1923, the need for a larger force led to the formation of the "Arab Legion," financed by Britain and under the command of British officers. Even as Britain incrementally granted sovereignty to Jordan, it retained this role in the Jordanian military until Sir John Glubb was relieved of command in 1956. The Arab Legion served as the Jordanian army, imposing both internal order in the realm, and defending it against incursions by Wahhabi tribesmen from the Najd (in what is today Saudi Arabia).

While the British provided security, forging a new national identity—in a land that never previously possessed one—fell to the Hashemites. These direct descendants of the Prophet, "placed Jordan into the continuum of Arab history, and then as a logical extension of it…the personal bridge connecting" the traditional narrative of Arab unity and the emerging reality of divided Arab states.[57] The establishment of borders, central administration, and nationality defined the new state. Membership in the Arab Legion was a critical unifying factor bringing together members of the many tribes that made up the Bedouin population. Emir Abdullah, not being native to the land of Transjordan, was able to convene and lead as a neutral arbiter over rival and often feuding chiefs. With no residual attachment to Ottoman identity nor competing political loyalty to local, familial, or tribal affiliations, and juxtaposed against colonial suzerainty, Abdullah was able to create a narrative of Jordanian citizenship, based on his own lineage, the army, and his deft balancing of local politics against external threats. As the nascent state evolved key government institutions "promoted the creation of a supra-tribal national culture that furthered Transjordanians' national identifications and instilled in them a sense of patriotism."[58]

There are important insights to draw from the consolidation of the state that is today Jordan, which are worthy of consideration even if not directly applicable or replicable in the 21st century. First, Jordanian sovereignty was achieved in stages, under the watchful eye of British senior management. The process of gaining full sovereignty took about 25 years—approximately one generation. Even today Jordan remains dependent on external financial support from the donor community, and security support from its allies, including the United States and Great Britain.[59] The principle of gradual or incremental sovereignty may no longer be politically acceptable, though Timor-Leste and Kosovo incubated under UN and North Atlantic Treaty Organization (NATO) protection for years. The process of state formation, however, must be allowed to unfold over whatever period of time is necessary to realize essential sovereign state functions (effective governance of territory and policing of borders). Premature sovereignty renders unready states vulnerable to the fissile dynamics of internal politics, the predations of external adversaries, and the corrosion of illicit networks and corrupt insiders. Permitting sovereignty to progress incrementally provides opportunities for local learning, institutional development, management and mitigation of internal conflicts, and gradual socialization of policies. The 25-year process of Jordanian sovereignty is not necessarily a guideline, but the generational nature of state consolidation must be acknowledged along with the bilateral or multilateral partnerships necessary to cultivate such a process. Any international actor, be it the United States, the UN, or an ad hoc coalition of the willing that wants to help build a state, must be in it for the long haul, with modest expectations for short-term benchmarks and no delusions about succeeding on the cheap.

Avoidance of conflict for at least five years following the cessation of hostilities is associated with dramatically greater likelihood of long-term security. It follows that taking the steps necessary to avoid conflict during the initial five-year window must be a high priority. The British, through the Mobile Force and then the Arab Legion, established and enforced security in Transjordan, suppressing internal fighting and external aggression. Like Jordan, even today in the midst of regional conflagration, both Kosovo and Timor-Leste are relatively peaceful, both having consolidated under the protective security shield of international partners. These cases, in addition to the cases of Afghanistan and Iraq, where international powers failed to achieve security and stabilization in the immediate postconflict period, suggest that security and stability must be a paramount priority. Without security and stability, no progress or consolidation is possible.

Singapore

The story of Singapore's journey from a corruption-plagued, politically unstable, resource-poor city-state populated largely by indentured Chinese and Indian laborers, to a top global economic power and standard-bearer for good governance less than four decades later provides several critical lessons in successful state-building. The East Asian island nation's remarkable transformation has been well-documented, not least by the man responsible for leading it "from third world to first," as he said himself—Singapore's founding father and first prime minister, Lee Kuan Yew.[60]

Singapore's achievement is even more impressive when you consider that it was unceremoniously asked to leave the Federation of Malaya in 1965, just two years after being accepted, because its prospects for independence were considered so unviable that it would later accept much humbler terms of membership.[61] Even Lee admitted, "We faced tremendous odds with an improbable chance of survival."[62]

Today, Singapore has the 10th highest gross domestic product (GDP) per capita in the world and stands as such an exemplary case of state turnaround and market building that countries around the world look to it regularly to provide examples and guidance for their own transformations.[63] Its success is widely acknowledged to be the result of effective policy planning, implementation, evaluation, and adaptation. Early on, its leaders made a series of bold decisions about what kind of country they wanted Singapore to become, and then systematically implemented policies designed to turn that vision into reality.

In 1965, unemployment in the newly independent country was at 14 percent and rising and its only major economic asset was its location alongside a high-volume shipping lane, which enabled it to act as a port between major international markets.[64] Understanding that ample employment opportunities were necessary for political and social stability, the government embraced an economic policy that prioritized substantially increasing sustainable job opportunities.[65]

This goal was part and parcel of Lee's central plan to bring Singapore to self-sufficiency, which led him to reject the familiar model of long-term foreign aid in favor of short-term aid to be used in service of the development of new industries. His philosophy is captured in his declaration, "The world does not owe us a living. We cannot live by the begging bowl."[66]

Within that self-sufficiency strategy were two unique decisions that guided Singapore's development. First, the government chose to vault over the prescribed stages of national economic planning theories of the time and instead look for opportunities to connect to the international economy via investment from multinational corporations (MNCs) that could bring skills-training and jobs to its workers.[67]

This decision was supported by a second transformational decision: rather than become another third-world factory in the region for Western goods, Singapore would make itself into a first-world oasis of service standards.

Before Singapore could attract MNCs, it needed an unimpeachable system of governance. The first plank of the economic policy's implementation strategy was a series of efforts to systematically curb corruption and institute integrity at the uppermost levels of what was then a very dishonest system. A former cabinet minister described how Lee, in his early days in office, would walk into offices and run his finger along bookshelves to check for dust, signaling his attention to detail. A parking attendant was fired for taking a bribe worth less than a dollar.[68] These new standards were gradually expanded throughout government over the following decades, and by 2013, Singapore was ranked fifth out of 179 countries in Transparency International's Corruption Perception Index.[69]

Once it had established a system for tackling corruption, the government implemented a highly successful combination of policies to attract foreign investment. These included: building transport infrastructure, creating industrial estates, offering equity participation in national industries, granting fiscal incentives (such as a 10-year tax-free status for investors) to promote exports or stabilize labor relations, implementing sound macroeconomic policies, engaging leading international businessmen, cultivating an investment compatible image through grooming public spaces, raising professional standards in the services industry, and generating positive publicity by sending officials to attend foundation-laying ceremonies and official openings of factories.[70] These combined efforts contributed towards a favorable investment climate with low transaction costs, low barriers to entry, and low risk of government or labor disruption to business operations.[71]

Singapore's first leaders were as bold in their approach to governance and public sector management as they were in their economic vision. On the eve of independence, they began an experiment in self-governance, with only the country's limited experience under the British colonial system and the Malaysian Federation as preparation. The People's Action Party (PAP) was Singapore's ruling party within the Malaysian Federation, and it won a vast majority of parliamentary seats in the first independent government. The PAP's top leadership was composed of a cadre of highly educated and technically capable professionals, with Lee at the helm. His team shared a strong work ethic, a common vision, and a mutual trust built through the shared experiences of World War II and the Japanese occupation, the end of colonialism, and the integration with and then secession from the Malayan Federation.[72]

That competent original team actively cultivated future leaders from the earliest days of self-governance. The government institutionalized several processes, including: efficient administrative procedures and processes for long-term planning; coordination between public, private, and civil actors; solicitation of input from stakeholders; implementation; monitoring; feedback flows; evaluation and revision of policy; creative adaptation of positive policy examples around the world; learning; and innovation.[73]

The administration avoided classic bureaucratic pitfalls by establishing clear purpose and principles that endure today: integrity, meritocracy, pragmatic orientation toward results, efficiency tempered by social equality goals, and socially inclusive stability.[74]

In 2004, then Prime Minister Lee Hsieng Loong articulated four principles of governance: leadership with vision, moral courage, and integrity; constant re-examination of old ideas and openness to new ones; self-reliance and individual responsibility, tempered with the provision of some social safety nets; and an inclusive society where citizens feel a sense of ownership and belonging.[75] In addition, Singapore has designated five National Values: nation before community and society above self, family as the basic unit of society, community support and respect for the individual, consensus above conflict, and racial and religious harmony.[76]

The established purposes of Singapore's government are: development of human capital as the country's main resource, encouragement of self-reliance and financial prudence, maintenance of stability to attract foreign direct investment (FDI) and international talent, fostering sustainable economic growth, achieving and maintaining global relevance, prioritization of long-term sustainability over short-term gain, and supporting proactive government intervention to improve the public welfare.[77]

The Singaporean public sector ranks well internationally and competes successfully against members of the private sector for the Singapore Quality Awards, which recognize outstanding achievement in organizational management.[78] Singaporeans enjoy high educational attainment and living standards in a country free of external debt, where government expenditure generally ranges between 14 and 18 percent of GDP.[79]

A strong testimony to Singapore's success in governance and public administration is its management of the nation's many transitions, including economic transitions, as described above, as well as transitions associated with social, political, international, and technological changes. For example, in 1989, Singapore launched TradeNet, the first nationwide electronic data interchange system in the world. Rather than simply digitizing existing processes, the Singaporean Trade Development Board used TradeNet to transform its organizational structure and business processes, network, and scope, resulting in productivity and competitiveness gains in both the public and private sectors.[80]

In 1995, the Singaporean government introduced the "Public Service for the 21st Century" initiative, in order to mold the public service into a body capable of undertaking "change as a permanent state," by optimizing each employee's potential, improving bureaucratic processes, building from coordinated action to a coordinated vision, empowerment of the ministries through decentralization of decisionmaking powers, and emphasis on superior leadership to counteract the public sector's lack of market competition. This program aimed to change organizational culture and processes by targeting employee well-being, continuous learning, high-quality customer service, and organizational reviews for the purpose of integrating new technologies, reducing inefficiencies, and enhancing innovation.[81]

The government's continuing investment in long-term planning and visioning is epitomized by the Centre for Strategic Futures (CSF), established in 2009, as part of the Strategic Policy Office within the Prime Minister's Office. The CSF develops tools and methodologies to promote strategic thinking and risk management throughout government, and develops collaborative networks between government agencies, international partners, and academic organizations.[82] The CSF hosted the 2013 Conference on Foresight and Public Policy, which identified four key issues in Singaporean strategic planning: the future of growth; the middle class; cities; and relations between citizens, corporations, and government.[83]

Singapore's experience has its critics and its limitations. Many point to its less than democratic practices, and Singapore itself in recent years has been looking at its model for citizen participation and enfranchisement. Some point to the issue of scale and argue that as Singapore is a city-state rather than a large country, its applicability may be limited. Still, countries across Asia and Africa continue to turn to Singapore to share its lessons and expertise for country and city transformations.

Colombia

Unlike Jordan or Singapore, which had minimal experience of self-government prior to the 20th century, Colombia has been a sovereign republic since the early 19th century. Political violence, however, has been constant throughout Colombia's history. The period 1948 through 1958, known historically as *la violencia*, was particularly wrenching, characterized by political assassinations, riots, uprisings, and extreme cruelty. As many as 200,000 Colombians lost their lives. The entente in the 1960s between the feuding Liberal and Conservative parties—representing different interests within the elite classes—that ended *la violencia* constituted a short-lived experiment in coalition government and conflict mitigation, but failed to protect the interests of the rural poor. The formation of rural-based and Marxist-influenced guerrilla movements, notably the National Liberation Army (ELN) in 1964, and the Revolutionary Armed Forces of Colombia (FARC) in 1966, marked the inauguration of another protracted period of political violence in Colombia.

The ELN and FARC insurgencies were primarily fought in Colombia's rural areas; however, during the 1980s, the war came to the cities. A new guerilla organization, "M-19," rose from the electoral controversy of 1970 and committed spectacular acts of terrorism, including the 1985 siege of the Ministry of Justice, resulting in over 100 casualties, and among the dead were 11 of the 21 justices of Colombia's Supreme Court. The growing cocaine trade added a further dimension of lawlessness, as the Medellín and Cali cartels both fought against, and corrupted from within, the Colombian state, armed forces, and police. By the late 1990s, many considered Colombia to be on the brink of state failure.[84]

In June 2016, the final details of a peace agreement that will end 50 years of violent conflict in Colombia were being negotiated.[85] The peace that is agreed could unravel—there remain significant challenges. However, the country "has been converging fast towards higher living standards since the early 2000's. Sound macroeconomic policy reforms— the adoption of an inflation targeting regime, a flexible exchange rate, a structural fiscal rule and solid financial regulation—have underpinned growth and reduced macroeconomic volatility."[86] From nationwide lawlessness in the late 1990s, when insurgent groups controlled as much as 40 to 50 percent of Colombian territory, the state has recaptured control, with "only six percent of municipalities...affected by terrorism by the end of 2014."[87] How did this dramatic reversal of national fortunes take place, and how was Colombia rescued from failure?

Some have argued that by the late 1990s, the Colombian people were simply fed up with conflict and insecurity, and demanded stabilization and peace. One seasoned observer wrote, "By 1999, Colombians had reached a collective conclusion that, if the deteriorating conditions remained unchecked, the viability of the nation was in question."[88] Perhaps it is true, but history indicates that popular will, though necessary, is not sufficient to either catalyze change or guarantee stability and peace. The formula for dramatic change in Colombia rested on four variables: determined political leadership, recapturing the military initiative, restoring law and order, and investing in social programs in marginalized areas both rural and urban.

Following the failed efforts of then President Andres Pastrana to negotiate an enduring peace with FARC and ELN (1998 to 2002), attitudes among the Colombian political leadership hardened. Pastrana initiated Plan Colombia with the support of the United States. An ambitious, long-term, and multifaceted plan for the recovery of sovereignty throughout Colombia, "At the core of the state-building, of course, was the modernization/ professionalization of the armed forces and police to gradually construct peace."[89] Though the Plan had significant economic and other components, Pastrana's successor, then President Alvaro Uribe, took office in 2002 "…with a platform based on establishing security," stating during his campaign, "If I am elected I will fight day and night, every minute during 24 hours a day, to restore security…."[90] Recognizing the complexity of Colombia's generations-old internal conflict and the socioeconomic root causes of the conflict, Uribe argued, "Of course we need to eliminate social injustice in Colombia but what is first? Peace. Without peace, there is no investment. Without investment, there are no fiscal resources for the government to invest in the welfare of the people."[91] Uribe ushered in a period of full commitment to defeating the insurgencies as well as the narco-traffickers, and held to that commitment for eight years. He imposed a national wealth tax to raise funds to complement foreign assistance.

From 2000 to 2010, the combined Colombian armed services' active and reserve components grew from 213,700 to 347,120, an increase of 62 percent. The defense budget trebled from $2 billion to $6.18 billion.[92] Over the 15 years, from 2000 to 2015, with substantial support from the United States and other allies, through high-value targeting and the deployment of joint task forces, among other initiatives and innovative tactics, the armed forces were able to reclaim lost territory, and degrade the FARC from over 20,000 to fewer than 8,000 combatants.

By circumscribing military operations and shrinking the space outside of the government's monopoly on the legitimate use of force, the Colombian authorities were able to focus on restoring law and order throughout the country. By 2007, all of Colombia's 1,099 municipalities had a formal state presence, up from 930 in 2002. The effort has been classic "clear, hold, build." According to Robert Killibrew:

> First, the military pushes the FARC out of a geographical space. Close behind the troops comes the National Police, who have evolved into a quasi-paramilitary force *acting under the rule of law* to secure the gains the military has just made, and courts to hear complaints—and the cops and the legal system stay permanently. Third, and with the cops and judges, comes economic assistance in the form of food grants, the making of truck farms, larger grants in in-kind assistance for economic development, electricity, email connectivity, roads, schools and all the trappings of good government.[93]

In 2002, when Uribe took office, the homicide rate was 69 per 100,000 annually, with over 28,000 homicides.[94] By 2013, the rate had dropped to 32 per 100,000.[95] In the same year, 132 Colombian fugitives were extradited to the United States. Kidnappings dropped from nearly 3,000 in 2000 to 288 in 2014. Increasing stability led to dramatic increases in economic growth with GDP increasing from approximately $100 billion in 2000 to $337.7 billion in 2014.[96]

Also critical to revitalizing the Colombian state was the emergence of a new narrative which cast the FARC as narco-traffickers, terrorists, and enemies of the state. This new narrative was reflected in Uribe's "Democratic Security Policy," emphasizing citizen rights, security, and the rule of law, which effectively discredited FARC claims to be the defender of people. Uribe's plan was partly underwritten by U.S. assistance, but also by over $1 billion raised through a "war tax" imposed on the wealthiest Colombians—that it was paid is an indicator of the success of the new narrative.[97] FARC's widely perceived failure to negotiate in good faith and use of the demilitarized zones granted to them by Pastrana as safe havens for regrouping and rearming further eroded their credibility. "The failed negotiations severely disillusioned the Colombian public and generated widespread support for adopting a hardline approach to security...."[98] The new narrative was helped by the dramatic terrorist attacks on the United States on September 11, 2001, which drew global attention to the brutality and inhumanity of terrorism. In 2004, the FARC was publicly condemned as a terrorist group by the UN High Commissioner for Human Rights.

With the FARC's military project at a dead end and the political space both within Colombia and globally contracting, in 2012, the FARC returned to the negotiating table a severely weakened force. Though the future, long-term success of Colombia's recovery over the last 15 years cannot be assumed, the country has made remarkable progress in rebuilding vital state security institutions, reinvigorating its economy, and creating a new social contract that has engaged the vast majority of Colombians. The Colombia-U.S. partnership has been a critical element in this historical process, and demonstrates that external support for state-building efforts, provided the political will and determined leadership is present on both sides, is not futile.

State-Builders' Toolbox

There is much we do not know about state-building. We do know from considerable experience that state-building is an arduous, labor-intensive, and time-consuming task. There is extensive literature on the subject and widely diverging views on how it should be done, but virtual unanimity regarding the intensiveness of the process.[99] What does state-building consist of? Though far from a science—still more alchemy than chemistry at this stage—there are a few principles that draw wide agreement.

In the litany of what must be done, arguably the most vital are the formation and consolidation of the security sector and the establishment of the rule of law.[100] The performance of the security sector, comprising those institutions and forces that guard sovereignty and ensure order, determines what civil, social, and economic activities may take place. The military and the police forces, along with their governing—preferably civilian—bodies, shape the environment for civic activity. The state must establish a secure and stable environment for both public and private life. Famously prescribed by Max Weber as the singular defining attribute of a state, "a state is a human community that (successfully) claims the monopoly of the legitimate use of physical force within a given territory."[101] More importantly, Weber specifies the legitimacy of the use of force. No state has or can enjoy a complete monopoly of the use of force—nor would we necessarily want it to do so. However, for the use of force to be legitimate, it must be sanctioned by the state. Historically, the state use of force has been conceived as a responsibility to protect citizens from external aggressors, though in many cases the state itself has been an aggressor. This unpleasant reality has recently been addressed by a growing acknowledgment that the state's responsibility for security extends beyond its own survival to its population; hence, the emerging concepts of "human security" and "citizen security."[102]

Establishing a monopoly of the legitimate use of force in a territory is no mean feat, and cannot be accomplished by brutal methods without sacrificing the legitimacy that is essential to effective governance. There are numerous U.S. government programs that provide assistance, training, equipping, mentoring, and other support to partner governments, both military and civilian agencies, for the purpose of building partner capacity (BPC). No amount of training and equipping, however, can substitute for the social contract between government and governed necessary to establish and sustain legitimacy. This must be achieved by our partners; and in this, our role can only be to help them identify methods, techniques, best practices (to the extent we know them), and lessons to enable their success. Controlling the use of force within sovereign territory, either directly or through delegation, is an essential function of a viable state.

The application of force in society must be bound by the rule of law, the establishment of which is another critical responsibility of the state. The state must establish the rule of law and mechanisms for articulation, adjudication, and redress of grievances. Doing so provides methods for the resolution of social and other disputes within society, provides predictability necessary for commerce, and ensures the security of citizens. The rule of law is not just a question of constitutions or statutes, though they form the legal framework in any country. It also requires that citizens have access to the law and the institutions of justice, and that they are not excluded from legal recourse by cost, language, distance, or identity. Under a genuine rule of law, the state itself is also subject to the law and cannot operate outside the law or the legal system. This includes prohibiting and punishing corruption, especially within government agencies both civilian and military.

In order to execute its required functions effectively and sustainably, the state must invest in and upgrade human capital; it must create a reliable and competent civil service to administer official state functions and manage state assets. The best-known method of succeeding in this is the application of merit-based recruitment, retention, and promotion. It must also provide a social and security environment conducive to education and public health. A professional civil service, including administrators, diplomats, and a range of support personnel necessary to operate complex systems is required to assume the full spectrum of governmental responsibilities.

The state must develop systems, human resources, and institutional mechanisms for raising revenue, securing state financing, and managing state assets and budgets. No state can operate effectively without a stable and predictable revenue stream sufficient to meet the costs of its obligations. There is controversy over what a state's obligations are and vast variation among states; but whatever obligations compose the social contract between government and governed must be within the financial means of the state. Taxation and regulation of commercial and financial activity is a responsibility only appropriate for the state. A banking system capable of interaction in the global financial network is beyond the capabilities of the private sector alone.

The degree to which the state must or should be involved in commercial activity is debatable, and indeed, the subject of wide historical debate that has generated large and powerful intellectual and political schools of thought on the subject. There is, though, a degree of consensus on the state's responsibility with respect to creating an environment in which citizens' economic needs are met.[103] The debate is over the balance of responsibility for meeting those needs within a conducive environment. Some argue the state should enable individual economic and commercial innovation, bearing only a modest responsibility beyond that through taxation, promotion of property rights, and limited redistributive programs. Historically, however, in many contexts of state formation and growth, the state has taken an active role in creating the institutions that underpin the market and establishing a pathway to industrial and agricultural development and technological innovation. The state has also intervened in times of acute market failure, including after the Asian financial crisis of 1998, and again after the global fiscal and financial crisis of 2008 to 2009.

Critical to state sustainability is an inclusive national narrative promoting citizenship. While not the exclusive responsibility of the state, typically only the state has access to the nationwide communication systems required to disseminate strategic messaging about national issues, though this is decreasingly true in the era of ubiquitous social media. Still, identity building and creating a national narrative remain governmental responsibilities; states must learn to utilize all emerging media to accomplish this. The drafting and adoption of a national constitution can contribute to an inclusive national narrative, as can elections. These can, of course, be divisive, but that is the craft of statesmanship: being able to manage and utilize such formal processes in support of the national interest. The willingness of citizens to pay taxes to the state is contingent upon an inclusive national narrative and a social contract that citizens accept. An effective state penetrates most aspects of public life, including education, public health, commerce, and dispute resolution, among many others. It can use those platforms to further the forging of an inclusive national narrative. This requires high standards of leadership, without which no state will succeed in any case.

The state must also create accessible mechanisms for interaction between citizens, civil society, and the state. Robust civil society encourages associative behavior conducive to social capital and enables citizens to pursue their interests equitably amidst the competing interests within any state. The state cannot form or create civil society, but it can communicate and interact transparently and responsively with civil society organizations. Furthermore, it can provide and secure the political space needed for their operations. Finding ways to consult and engage citizens, particularly in active and decisionmaking roles, can help secure their participation in and loyalty to the state. Many governmental functions and social services do not need to be carried out by government entities, but if the government can set the policy frameworks and regulations, citizens, nongovernmental actors, and the private sector can be involved in different ways in implementing policies.

The most critical endowment of any state is its human capital. The state is a bureaucratic abstraction composed of people, and can be no better than the people who populate it. This is no less true of government institutions and agencies than of any other organization. An educated and healthy population base is the critical ingredient at the heart of a successful state, as it forms the reservoir from which good leadership is drawn. How does a state invest in human capital? "Singapore provides better schools and hospitals and safer streets than most Western countries—and all with a state that consumes only 19 percent of GDP."[104] Obviously, through budgetary allocations to education and public health programs and institutions, but less obviously by creating a culture of merit-based incentives and disincentives. One of the many successful innovations in Singapore has been its well-compensated, meritocratic public service professions, including, significantly, the Singapore Civil Service. "In Singapore meritocracy reigns all the way down the system."[105]

These approaches reflect experience from a diverse array of geographic, cultural, and economic environments, but must not be considered as a sequential or structural template for building effective states. While some generalizations may be widely transferable (i.e., the primacy of security) context-specificity is paramount and several additional factors must be recognized and respected in any effort to support effective state-building. Perhaps most important is host country ownership. As stated in the Paris Declaration on Aid Effectiveness, "Partner countries [must] exercise effective leadership over their development policies, and strategies and co-ordinate development actions."[106] This is to say simply that the recipient or host country (i.e., the assistance receiver) must not only fully embrace the proposed development effort, but be full party to its design and implementation. The host country must "take the lead in coordinating aid at all levels in conjunction with other development resources in dialogue with donors and encouraging the participation of civil society and the private sector."[107] Only then is it likely the country in question will develop the self-organizing dynamics required for the long-term sustainability of institutional development.[108] They must "want it" at least as much as we want it.

There is no substitute for good leadership. Our brief examinations of Jordan, Singapore, and Colombia provide evidence that enlightened leaders can, with external support when necessary, overcome the many challenges to building an effective state. "Political leadership is crucial because of the way it affects the quality and autonomy of the bureaucracy."[109] In Jordan, King Abdullah and his successor King Hussein carefully and skillfully navigated the violent tides of conflict and violence in the Middle East, maintaining critical alliances with powerful external allies, and balancing the interests of competing domestic interest groups. Singapore's Lee Kwan Yew established and enforced a set of standards for ethical behavior in governance that has made Singapore one of the most successful states in the world today. In Colombia, then President Alvaro Uribe and President Juan Manuel Santos, though taking diametrically opposed political stances toward the current peace process, both acted strongly in support of a vision of ending a generations-long war in Colombia.

Good leadership is anchored by the indivertible will to accomplish the necessary objectives and goals of good governance. In the words of Verena Fritz and Alina Rocha Menocal, what is needed is "a political leadership…committed to development and, in most cases, the uprooting of traditional elites." In the most successful cases, such as the Asian Tigers, "Development was regarded as a 'national project' of the first priority."[110] Such will—sometimes referred to as political will—goes far beyond rhetoric, and often requires great compromise, and even self-sacrifice. It often means taking on deeply entrenched and self-defending interest groups. In the absence of good leaders, no effort or commitment on the part of the international community is likely to succeed in building an effective state.

Ultimately, the factor perhaps most underestimated, but of critical importance is time—the amount of time necessary for any development project, let alone a state-building project. It may be advantageous to think of state-building as a process rather than as a project in order to avoid the anticipation of completion that a project suggests. We look forward to the completion of a project; a process may endure indefinitely. We believe state-building is an indefinite process, with no endpoint. The quest for answers to fundamental questions regarding the relation of citizens to the state, and to each other, is eternal. The so-called "developed states" are still building their institutions and governance processes, as is clear from recent developments in Europe. The earliest stages of state-building are especially fraught, and less-developed states must be understood within their evolutionary context. According to the World Bank, "International assistance needs to be sustained for a minimum of 15 years to support most long-term institutional transformations."[111] Considering this estimate must include states not in the midst of conflict, we should assume significantly more time for those suffering from conflict, such as Afghanistan or Iraq; with very limited existing infrastructure or resources like Mali or Guatemala; or with highly complex compositions, for example, the DRC. Imposing unrealistic timeframes onto such polities can result in "premature load-bearing," where new processes and practices are forced onto institutions before they are tested, and the environment is prepared.[112]

Development organizations often refer to 5- or 10-year plans; perhaps it would be more appropriate to consider realistic expectations of a country in 25 or even 30 years, and work backwards from there, establishing realistic benchmarks along the way. In that longer trajectory, rational sequencing of priorities may be more strategic. Consider, for example, that many of today's strong democratic states began their trajectory toward democracy with periods, even prolonged periods of autocratic rule as they built the bureaucratic and economic architecture of the state—a fact we must seriously consider. Research suggests that at least for postconflict states the highest immediate priority—often requiring as long as a decade—must be avoiding a conflict relapse, which may require the prolonged prioritization of security and stability.[113] In recent decades, the western emphasis on democratization has, on occasion, hurried processes that historically have taken decades, if not generations. By now, we should have learned the lesson that "democratization before economic take-off, and incomplete democratizations, also entails important risks. In particular, expectations are raised that cannot be satisfied, and clientelistic systems continue or even intensify where the potential authoritarian top-down control is not replaced by effective accountability to citizens."[114] In state-building, patience is not only a virtue, it is a necessity.

Of near if not equal importance to patience, is the need to embrace evidence-based adaptation as the state-formation process proceeds. As execution begins to show results revision of the original plan may be appropriate or required. In state formation, so much is subject to unpredictable and nonlinear developments.[115] Patient adherence to a well-conceived plan can easily slide into stubborn insistence if evidence of failure begins to cascade; in which case not only is flexibility required, but rapid adaptation to emerging circumstances.

These reflections provide merely a snapshot of what is entailed in the epic task of state-building. While we offer merely a notional short list of state responsibilities and functions (of which there are many more), described in a summary manner, we emphasize the centrality of the state to sustaining the rule-based world order. A world without order is a frightening prospect that recalls Hobbes's characterization of the natural state of mankind before government is established as "every man against every man," and a life that is "solitary, poor, nasty, brutish, and short."[116]

Notes

[1] Randall Schweller, *Maxwell's Demon and the Golden Apple: Global Discord in the New Millennium* (Baltimore, MD: Johns Hopkins University Press, April 2014).

[2] Robert Rotberg, "Failed States in a World of Terror," *Foreign Affairs*, July/August 2002.

[3] The White House, *The National Security Strategy of the United States*, September 2002, accessed at <http://georgewbush-whitehouse.archives.gov/nsc/nss/2002/>.

[4] Alexander Woodside, *Lost Modernities: China, Vietnam, Korea, and the Hazards of World History* (Cambridge, MA: Harvard University Press, 2006).

[5] Max Roser, "Life Expectancy," *OurWorldinData.org*, available at <https://ourworldindata.org/life-expectancy/>; United Nations Development Programme, "Life expectancy at birth (years)," last modified November 15, 2013.

[6] Angus Maddison, *The World Economy: A Millennial Perspective* (Paris: Development Centre of the Organization for Economic Co-operation and Development, 2001).

[7] The rate of 35 percent refers solely to the average literacy of six European countries in 1650, due to lack of data from other countries and regions. Max Roser, "Literacy," *OurWorldinData.org*, available at <https://ourworldindata.org/literacy/>.

[8] United Nations Development Programme, *Human Development Report* (New York, NY: Oxford University Press, 1999).

[9] David Ronfeldt, *In Search of How Societies Work: Tribes—the First and Forever Form*, RAND Working Paper (Arlington, VA: RAND Pardee Center, December 2006). For an interesting reflection on the socializing advantages of tribal communities, see Sebastian Junger, *Tribe: On Homecoming and Belonging* (Lebanon, IN: Twelve, May 2016).

[10] For a more optimistic view, see Kishore Mahbubani and Lawrence H. Summers, "The Fusion of Civilization," *Foreign Affairs*, May/June 2016.

[11] Thomas Hobbes, *Leviathan*, 1651.

[12] Stephen D. Krasner, *Governance Failures and Alternatives to Sovereignty*, CDDRL Working Papers (Stanford, CA: Stanford Institute on International Studies, November 2, 2004), available at <http://fsi.stanford.edu/sites/default/files/enhanced_sov_krasner_Aug_1_04.pdf>.

[13] Robert D. Kaplan, "The Coming Anarchy," *The Atlantic*, February 2004, available at <http://www.theatlantic.com/magazine/archive/1994/02/the-coming-anarchy/304670/>.

[14] For a discussion of the attributes of sovereignty, see Dan Philpott, "Sovereignty," in the *Stanford Encyclopedia of Philosophy*, last modified March 25, 2016, available at <http://plato.stanford.edu/entries/sovereignty/>.

[15] Laza Kekic, "Ripe for Rebellion? Where Protest is Likeliest to Break Out," The World in 2014, *The Economist*, November 18, 2013.

[16] The Fund for Peace's "Fragile States Index" classifies 107 of 178 states (60 percent) as in the "warning" through "very high alert" categories. See Fund for Peace, "Fragile States Index 2015," available at <http://fsi.fundforpeace.org/>.

[17] Anne L. Clunan and Harold A. Trinkunas, ed., *Ungoverned Spaces: Alternatives to State Authority in an Era of Softened Sovereignty* (Stanford, CA: Stanford University Press, 2010); Angel Rabasa, Steven Boraz, Peter Chalk, Kim Cragin, Theodore W. Karasik, Jennifer D.P. Moroney, Kevin A. O'Brien, and John E. Peters, *Ungoverned Territories: Understanding and Reducing Terrorism Risks* (Arlington, VA: RAND Corporation, 2007); M.D. Phillips and E.A. Kamen, "Entering the Black Hole: The Taliban, Terrorism, and Organised Crime," *Journal of Terrorism Research* 5, no. 3 (2014); David H. Gray and Kristina LaTour, "Terrorist Black Holes: Global Regions Shrouded in Lawlessness," *Global Security Studies* l, no. 3 (Fall 2010), available at <http://globalsecuritystudies.com/LaTour%20Black%20Holes.pdf>.

[18] Dominic Tierney, "The Backlash Against Nation-Building," *PRISM* 5, no. 3 (Summer 2015), available at <http://cco.ndu.edu/Publications/PRISM/PRISMVolume5,no3.aspx>.

[19] Anne Evans, Nick Manning, Yasin Osmani, Anne Tully, and Andrew Wilder, *A Guide to Government in Afghanistan* (Washington, DC: World Bank, 2004), 38, available at <https://openknowledge.worldbank.org/handle/10986/14937>.

[20] Statement of Clare Lockhart, testimony before the Senate Committee on Foreign Relations, U.S. Senate, September 17, 2009, available at <http://www.foreign.senate.gov/imo/media/doc/LockhartTestimony090917a1.pdf>.

[21] Commission on Presidential Debates, "The Second Gore-Bush Presidential Debate," transcript, October 11, 2000, available at <http://www.debates.org/?page=october-11-2000-debate-transcript>.

[22] Condoleezza Rice, "Promoting the National Interest," *Foreign Affairs,* January/February 2000.

[23] Michael R. Gordon, "The 2000 Campaign: The Military; Bush Would Stop U.S. Peacekeeping in Balkan Fights, *The New York Times*, October 21, 2000, available at <http://www.nytimes.com/2000/10/21/us/the-2000-campaign-the-military-bush-would-stop-us-peacekeeping-in-balkan-fights.html?pagewanted=all>.

[24] "An Interview with Richard L. Armitage," *PRISM* 1, no. 1 (December 2009); see also Dov Zakheim, *A Vulcan's Tale: How the Bush Administration Mismanaged the Reconstruction of Afghanistan* (Washington, DC: Brookings Institution Press, May 2011).

[25] James Risen, *State of War: The Secret History of the CIA and the Bush Administration* (New York, NY: Free Press, 2006), 77.

[26] U.S. Department of Defense, Office of the Under Secretary for Policy Notes from Stephen Cambone [Rumsfeld's Comments], September 11, 2001. This indicates that a few hours after the 9/11 attacks, Rumsfeld spoke of attacking Iraq as well as Osama bin Laden and directed Defense Department lawyer Jim Hayes to get "support" for a supposed link between Iraq and Osama bin Laden from Paul Wolfowitz.

[27] John Esterbrook, "Rumsfeld: It would be a short war," *CBS*, November 15, 2002.

[28] Interview with Dick Cheney, "Meet the Press," *NBC*, March 16, 2003.

[29] Statement of Paul Wolfowitz, testimony before the House Appropriations Committee, U.S. House of Representatives, March 27, 2003.

[30] Interview with Dick Cheney, "Meet the Press."

[31] Nora Bensahel, Olga Oliker, Keith Crane, Richard R. Brennan, Jr., Heather S. Gregg, Thomas Sullivan, and Andrew Rathmell, *After Saddam: Prewar Planning and the Occupation of Iraq* (Arlington, VA: RAND Corporation, 2008), available at <http://www.rand.org/content/dam/rand/pubs/monographs/2008/RAND_MG642.pdf>.

[32] News Services, "Rumsfeld Forbade Planning for Postwar Iraq, General Says," Washington in Brief, *The Washington Post*, September 9, 2006.

[33] The most substantive and comprehensive effort addressing the governance challenges of post-Saddam Iraq was the Future of Iraq Project, organized by the Department of State. Begun in October 2001, the project generated a 13-volume report that included both strategies and recommendations. This effort was dismissed by the Department of Defense leadership, and those involved in the Future of Iraq Project were prevented from participating in invasion and postinvasion planning. "New State Department Releases on the 'Future of Iraq' Project," *National Security Archive*, September 1, 2006, available at <http://nsarchive.gwu.edu/NSAEBB/NSAEBB198/>.

[34] L. Paul Bremer, "Lessons Learned in the Fog of Peace," in *Commanding Heights: Strategic Lessons from Complex Operations,* ed. Michael Miklaucic (Washington, DC: National Defense University Press, 2010).

[35] "Coaliton [sic] Provisional Authority Order Number 1: De-Ba`athification of Iraqi Society," *National Security Archive*, available at <http://nsarchive.gwu.edu/NSAEBB/NSAEBB418/docs/9a%20-%20Coalition%20Provisional%20Authority%20Order%20No%201%20-%205-16-03.pdf>; "Coalition Provisional Suthority [sic] Order Number 2: Dissolution [sic] of Entities," *National Security Archive*, available at <http://www.iraqcoalition.org/regulations/20030823_CPAORD_2_Dissolution_of_Entities_with_Annex_A.pdf>.

[36] Samir al-Khalil, *Republic of Fear* (New York, NY: Pantheon Books, 1989).

[37] L. Paul Bremer III, "Iraq's Path to Sovereignty," *The Washington Post*, September 8, 2003.

[38] *The Paris Declaration on Aid Effectiveness and the Accra Agenda for Action* (Paris: OECD, 2005/2008), available at <http://www.oecd.org/dac/effectiveness/34428351.pdf>.

[39] Bremer III, "Iraq's Path to Sovereignty."

[40] "Sistani for Early Iraq Polls at All Levels," *AlJazeera*, November 27, 2003.

[41] Robert M. Perito, "Provincial Reconstruction Teams in Iraq," *United States Institute of Peace*, March 1, 2007.

[42] Bensahel et al., *After Saddam*.

[43] Khaled Yacoub Oweiss, "Mosul Falls—Chaos Takes Hold in Iraqi Cities," *Reuters*, April 11, 2003.

[44] Sean Loughlin, "Rumsfeld on Looting in Iraq: 'Stuff Happens,'" *CNN Washington*, April 12, 2003.

[45] Gilbert Burnham, Shannon Doocy, Elizabeth Dzeng, Riyadh Lafta, and Lee Roberts, The Human Cost of the War in Iraq: A Mortality Study, 2002-2006, (Baltimore, MD: Johns Hopkins University, October 2006).

[46] David Petraeus, "How We Won in Iraq," *Foreign Policy*, October 29, 2013.

[47] George W. Bush, "President's Address to the Nation," *Office of the Press Secretary*, January 13, 2007.

[48] Jason H. Campbell and Michael E. O'Hanlon, "The State of Iraq: An Update," *Brookings Institution*, December 22, 2007.

[49] Petraeus, "How We Won in Iraq."

[50] "Agreement Between the United States of America and the Republic of Iraq on the Withdrawal of United States Forces from Iraq and the Organization of Their Activities During Their Temporary Presence in Iraq," *ProCon.org*, available at <http://usiraq.procon.org/sourcefiles/SOFA-11-19-08.pdf>.

[51] Barack Obama, "My Plan for Iraq," *The New York Times*, July 14, 2008.

[52] Charles Tilly, *Coercion, Capital and European States, A.D. 990-1990* (Oxford: Basil Blackwell Ltd., 1990).

[53] Sykes-Picot Agreement (officially the Asia Minor Agreement), May 15-16, 1916, available at <http://wwi.lib.byu.edu/index.php/Sykes-Picot_Agreement>.

[54] Elizabeth Monroe, *Britain's Moment in the Middle East, 1914-1971* (Baltimore, MD: Johns Hopkins University Press, 1981).

[55] Philip Robins, "Jordan: Among Three Nationalisms," in *Nation Building, State Building, and Economic Development: Case Studies and Comparisons*, ed. S.C.M. Paine and M.E. Sharpe (New York, NY: Routledge, 2010).

[56] Joseph A. Massad, *Colonial Effects: The Making of National Identity in Jordan* (New York, NY: Columbia University Press, 2001).

[57] Betty S. Anderson, *Nationalist Voice in Jordan: The Street and the State* (Austin, TX: University of Texas Press, 2005).

[58] Matthew Liston, "Nation-Building in Jordan: The Hashemite Use of Educational Reform as a Means to Legitimate their Rule," (senior thesis, Colorado College, 2013), available at <https://digitalcc.coloradocollege.edu/islandora/object/coccc%3A8208>.

[59] Few contemporary states have the luxury of such consistent historical "sponsor" countries. Few would accept the conditional nature of early Jordanian sovereignty. Both Timor-Leste and Kosovo had prolonged "trusteeship" periods under United Nations auspices, which enabled them to obtain broad international recognition while developing critical governance capacities. Both continue to receive considerable external financial support to this day, and Kosovo continues to host a NATO security force for stability.

[60] Lee Kuan Yew, *From Third World to First: The Singapore Story 1965-2000* (New York, NY: Harper Collins, 2000).

[61] Clare Lockhart and Ashraf Ghani, *Fixing Failed States* (Oxford: Oxford University Press, 2008), 36.

[62] Yew, *From Third World to First*, 73

[63] "The 20 countries with the largest gross domestic product (GDP) per capita in 2016 (in U.S. dollars)," *statista*, available at <http://www.statista.com/statistics/270180/countries-with-the-largest-gross-domestic-product-gdp-per-capita/>.

[64] Yew, *From Third World to First*.

[65] Ngiam Tong Dow, former chairman of the Economic Development Board, interview by Institute for State Effectiveness (ISE) staff, 2009.

[66] Yew, *From Third World to First*.

[67] W.W. Rostow, "The Stages of Economic Growth," *The Economic History Review* 12, no. 1 (1959).

[68] Former senior officials, interview by ISE staff, Singapore, 2009.

[69] Transparency International, "Corruption Perceptions Index 2013," available at <http://cpi.transparency.org/cpi2013/results/>.

[70] The industrial estates mentioned includes the Juron Industrial Estate. Yew, *From Third World to First*.

[71] Shahid Yusuf and Kaoru Nabeshima, *Some Small Countries Do It Better: Rapid Growth and Its Causes in Singapore, Finland, and Ireland* (Washington, DC: World Bank, 2012), 14-15.

[72] Yew, *From Third World to First*.

[73] Boon Siong Neo and Geraldine Chen, *Dynamic Governance: Embedding Culture, Capabilities, and Change in Singapore* (Singapore: World Scientific, 2007), 17.

[74] Ibid., 26-27.

[75] Siong Guan Lim and Joanne H. Lim, *The Leader, the Teacher, and You: Leadership through the Third Generation* (Singapore: World Scientific, 2013).

[76] Lim and Lim, *The Leader, the Teacher, and You.*

[77] Neo and Chen, *Dynamic Governance*, 26-27.

[78] Ibid., 11.

[79] Ibid., 9.

[80] Hock-Hai Teo, Bernard C.Y. Tan, and Kwok-Kee Wei, "Organizational Transformation Using Electronic Data Interchange: The Case of TradeNet in Singapore," (Singapore: National University of Singapore), available at <http://www.comp.nus.edu.sg/~teohh/tenure/research_port/JMIS.PDF>.

[81] Lim and Lim, *The Leader, the Teacher, and You.*

[82] Centre for Strategic Futures, "Vision, Mission, and Key Roles," *Government of Singapore*, last modified February 14.

[83] Centre for Strategic Futures, "Foresight and Public Policy," Government of Singapore, 2013.

[84] Phillip McLean, "Colombia: Failed, failing, or just weak?" *The Washington Quarterly* 25, no. 3 (2002), 123-134. Also, Independent Task Force, *Toward Greater Peace and Security in Colombia: Forging a Constructive U.S. Policy* (Washington, DC: Council on Foreign Relations Press, October 2000).

[85] Ed Vulliamy, "Colombia peace deal with Farc is hailed as new model for ending conflicts," *The Guardian*, September 26, 2015, available at <http://www.theguardian.com/world/2015/sep/26/colombia-farc-peace-santos>.

[86] OECD Economic Surveys, "Colombia," January 2015, available at <https://www.oecd.org/eco/surveys/Overview_Colombia_ENG.pdf>.

[87] June S. Beittel, *Peace Talks in Colombia* (Washington, DC: Congressional Research Service, March 31, 2015); Mauricio Solaún, *U.S. Interventions in Latin America: "Plan Colombia"* (Champaign, IL: University of Illinois, Urbana-Champaign, April 24, 2002); Juan Carlos Pinzon Bueno, "Preface," in *A Great Perhaps: Colombia, Conflict and Convergence*, ed. Dickie Davis, David Kilcullen, Greg Mills, and David Spencer (London: Hurst and Company, 2016).

[88] Michael Shifter, "Plan Colombia: A Retrospective," *Americas Quarterly*, Summer 2012.

[89] Solaún, *U.S. Interventions in Latin America.*

[90] "Interview with Alvaro Uribe," *PRISM* 3, no. 3 (June 2012), available at <http://cco.ndu.edu/Portals/96/Documents/prism/prism_3-3/prism139-144_velez.pdf>.

[91] "Uribe Defends Security Policies," *BBC News*, November 18, 2004, available at <http://news.bbc.co.uk/2/hi/americas/4021213.stm>.

[92] International Institute for Strategic Studies, *The Military Balance*, 2000, 2010, 2011.

[93] U.S. Army Colonel Robert Killebrew (Ret.), "Colombia's Neglected Success Story," *Foreign Policy*, July 29, 2010, available at <http://foreignpolicy.com/2010/07/29/colombias-neglected-success-story/>.

[94] United Nations Office of Drugs and Crime, "Global Study on Homicide 2013," March 2014.

[95] World Bank, "Intentional Homicides," available at <http://data.worldbank.org/indicator/VC.IHR.PSRC.P5>.

[96] Dan Restrepo, Frank O. Mora, Brian Fonseca, and Jonathan D. Rosen, *The United States and Colombia: From Security Partners to Global Partners in Peace* (Washington, DC: Center for American Progress, February 2, 2016), available at <https://www.americanprogress.org/issues/security/report/2016/02/02/130251/the-united-states-and-colombia-from-security-partners-to-global-partners-in-peace/>.

[97] Peter DeShazo, Tanya Primiani, and Phillip McLean, *Back from the Brink: Evaluating Progress in Colombia, 1999-2007* (Washington, DC: Center for Strategic and International Studies, November 2007), available at <https://books.google.com/books?hl=en&lr=&id=F9guVNfWSpsC&oi=fnd&pg=PR5&dq=plan+colombia+and+farc&ots=pa9hBDJFWl&sig=KQ1lScT4e7FdiM5FLr9w3w-IE00#v=onepage&q=plan%20colombia%20and%20farc&f=false>.

[98] Beittel, *Peace Talks in Colombia.*

[99] For a sampling of this literature, see Lockhart and Ghani, *Fixing Failed States*; Francis Fukuyama, *State-Building Governance and World Order in the 21st Century* (Ithaca, NY: Cornell University Press, 2004); and James Dobbins, Seth G. Jones, Keith Crane, and Beth Cole DeGrasse, *The Beginner's Guide to Nation Building* (Arlington, VA: RAND Corporation, 2007).

[100] In *Fixing Failed States*, Ghani and Lockhart set out a framework for the 10 functions the state must perform in the contemporary era. Those functions build on this list.

[101] Max Weber, "Politics as a Vocation," 1919.

[102] S. Neil MacFarlane and Yuen Foong Khong, *Human Security and the UN: A Critical History*

(Bloomington, IN: Indiana University Press, January 2006); Roland Paris, "Human Security - Paradigm Shift or Hot Air?" *International Security* 26, no. 2 (2001): 87-102; United Nations Development Programme, *Human Development Report for Latin America 2013-2014*, November 12, 2013.

[103] James Dobbins et al., *The Beginner's Guide to Nation-Building*.

[104] "Go East Young Bureaucrat," *The Economist*, March 17, 2011.

[105] Ibid.

[106] *The Paris Declaration on Aid Effectiveness*.

[107] Ibid.

[108] Cedric de Coning, "From Peacebuilding to Sustaining Peace: Implications of Complexity for Resilience and Sustainability," *Resilience* (2016), available at <http://dx.doi.org/10.1080/21693293.2016.1153773>.

[109] V. Fritz and A. Rocha Menocal, "(Re)building Developmental States: From Theory to Practice," Working Paper 274, *Overseas Development Institute*, September 2006.

[110] Ibid.

[111] Conflict, Security, and Development, "World Development Report 2011," *World Bank*.

[112] Lant Pritchett, Michael Woolcock, and Matt Andrews, *Capability Traps? The Mechanisms of Persistent Implementation Failure*, Working Paper 234 (Washington, DC: Center for Global Development, December 2010).

[113] Paul Collier, *Development and Conflict* (Oxford: Centre for the Study of African Economies, Oxford University, October 1, 2004).

[114] Fritz and Menocal, "(Re)building Developmental States."

[115] de Coning, "From Peacebuilding to Sustaining Peace."

[116] Hobbes, *Leviathan.*

15

Communicate, Cooperate, and Collaborate (C³) Through Public-Private Partnerships (P³) to Counter the Convergence of Illicit Networks

Celina Realuyo

Today, we face a broad spectrum of security threats, such as global terrorism, transnational organized crime, economic crises, cyber-attacks, extreme natural disasters, and revisionist states that have made national security more challenging than ever before. The complexity of these security threats, particularly from illicit networks, like terrorists, criminals, and proliferators, requires a multidisciplinary approach to comprehend and counter. The convergence of these illicit networks, and the magnitude, velocity, and violence associated with their illicit activities are overwhelming governments and threatening state sovereignty and prosperity. Governments are no longer able to guarantee the security, prosperity, rule of law, and governance their people expect and deserve. Oftentimes, average citizens who "see something and say something" are the first to recognize anomalies and identify threats; they know their sector, workplace, or community best. Therefore, governments need to actively identify and engage partners in the private and civic sectors to better detect, dismantle, and deter the illicit networks that undermine our security and prosperity.

This chapter will emphasize the need to foster robust partnerships among the public, private, and civic sectors of society to counter the convergence of illicit networks at home and abroad. It will examine three specific areas where multi-sector collaborations are underway and have experienced some success: (1) protecting privacy, cyberspace, and critical infrastructure with the private sector through formal information sharing mechanisms; (2) combating threats to the international banking system, such as terrorist financing, money laundering, and cybercrime, in partnership with the financial services sector; and (3) countering violent extremism and foreign fighter recruitment by the Islamic State of Iraq and the Levant (ISIL) with local communities and civil society organizations in the United States. These examples illustrate the challenging nature of these security threats and the importance of harnessing resources across various sectors to counter them. Since the government no longer enjoys a monopoly on national security or the use of force as it did in the past, it must adopt a "whole of society" approach to understand and address the evolving threats to our security and prosperity in this globalized world. In the 21st century, all sectors need to communicate, cooperate, and collaborate (C³) through public-private partnerships (P³) to counter the convergence of illicit networks and ensure national security.

Protecting Cyberspace

Cybersecurity is considered one of the most daunting security challenges of the 21st century. It has become a leading security concern at the individual, enterprise, national, and international levels. Not a day goes by without some headline news item featuring the latest cyber-hack or cyber espionage case around the world. Whether it be the Target data breach with 110 million records compromised in 2013 or the massive disclosure of up to 21.5 million U.S. government employees' personal and biometric data through the Office of Personnel Management (OPM) hack revealed in 2015, cyber vulnerability affects each of us personally and professionally.[1] Privacy and civil liberties are at the center of the debate over data breaches through cyberspace.

Perhaps the most publicized cyber-attack to date was the one against Sony Pictures Entertainment on November 24, 2014, which crippled its movie and television studio. The cyber intrusion compromised data including personal information, such as the social security numbers of Sony Pictures' employees and their families, sensitive emails between employees, executive compensation information, and copies of (previously) unreleased Sony films. Since the discovery of the attack, Sony has been working to repair the damage caused by a group calling itself "Guardians of Peace," in an assault that the U.S. government has blamed on North Korea, spending an estimated $15 million.[2]

Sony Pictures is not the only corporate victim of cyber insecurity. Pricewaterhouse Coopers' Global State of Information Security 2015 Survey reported that the number of detected data security incidents among companies queried soared to a total of 42.8 million in 2014, a 48 percent leap over the number in 2013. This increase comes at great cost, with total financial losses attributed to security compromises increasing 34 percent over 2013.[3] For the past decade, the United States and many other countries have been grappling with how to define cybersecurity and how to protect against the broad spectrum of security threats that emanate from the cyber domain. Cyber threats can emerge from a number of different sources: nation-states that might engage in cyber warfare, foreign intelligence services, corporate espionage, terrorists, criminals, hackers, hacktivists, and insider threats.[4]

U.S. Cybersecurity Efforts

The Obama Administration considers cybersecurity one of the most important challenges facing the nation today. President Obama has said:

> America's economic prosperity, national security, and our individual liberties depend on our commitment to securing cyberspace and maintaining an open, interoperable, secure, and reliable internet. Our critical infrastructure continues to be at risk from threats in cyberspace, and our economy is harmed by the theft of our intellectual property. Although the threats are serious and they constantly evolve, I believe that if we address

them effectively, we can ensure that the internet remains an engine for economic growth and a platform for the free exchange of ideas.[5]

To secure cyberspace, the U.S. government has implemented a wide range of policies, both domestic and international, to improve our cyber defenses, enhance our response capabilities, and upgrade our incident management tools. Several of these initiatives entail close collaboration between the public and private sectors to protect our critical infrastructure and safeguard our privacy. The Obama Administration's Priorities on Cybersecurity are:

1. Protecting the country's critical infrastructure—our most important information systems—from cyber threats;
2. Improving our ability to identify and report cyber incidents so that we can respond in a timely manner;
3. Engaging with international partners to promote internet freedom and build support for an open, interoperable, secure, and reliable cyberspace;
4. Securing federal networks by setting clear security targets and holding agencies accountable for meeting those targets; and
5. Shaping a cyber-savvy workforce and moving beyond passwords in partnership with the private sector.

The Administration is employing the following principles in its effort to strengthen cybersecurity: a whole-of-government approach, network defense first, protection of privacy and civil liberties, public-private collaboration, and international cooperation and engagement.[6] On February 12, 2013, President Obama signed *Executive Order (EO) 13636: Improving Critical Infrastructure Cybersecurity*, a new approach to critical infrastructure cybersecurity, and released Presidential Policy Directive (PPD) 21, which seeks to increase the overall resilience of our national critical infrastructure.[7] Together, these measures drive action toward a whole of community approach to risk management, security, and resilience.

The Cybersecurity Framework developed by the National Institute of Standards and Technology (NIST) working with industry, consists of standards, guidelines, and best practices, to promote the protection of critical infrastructure through cyber risk management.[8] To support these goals, the Department of Homeland Security (DHS) created the Critical Infrastructure Cyber Community C³ (pronounced "C-cubed") Voluntary Program in February 2014. This program is an innovative P³ designed to help align critical infrastructure owners and operators with existing resources that will assist their efforts to adopt the Cybersecurity Framework and manage their cyber risks.[9]

The C³ Voluntary Program emphasizes three C's:

1. *Converging* critical infrastructure community resources to support cybersecurity risk management and resilience through use of the Framework;

2. *Connecting* critical infrastructure stakeholders to the national resilience effort through cybersecurity resilience advocacy, engagement, and awareness; and

3. *Coordinating* critical infrastructure cross-sector efforts to maximize national cybersecurity resilience.[10]

The primary goals of the C³ Voluntary Program are to support industry in increasing cyber resilience, to increase awareness and use of the Cybersecurity Framework, and encourage organizations to manage cybersecurity as part of an all-hazards approach to enterprise risk management. With over 85 percent of U.S. critical infrastructure in the hands of the private sector, government programs like C³ are fundamental to promote P³ in the cyber arena.[11] Homeland Security Secretary Jeh Johnson considers counterterrorism and cybersecurity the top priorities for his agency and emphasizes that his team cannot execute these missions without working hand in hand with the private and civic sectors.[12]

2015 White House Cybersecurity Summit

Highlighting the importance of collaborating with the private and civic sectors, the Obama Administration held a White House Summit on Cybersecurity and Consumer Protection on February 13, 2015 at Stanford University. The Summit convened leaders from businesses throughout the economy, consumer and privacy groups, educators, students, law enforcement, and other government agencies. Private sector participants included the CEOs of Apple Inc., American International Group Inc., American Express Corp., Bank of America Corp., Pacific Gas & Electric Co., Kaiser Permanente, and MasterCard.[13] At the event, over two dozen companies made commitments to share best practices, adhere to stronger security standards, use the Cybersecurity Framework of Standards and Best Practices to manage their cyber risk, share cyber threat information, and adopt more secure payment technologies.[14]

On the occasion of the Summit, President Obama issued an *Executive Order (EO) on Promoting Private Sector Cybersecurity Information Sharing* to encourage the development of Information Sharing and Analysis Organizations (ISAOs). These ISAOs would serve as the hubs for sharing critical cybersecurity information and promoting collaboration for analyzing this information both within and across industry sectors; the private sector has been organizing its communities of interest and forming ISAOs. However, some of the corporate executives at the Summit said the government needed to do more, including passing legislation that offers liability protection to companies participating in sharing programs.[15] The EO clarifies DHS' authority to enter into agreements with ISAOs, streamlines private sector companies' ability to access classified cybersecurity threat information, and provides strong privacy and civil liberties protections.[16]

To enhance collaboration, DHS has developed a system for the automated sharing of cyber threat indicators with the private sector and government that includes privacy and civil liberties protections. Interested companies can work with the National Cybersecurity and Communications Integration Center (NCCIC) to prepare their networks for the automated sharing of cyber threat indicators. In addition, NIST has created the National Cybersecurity Center of Excellence to partner with the private sector, academia, and other government agencies in order to find solutions to security problems inherent in technology. The center will produce generally available standards-based reference designs, templates, and example "builds," in order to reduce costs and complexities and enable companies in all sectors to use more secure technology.[17]

International Cooperation

Since cyberspace knows no borders and cyber threats can have worldwide impact, the Obama Administration has taken several steps to strengthen U.S. global leadership on the cyber front through bilateral and multilateral engagements. In 2011, the White House issued the "International Strategy for Cyberspace: Prosperity, Security, and Openness in a Networked World" that seeks a future of cyberspace open to innovation, interoperable the world over, secure enough to earn people's trust, and reliable enough to support their work. The strategy focuses on the following policy priorities for securing cyberspace:

- Economy: Promoting International Standards and Innovative, Open Markets; Protecting Our Networks; Enhancing Security, Reliability, and Resiliency.
- Law Enforcement: Extending Collaboration and the Rule of Law.
- Military: Preparing for 21st-Century Security Challenges.
- Internet Governance: Promoting Effective and Inclusive Structures.
- International Development: Building Capacity, Security, and Prosperity.
- Internet Freedom: Supporting Fundamental Freedoms and Privacy.[18]

Cybersecurity is becoming a regular agenda item in diplomatic engagements and international security conferences. During Chinese President Xi Jinping's state visit to Washington in September 2015, cybersecurity was one of the most prominent and controversial issues that resulted in a groundbreaking bilateral agreement on corporate espionage. After intense discussions, President Obama said the United States and China had reached a "common understanding" that neither country's government will conduct or knowingly support cyber-enabled theft of intellectual property, including trade secrets or other confidential business information, with the intent of providing competitive advantages to companies or commercial sectors. President Obama said that he had told President Xi during two hours of meetings at the White House that the escalating cycle of cyber-attacks against American targets "has to stop," and warned his Chinese counterpart that the United States would go after and punish perpetrators of those offenses through traditional law enforcement tools and, potentially, with sanctions.[19]

While the agreement marks progress in addressing the sensitive subject of cyber espionage and intellectual property theft with China, many observers, including President Obama, are taking a "trust but verify" approach to see what actual impact this pact will have on cyber intrusions. Director of National Intelligence James Clapper and other top U.S.

military officials have said cyber threats are increasing in frequency, scale, sophistication, and severity, and the United States needs the same kind of deterrent capability in cyberspace that it maintains for nuclear weapons. He told the Senate Armed Services Committee that the U.S.-China agreement did not include specific penalties for violations, but that the U.S. government could use economic sanctions and other tools to respond if needed. He viewed the cyber agreement between China and the United States on curbing economic cyber espionage as a "good first step," but noted it was not clear how effective the pact would be.[20] It is not yet clear if P³ can help bridge that gap, as private companies are often the targets of Chinese attacks.

Cyberspace is the new operating environment for consumers, producers, private enterprise, government agencies, and nongovernmental organizations (NGOs). While the internet has engendered many of the positive aspects of globalization, like better access to goods, services, capital, and information, cyberspace and the critical infrastructure it supports are vulnerable to a broad spectrum of threats. As all countries are grappling with the evolving challenges of cybersecurity, there are calls for the establishment of international norms to assist with common definitions regarding cyber threats and responses to attacks. More confidence-building measures and capacity building in cybersecurity are needed, especially in emerging markets, to promote information sharing across sectors and borders, according to Ambassador Makita Shimokawa, Japanese Foreign Ministry deputy director of UN Affairs and Cyber Policy.[21] Cyber actors, including nation-states, terrorists, criminals, and hackers, capitalize on vulnerabilities to steal information and money and are developing capabilities to disrupt, destroy, or threaten the delivery of essential services. To keep up with technological advances and protect cyberspace, governments and the private sector need to deepen their collaborations on cybersecurity to complement the initiatives described above.

Safeguarding the International Financial System

After the September 11 attacks, there was intense focus on stemming the flow of funding to al-Qaeda. This financial front of the war on terror complemented the formidable military and intelligence campaign in Afghanistan, known as Operation *Enduring Freedom*. This multifaceted approach recognized that financing is perhaps the most critical enabler for any terrorist organization to proselytize, plot, and execute their agenda. Financial forensic analysis of the 9/11 attacks demonstrated how the al-Qaeda operatives used the international financial system, through prominent banks and money services businesses, to fund the planning and execution of the largest terrorist attack on American soil in our country's history.[22] The 9/11 Commission estimated that the September 11 attacks cost al-Qaeda about $500,000, killed 19 hijackers and nearly 3,000 innocent victims, and changed the national security landscape of the U.S. forever.

Since the 1970s, the U.S. government has been working with the private sector to pursue financial crimes like fraud, tax evasion, and money laundering. Under the Bank Secrecy Act (BSA) of 1970, U.S. financial institutions are required to assist U.S. government agencies to detect and prevent money laundering. Specifically, the BSA requires banks to keep records of cash purchases, file reports of cash transactions exceeding $10,000, and report suspicious activity that might indicate money laundering, tax evasion, or other criminal activities.[23] Financial institutions are required to know their clients (and their clients' clients), monitor their transactions for anomalies, and report them to the authorities. In the United States, the Financial Crimes Enforcement Network (FinCEN) serves as the country's financial intelligence unit (FIU), collecting and analyzing suspicious transaction reports. FinCEN has global counterparts, providing the opportunity for global cooperation. Once it came to light that al-Qaeda used the formal banking system to finance the 9/11 attacks, banks and other financial institutions, concerned with reputational risk, realized they had a new and vital task—to detect and report possible cases of terrorist financing. Now bankers needed to understand and identify how terrorist groups raise, move, store, and use money, and what vulnerabilities exist in the banking system to prevent future cases of terrorist financing.

U.S. Counterterrorism Financing and Anti-Money Laundering Efforts

The U.S. law enforcement and intelligence communities work closely with officials at various financial institutions, who have been vetted and hold an active security clearance, to investigate and prosecute specific cases of terrorist financing and money laundering. In many instances, bank officials are former law enforcement agents or bank regulators. The financial sector has invested billions in human, technological, and financial resources to enhance their anti-money laundering/counterterrorist financing (AML/CTF) compliance capabilities. While these relationships between the public and private sectors have been quite productive, particularly in detecting Iranian sanctions violations, some financial institutions have expressed frustration regarding the lack of information flow from the government on the impact this cooperation has had on actual cases. The private sector has called for better two-way communications and more information to justify the immense investment in AML/CTF programs to their shareholders; they want to see how "following the money trail" has fought terrorism and crime.

In 2010, the Financial Intelligence and Information Sharing Working Group (FIIS WG) was established, following the completion of a P³ pilot project under the auspices of the Office of the Director of National Intelligence's Office of Private Sector Partnerships. The project's content aside, both the analysts and the business people involved found that the shared communication about threat-finance typologies was productive. To keep the dialogue going, they started a freestanding group, which organically blossomed due to the information sharing gaps in this area. While the group has no affiliation with the government, the FIIS WG is intended to provide experts in the financial services industry and the U.S. government with a forum to informally discuss relevant topics, including protection of critical financial infrastructure, prevention of fraud, terrorist financing, and money laundering.

FIIS WG meetings and the relationships formed at those events facilitate information flow and bridge cultural gaps between government and industry. The FIIS WG eventually found a home with the American Security Project, and its members include hundreds of representatives of both public and private sector entities, including regulatory, intelligence, defense, and law enforcement agencies, financial institutions, think tanks, consultancies, and academia.[24] The FIIS WG is considered a peer-to-peer community of practice and useful forum to discuss red flags, trends, emerging financial technologies, new payment systems, virtual currencies, alternative value systems, and the threats and vulnerabilities that accompany them.

Besides the banks themselves, several trade associations and NGOs have become actively involved in raising awareness, training, and educating the financial industry on the threats to the international financial system from financial crimes. One such example is the Association of Certified Anti-Money Laundering Specialists (ACAMS). It is the largest international membership organization dedicated to enhancing the knowledge, skills, and expertise in AML/CTF, as well as financial crime detection and prevention. Members represent various financial institutions, regulatory bodies, law enforcement agencies, and industry sectors. ACAMS circulates and discusses the latest trends and case studies in money laundering and terrorist financing through seminars, forums, international conferences, and local chapters.[25] The participation of senior U.S. government officials from the Departments of Treasury, Justice, and Homeland Security, the bank regulators, and law enforcement agencies responsible for combating terrorist financing and money laundering, at ACAMS events, demonstrates the active outreach conducted by the U.S. government to promote P[3].

Countering Emerging Threats to Financial Sector

Another example of cross-sector collaboration to protect the international financial system is the Financial Services Information Sharing and Analysis Center (FS-ISAC). It serves as the global financial industry's "go-to" resource for cyber and physical threat intelligence analysis and sharing. FS-ISAC is unique in that it was created by and for members in 1999 to prepare for Y2K and operates as a member-owned nonprofit entity. It was established by the financial services sector in response to the 1998 Presidential Decision Directive 63 (later updated by the 2003 Homeland Security Presidential Directive 7) that mandated that the public and private sectors share information about physical and cybersecurity threats and vulnerabilities to help protect U.S. critical infrastructure, of which the financial sector is a vital component.[26]

In response to emerging global threats in cyberspace to the financial sector, FS-ISAC's board extended its charter in 2013, to share information between financial services firms around the world. Constantly gathering reliable and timely information from financial services providers, commercial security firms, federal/national, state and local government agencies, law enforcement, and other trusted resources, the FS-ISAC can quickly disseminate physical and cyber threat alerts and other critical information to other organizations. This information includes analysis and recommended solutions from leading industry experts.

The Center's Critical Infrastructure Notification System (CINS) allows the FS-ISAC to send security alerts to multiple recipients around the globe almost simultaneously, while providing for user authentication and delivery confirmation. The system also provides an anonymous information sharing capability across the entire financial services industry. This protects members' proprietary information and client confidentiality. When they receive a submission, industry experts verify and analyze the threat and identify any recommended solutions before alerting FS-ISAC members. This procedure assures that member firms receive the latest tried-and-true procedures and best practices for guarding against known and emerging security threats.[27] Peer-to-peer collaboration brokered by organizations like FS-ISAC, combined with notifications to the appropriate government officials in the United States and elsewhere, is an example of timely and effective mechanisms to detect, address, and prevent threats to the financial system in the traditional and cyber domains.

International Cooperation

However, unilateral, single-country efforts will not be sufficient to address this threat. Our international financial system is far more interconnected and interdependent than ever before. International cooperation between the public and private sectors is, therefore, paramount. The Financial Action Task Force (FATF) is an intergovernmental body established in 1989, by the ministers of its 34 member jurisdictions. The FATF sets standards and promotes effective implementation of legal, regulatory, and operational measures for combating money laundering, terrorist financing, and other related threats to the integrity of the international financial system. It serves as a "policy-making [sic] body" working to generate the necessary political will to bring about national legislative and regulatory reforms to protect the global financial system. The FATF has developed a series of recommendations that are recognized as the international standard for combating money laundering, the financing of terrorism, and the proliferation of weapons of mass destruction.[28]

The FATF values private sector expertise and operational knowledge, as essential resources to evaluate the application of the AML/CFT requirements to business practices, and to encourage the practical adoption of the standards. The private sector can serve as a helpful sounding board to test or assess the potential impact of measures under consideration, or brainstorm on possible technical solutions in a specific field affecting the financial industry. It is also an important way to learn about market developments and access new information regarding emerging threats and vulnerabilities to the global financial system.[29] The FATF Private Sector Consultative Forum is the formal means to reach out to and cooperate with private sector stakeholders. It holds open consultations with interested stakeholders and sets up working groups to examine specific issues including financial innovations like mobile payments, virtual currencies, and store of value cards that could be vulnerable to money laundering or terrorist financing.[30] These examples of domestic and international P³ have strengthened the international financial system's ability to detect and deter terrorist financing, money laundering, and other financial crimes. According to declassified intelligence reports, groups like al-Qaeda and Mexican drug cartels decided to refrain from using the formal banking sector due to the enhanced compliance and monitoring measures adopted by the private sector.

Countering Violent Extremism

Terrorism and political violence have existed since the dawn of civilization. Unfortunately, the tragic attacks of September 11 perpetrated by the Islamic terrorist group al-Qaeda turned the threat of violent extremism into stark reality with 19 terrorists using airplanes as weapons of mass destruction to kill 2,977 innocent victims in New York, the Pentagon, and Shanksville, Pennsylvania. Since then, the United States and other governments have been trying to devise effective strategies to counter terrorism. The calls to counter violent extremism (CVE) have grown louder lately, since the April 2013 Boston Marathon bombing and the rapid rise of the Islamic State of Iraq and the Levant (ISIL) and its ability to recruit tens of thousands of sympathizers and foreign fighters from around the world. Fears that an ISIL foreign fighter could return to his or her country of origin and commit acts of terror have heightened homeland security concerns and redoubled CVE efforts in Western countries.

The Threat of Violent Extremism in the United States and Foreign Fighter Flow

With the rise of ISIL, law enforcement and intelligence officials at the federal, state, and local levels around the country have been on high alert for possible terrorist attacks at levels not seen since the immediate post-9/11 period. Counterterrorism analysts are hard at work monitoring all three phases of radicalization: online radicalization, recruitment, and violent action. Federal Bureau of Investigation (FBI) Director James Comey told the Senate in July 2015 that more than 200 Americans have tried to join Islamic extremists in Iraq and Syria. "Whether or not the individuals are affiliated with a foreign terrorist organization and are willing to travel abroad to fight or are inspired by the call to arms to act in their communities," he said, "they potentially pose a significant threat to the safety of the United States and its citizens."[31] As of September 1, 2015, U.S. authorities have charged 64 men and women in 20 states with alleged ISIL activities; men outnumber women in those cases by about 5 to 1, with an average age of 25. The FBI reports that, in a handful of cases, it has disrupted plots targeting U.S. military or law enforcement personnel.[32]

More recently, on October 8, 2015, Comey testified before Congress that while counterterrorism remains the FBI's top priority, the threat itself has changed in two significant ways. First, the "progeny of al-Qaeda"—including ISIL, al-Qaeda in the Arabian Peninsula (AQAP), and al-Qaeda in the Islamic Maghreb (AQIM)—have become our focus, rather than al-Qaeda itself. Second, we are faced with an explosion of terrorist propaganda and training on the internet, in particular social media, that has made recruitment possible even without operatives within the United States.[33] Dozens of people in the United States are engaged in conversations with overseas supporters of ISIL that the FBI cannot monitor.

Director Comey has been warning for months that when ISIL supporters find someone in the United States through messages on social media, they then employ software that encrypts their communications, making it impossible for the FBI to follow them, even with a court order. This is known as the challenge of "going dark."[34] ISIL has demonstrated a surprising level of social media sophistication in its ability to attract sympathizers and recruit foreign fighters. The evolution of terrorist recruitment through the use of the internet has underscored the urgent need for better CVE strategies; ultimately, this will require robust partnerships between governments and civil societies.

According to a September 2015 Homeland Security Committee's Bipartisan Foreign Fighter Task Force Report, ISIL has recruited some 25,000 foreign fighters since 2011. The report found that despite concerted efforts to stem the flow, the United States has largely failed to stop Americans from traveling overseas to join jihadists. Of the hundreds of Americans who have sought to travel to the conflict zones in Syria and Iraq, authorities have only interdicted a fraction of them. The report concluded that:

- The U.S. government lacks a national strategy for combating terrorist travel and has not produced one in nearly a decade;
- To catch foreign fighters, intelligence needs to be exchanged quickly, but at home and abroad, we still face barriers to information sharing;
- Despite improvements since 9/11, some foreign partners are still sharing information with us about terrorist suspects in a manner that is ad hoc, intermittent, and often incomplete;
- Since 9/11, America has gotten better at keeping terrorists out of the country, but we are doing far too little to keep Americans from leaving to train in terrorist safe havens;
- Few initiatives exist nationwide to raise awareness about foreign fighter recruitment; and
- Gaping security weaknesses overseas—especially in Europe—are putting the U.S. homeland in danger by making it easier for aspiring foreign fighters to migrate to terrorist hot spots and for jihadists to return to the West.[35]

"The alarming threat of extremist ideology possibly influencing foreign fighters is very apparent in today's world," said Ranking Member Bennie Thompson (D-MS). "The Task Force has found that although there are serious government efforts to address the radicalization of foreign fighters, there is much more we can do in terms of sharing information with our international partners, assisting law enforcement, and bolstering community awareness."[36] These conclusions underscore the continued importance of pursuing vigorous CVE strategies that engage all sectors of society and our foreign counterparts.

The Threat of Homegrown Terrorism and ISIL-Inspired Attacks in the United States

FBI Director Comey flagged the threat of homegrown terrorism when he stated in October 2015 that over 900 FBI investigations of ISIL suspects were underway in 50 states; these are active, open investigations and do not include the 66 Muslim men and women around the country who have already been charged with alleged ISIL activities over the past 2 years.[37] While ISIL initially focused on consolidating its power base in Syria and Iraq, the Paris attacks on November 13, 2015, directed by ISIL and perpetrated by radicalized French and Belgian citizens, and the ISIL-inspired attack in San Bernardino, California, demonstrate the global aspirations and reach of ISIL and turned terrorism into the leading national security concern in the 2016 U.S. presidential election debates.

The San Bernardino attack killed 14 people and seriously injured 22 at a San Bernardino County Department of Public Health training event and holiday party on December 2, 2015. Syed Rizwan Farook, an American-born U.S. citizen of Pakistani descent, who worked as a health department employee and his Pakistani-born wife, Malik, who came to the United States on a fiancée visa, carried out the attack against his coworkers. The FBI considered the couple "homegrown violent extremists" (HVEs) inspired by foreign terrorist groups, who had become radicalized over several years prior to the attack. Farook and Malik had traveled to Saudi Arabia in the years before the attack and had amassed a large stockpile of weapons, ammunition, and bomb-making equipment in their home. They jointly pledged allegiance to ISIL on social media shortly before they were killed in a shootout with police, according to the FBI.[38] In light of this attack in the homeland, the U.S. strategy to degrade and destroy ISIL and programs to counter violent extremism have come under close scrutiny and criticism for their lack of effectiveness as ISIL expanded its influence in 2015.

U.S. Efforts to Counter Violent Extremism

CVE includes the preventative aspects of counterterrorism, as well as interventions to undermine the attraction of extremist movements and ideologies that seek to promote violence. CVE efforts by the U.S. government seek to address the root causes of extremism through community engagement, including the following programs:

- *Building Awareness* - including briefings on the drivers and indicators of radicalization and recruitment to violence;
- *Countering Extremist Narratives* - directly addressing and countering violent extremist recruitment narratives, such as encouraging civil society-led counter-narratives online; and
- *Emphasizing Community-Led Intervention* - empowering community efforts to disrupt the radicalization process before an individual engages in criminal activity.[39]

In August 2011, the White House released "Empowering Local Partners to Prevent Violent Extremism in the United States," the first national strategy to prevent violent extremism domestically. The U.S. strategy is based on two premises: that communities provide the solution to violent extremism; and CVE efforts are best pursued at the local level, tailored to local dynamics, where local officials continue to build relationships within their communities through community policing and outreach mechanisms. It considers the federal government's most effective role in strengthening community partnerships and preventing violent extremism is as a facilitator, convener, and source of research and findings.[40]

Since the release of the 2011 Strategy, local governments and communities around the country have developed prevention frameworks that address the unique issues facing their respective communities. Three cities—Boston, Los Angeles, and the Twin Cities—with help from the federal government, have created pilot programs to foster partnerships between local government, law enforcement, mayors' offices, the private sector, local service providers, academia, and many others who can help prevent violent extremism. Each city has created an action plan addressing the root causes and community needs they identified. The pilot framework developed by these three cities emphasizes the strength of local communities with the premise that well-informed and well-equipped families, communities, and local institutions represent the best defense against violent extremist ideologies.

The Role of the Department of Homeland Security in CVE

With its counterterrorism mission and a full-time CVE coordinator, DHS serves as the de facto lead agency in CVE in the United States. According to DHS, violent extremists are defined as "individuals who support or commit ideologically motivated violence to further political goals." DHS recognizes that the threat posed by violent extremism is neither constrained by international borders nor limited to a single ideology. Violent extremist threats within the United States can come from a range of violent extremist groups and individuals, including domestic terrorists and HVEs. A domestic terrorist differs from an HVE in that the former is not inspired by, and does not take direction from, a foreign terrorist group or other foreign power.[41]

The increasingly innovative use of the internet, social media, and information technology by violent extremists is making CVE efforts more difficult and complex. Accordingly, DHS has designed a CVE approach that addresses all forms of violent extremism, regardless of ideology, and that focuses not on radical thought or speech, but instead on preventing violent attacks. This approach provides numerous physical and virtual environments to promote information sharing and collaboration between federal, state, local, territorial, tribal, private, civilian, and international entities working to counter the threat of violent extremism.[42]

The DHS' CVE efforts are focused on three broad objectives:

1. *Understand Violent Extremism* - Support and coordinate efforts to better understand the phenomenon of violent extremism, including assessing the threat it poses to the nation as a whole and within specific communities;

2. *Support Local Communities* - Bolster efforts to catalyze and support community-based programs, and strengthen relationships with communities that may be targeted for recruitment by violent extremists; and

3. *Support Local Law Enforcement* - Deter and disrupt recruitment or individual mobilization through support for local law enforcement programs, including information-driven, community-oriented policing efforts, which for decades have proven effective in preventing violent crime.[43]

The DHS Building Communities of Trust (BCOT) initiative focuses on developing relationships of trust between law enforcement, fusion centers, and the communities they serve, particularly immigrant and minority communities, so that the challenges of crime control and prevention of terrorism can be addressed. In coordination with federal partners, DHS hosts conferences, workshops, and online forums for federal, state, local, territorial, tribal, private sector, civilian community, and international partners in order to share information about CVE in the form of outreach.

In collaboration with the Department of Justice and state and local law enforcement partners, DHS has trained thousands of frontline officers, first responders, and community leaders, and continues to provide CVE training to interested communities. These efforts work to improve communication, build trust, and encourage collaboration between officers and the communities they serve and protect. Training topics include effective policing without the use of ethnic or racial profiling, and best practices in community outreach.

DHS prioritizes CVE activities through grants that directly support state and local partners and community outreach efforts to understand, recognize, report, and respond to potential indicators of terrorist activity. In addition, DHS produces substantial analysis and research on trends in HVE, domestic terrorism, and terrorist propaganda to support federal, state, local, territorial, and tribal officials in identifying and mitigating violent extremist threats to the homeland.[44] These projects are too new to provide robust evidence of their successes; however, even in their nascent stage they have shown promise.

International CVE Efforts

The disturbing beheadings of Western hostages, remarkable military offensives in Iraq and Syria, the persecution of religious minorities, and compelling foreign fighter recruitment campaign by ISIL in the summer of 2014 directed the world's attention to countering terrorism once again. The United Nations designated ISIL as a terrorist group on June 2, 2014 under the UN al-Qaeda Sanctions List. President Obama, addressing the UN General Assembly in September 2014, called on member nations to do more to address violent extremism within their regions. He chaired a UN Security Council summit that unanimously adopted UNSC Resolution 2178, condemning violent extremism and underscoring the need to prevent travel and curb support for foreign terrorist fighters (FTFs) destined for ISIL.

The United States established and leads a coalition of more than 60 partners committed to degrading and ultimately destroying ISIL. Besides the military, intelligence, and financial lines of effort of the campaign to counter ISIL, the Coalition is working to erode ISIL's appeal by strengthening capabilities to counter the group's messages of hate. The State Department's Center for Strategic Counterterrorism Communications (CSCC) was created to coordinate, orient, and inform government-wide foreign communications activities targeted against terrorism and violent extremism, particularly al-Qaeda and its affiliates and adherents.[45] It operationalized an Interagency Counterterrorism Communications (ICC) cell to improve cross-government collaboration on countering ISIL's online messaging.[46] The ICC plans to enlist U.S. Embassies, military leaders, and regional allies in a global messaging campaign to discredit groups such as ISIL. The plan is to be "more factual and testimonial," said Rashad Hussain, 36, a former White House adviser brought in to lead this effort. It will seek to highlight ISIL hypocrisy, emphasize accounts of its defectors, and document its losses on the battlefield—without recirculating its gruesome images or matching its snide tone.[47] With an annual budget of less than $6 million though, the CSCC is still struggling to effectively stem the spread of al-Qaeda and ISIL ideology.

White House 2015 CVE Summit

From February 17 to 19, 2015, the White House hosted the Summit on CVE to highlight domestic and international efforts to prevent violent extremists from radicalizing, recruiting, or inspiring individuals or groups in the United States and abroad to commit acts of violence. Foreign leaders from over 65 countries, senior officials from the UN and regional organizations, and private and civil society representatives convened at the Department of State to discuss a broad range of challenges facing nations working to prevent and counter violent extremism.[48]

At the Summit, President Obama asked participants to focus on how to empower communities to protect their families, friends, and neighbors from violent ideologies and recruitment. He suggested the following specific areas for international cooperation on CVE:

- First, we must remain unwavering in our fight against terrorist organizations and continue our mission to degrade and ultimately destroy ISIL.
- Second, we have to confront the warped ideologies espoused by terrorists like al-Qaeda and ISIL, especially their attempt to use Islam to justify their violence.
- Third, we must address the grievances that terrorists exploit, including economic grievances.
- Fourth, we have to address the political grievances that terrorists exploit and protect human rights.[49]

Continuing the work of the February 2015 CVE Summit, President Obama convened another summit on September 29, 2015 at the UN General Assembly, highlighting the international community's efforts to counter ISIL, address FTFs, and CVE. The president made it clear that ISIL poses a threat to the United States and the international community, and that we will use all instruments of power to defeat it. He has also emphasized that this fight will not be won quickly, or solely through military means. This is a long-term struggle that will be won with a comprehensive approach in concert with state and nongovernmental actors across the globe—which is exactly what we are doing.

The Summit showcased some new CVE initiatives like the Strong Cities Network to support the development of effective rights-based community focused programs and training to build resilience against violent extremism; the Global Youth Summit to Counter Violent Extremism that brought together more than 80 global youth leaders and organizations from more than 45 countries to build support for innovative youth-led initiatives; and the Peer-to-Peer Global University Challenge.[50]

The Peer-to-Peer Global University Challenge is a program within the State Department geared towards millennials to counter extremism among their peers and in their communities around the world. The objective is to design and implement a social or digital initiative, product, or tool to motivate and empower their peers to join the movement in CVE. Teams of students at 23 universities in the United States, Canada, North Africa, the Middle East, Europe, Australia, and Asia started the competition in January 2015. Participants were encouraged to use creative messaging and tactics to best reach their peers such as websites, campus and community events, social campaigns, viral videos, social movements, mobile apps, blogs, and education tool kits.[51]

The top three teams were invited to the U.S. Department of State to present their campaigns in June 2015. "Millennials can speak better to millennials, there's no question about that," State Department Principal Deputy Assistant Secretary Kelly Keiderling said, who was a judge in the competition. Missouri State University (Springfield, Missouri, U.S.) won the competition with their campaign, One95, a virtual collaboration platform targeting middle school students with a social media component, curriculum for teachers, and created the hashtag #EndViolentExtremism. Curtin University (Perth, Australia) took second place with their mobile application, 52Jumaa, that sends daily positive affirmations about Islam to users' smartphones, allows them to connect with other Muslims and asks them to complete a selfless act of kindness every Friday, the day of prayer. Mount Royal University (Calgary, Alberta, Canada) launched a prevention campaign called the WANT Movement ("We Are Not Them"). The team hosted a series of workshops and seminars that taught the differences between the Islamic faith and the belief systems embraced by terrorist organizations, and won third place.[52]

Reorganization of U.S. CVE Efforts

On January 8, 2016, the Obama Administration announced it was revamping its CVE efforts. It would establish a new counterterrorism task force, based at the DHS, to coordinate the government's domestic counter-radicalization efforts and serve as a conduit for ideas, grants, and other resources to community groups across the country. The task force will be led by George Selim, a Homeland Security official, who previously served at the White House as director for Community Partnerships, where he was in regular contact with local law enforcement agencies and Muslim communities. U.S. officials said that the new unit will be made up of representatives from at least 11 departments or agencies and that its mission will involve using data to find better ways to combat radicalization, as well as funding and supporting intervention efforts.[53]

In addition, the State Department is refocusing its CVE communications efforts through a new Global Engagement Center. This center will more effectively coordinate, integrate, and synchronize messaging to foreign audiences that undermines the disinformation espoused by violent extremist groups, including ISIL and al-Qaeda, and that offers positive alternatives. The Center will focus more on empowering and enabling partners, governmental and nongovernmental, who are able to speak out against these groups and provide an alternative to ISIL's nihilistic vision. To that end, the Center will offer services ranging from planning thematic social media campaigns to providing factual information that counters disinformation, to building capacity for third-parties to effectively utilize social media, to research and evaluation.

According to the State Department, the Global Engagement Center will employ a strategy defined by:

- Drawing upon data and metrics to develop, test, and evaluate themes, messages, and messengers;
- Building narratives around thematic campaigns on the misdeeds of our enemy (e.g., poor governance, abuse of women, narratives of defectors), not the daily news cycle;
- Focusing on driving third-party content, in addition to our own; and
- Nurturing and empowering a global network of positive messengers.

The Center will implement its strategy by:

- Seeking out and engaging the best talent, within the technology sector, government and beyond;
- Engaging across our government to coordinate, integrate, and synchronize counterterrorism communications directed toward foreign audiences;
- Identifying and enabling international partners with credibility and expertise;
- Establishing and implementing a campaign-focused culture;
- Scaling up data science and analytics and using both throughout the design, implementation and evaluation phases of these campaigns;
- Providing seed funding and other support to NGOs and media startups focused on CVE messaging;

- Identifying gaps in U.S. government messaging and counter-messaging capabilities directed toward foreign audiences, and recommending steps to resolve them;
- Sharing information and best practices with U.S. government agencies focused on the challenge of HVE; and
- Amplifying the successes of the Counter-ISIL Coalition in defeating ISIL on both the military and information battlefield.

The Center will continue to be housed within the Department of State and staffed by experts from the private sector and U.S. government agencies charged with protecting our national interests and security—as well as the security of our allies—against the threat of international terrorism.[54]

These new initiatives based at the Departments of Homeland Security and State are aimed at disrupting recruitment and radicalization efforts by terrorist groups that increasingly exploit social media platforms and encrypted communications technologies, often developed in the United States but beyond the reach of law enforcement. The Administration hopes this streamlining of CVE efforts at the DHS, complemented by the strategic communications efforts at the State Department will have more impact than other CVE programs to date.

CVE initiatives require close collaboration between the public, private, and civic sectors and cross-border cooperation as well to have real impact. Messaging to CVE must come from civil society rather than the government. Government needs to serve as a more engaged facilitator and enabler to get that messaging out into the respective communities.

Conclusion

The complexity of 21st-century security threats reflects the dark side of globalization that has empowered the convergence of illicit networks like terrorists, criminals, and cyber actors. Governments have been hard-pressed to counter these emerging threats with traditional security strategies and policies. They can no longer go at it alone to ensure our prosperity and security. A whole-of-society approach that employs the vast knowledge and resources of the public, private, and civic sectors is essential to devise innovative responses to address these formidable threats. The examples described above from the cybersecurity, financial services, and CVE arenas illustrate some productive initiatives of communication, cooperation, and collaboration through cross-sector partnerships at the national and international levels.

Governments need to institutionalize more mechanisms for P[3] to address emerging security issues. In addition, they should dedicate funding to support these initiatives and measure their effectiveness. Governments should document and publicize success stories of whole-of-society solutions to national security issues the way the "see something, say something" campaign helped two street vendors alert the police to avert a car bomb attack at New York's Times Square in 2010. Senior leaders across the governments, private enterprise, and civil society organizations must also advocate and support efforts towards collaborative problem-solving.

Governments should actually go one step further and deputize their citizens as proactive contributors to their national security environment. As those closest to their community or industry know their environment best, they are best suited to contribute as first alerts, if not actual first responders, in the face of an emerging threat, whether it be criminal, terrorist, cyber, or critical infrastructure in nature. Governments need to make it easy for citizens to report security anomalies and share information effectively without compromising their privacy. Educating and sensitizing our citizenry to emerging security threats in the 21st century should be a basic element of government outreach programs. Communicating, cooperating, and collaborating (C³) by promoting public-private partnerships (P³) is just good public policy to enhance our national security at home and abroad.

Notes

[1] Elizabeth Palermo, "The 10 Worst Data Breeches of All Time," *Tom's Guide*, February 6, 2015, available at <http://www.tomsguide.com/us/biggest-data-breaches,news-19083.html>.

[2] Ryan Faughnder, "Sony says studio hack cost it $15 million in fiscal third quarter," *Los Angeles Times*, February 4, 2015, available at <http://www.latimes.com/entertainment/envelope/cotown/la-et-ct-sony-hack-cost-20150204-story.html>.

[3] "Global State of Information Security® 2015 Survey," *PwC*, available at <http://www.pwc.com/gx/en/issues/cyber-security/information-security-survey/key-findings.html>.

[4] For more details, see Industrial Control Systems Cyber Emergency Response Team (ICS-CERT), "Cyber Threat Source Descriptions," *Department of Homeland Security*, available at <https://ics-cert.us-cert.gov/content/cyber-threat-source-descriptions>.

[5] "Cybersecurity," *White House*, available at <https://www.whitehouse.gov/issues/foreign-policy/cybersecurity>.

[6] Ibid.

[7] Office of the Press Secretary, "FACT SHEET: Administration Cybersecurity Efforts 2015," press release, July 9, 2015, available at <https://www.whitehouse.gov/the-press-office/2015/07/09/fact-sheet-administration-cybersecurity-efforts-2015>.

[8] "Executive Order 13636: Cybersecurity Framework," *NIST Cybersecurity Framework*, available at <http://www.nist.gov/cyberframework/>.

[9] Suzanne Spaulding, "DHS Launches the C³ Voluntary Program, A Public-Private Partnership to Strengthen Critical Infrastructure Cybersecurity," *Department of Homeland Security*, February 12, 2014, available at <http://www.dhs.gov/blog/2014/02/12/dhs-launches-c%C2%B3-voluntary-program>.

[10] Ibid.

[11] U.S. Government Accountability Office, *Critical Infrastructure Protection: Progress Coordinating Government and Private Sector Efforts Varies by Sectors' Characteristics*, GAO-07-39 (Washington, DC: U.S. Government Accountability Office, October 16, 2006), available at <http://www.gao.gov/products/GAO-07-39>.

[12] Secretary of Homeland Security Jeh Johnson, "Jeh Johnson on U.S. Cybersecurity Readiness" (remarks at the Council on Foreign Relations, Washington, DC, November 4, 2015), available at <http://www.cfr.org/homeland-security/jeh-johnson-us-cybersecurity-readiness/p37196>.

[13] Rachel King, "Obama Signs Info Sharing Executive Order, But Concerns Remain," *Wall Street Journal* CIO Report, February 13, 2015, available at <http://blogs.wsj.com/cio/2015/02/13/obama-signs-info-sharing-executive-order-but-concerns-remain/>.

[14] Office of the Press Secretary, "FACT SHEET: Administration Cybersecurity Efforts 2015."

[15] King, "Obama Signs Info Sharing Executive Order."

[16] Office of the Press Secretary, "FACT SHEET: Executive Order Promoting Private Sector Cybersecurity Information Sharing," press release, February 12, 2015, available at <https://www.whitehouse.gov/the-press-office/2015/02/12/fact-sheet-executive-order-promoting-private-sector-cybersecurity-inform>.

[17] Office of the Press Secretary, "FACT SHEET: Administration Cybersecurity Efforts 2015."

[18] The White House, "International Strategy for Cyberspace," *White House*, available at <https://www.whitehouse.gov/sites/default/files/rss_viewer/International_Strategy_Cyberspace_Factsheet.pdf>.

[19] Julie Hirschfeld Davis and David E. Sanger, "Obama and Xi Jinping of China Agree to Steps on Cybertheft," *New York Times*, September 25, 2015, available at <http://www.nytimes.com/2015/09/26/world/asia/xi-jinping-white-house.html?_r=0>.

[20] Andrea Shalal, "Top U.S. Spy Says Skeptical sbout U.S.-China Cyber Agreement," *Reuters*, September 30, 2015, available at <http://www.reuters.com/article/2015/09/30/us-usa-cybersecurity-idUSKC-N0RT1Q820150930>.

[21] Ambassador Makita Shimokawa, Japanese Foreign Ministry Deputy Director of UN Affairs and Cyber Policy, "Global Approaches to Cybersecurity" (remarks at the Council on Foreign Relations, Washington, DC, November 4, 2015).

[22] National Commission on Terrorist Attacks Upon the United States, *9/11 Commission Report*, 2005, available at <http://www.9-11commission.gov/report/>.

[23] "Bank Secrecy Act," *FinCEN*, available at <http://www.fincen.gov/statutes_regs/bsa/>.

[24] "Threat Finance and Financial Intelligence," American Security Project, available at <http://www.americansecurityproject.org/asymmetric-operations/threat-finance-and-financial-intelligence/>.

[25] "What is ACAMS?" ACAMS, available at <http://www.acams.org/join-acams/#tabbed-nav=what-is-acams>.

[26] "About FS-ISAC," *Financial Services Information Sharing and Analysis Center*, available at <https://www.fsisac.com/about>.

[27] Ibid.

[28] "Who we are," *FATF*, available at <http://www.fatf-gafi.org/about/>.

[29] FATF, *Public and private sector partnership in fighting financial crime: The FATF Recommendation* (Paris: FATF, February 2012), available at <http://www.fatf-gafi.org/documents/documents/publicandprivatesectorpartnershipinfightingfinancialcrime.html>.

[30] FATF President, "G8 Public-Private Sector Dialogue on anti-money laundering and countering the financing of terrorism (AML/CFT)," *FATF*, available at <http://www.fatf-gafi.org/publications/fatfgeneral/documents/ppsdsept13.html>.

[31] Julian Hattem, "FBI: More than 200 Americans have tried to fight for ISIS," *The Hill*, July 8, 2015, available at <http://thehill.com/policy/national-security/247256-more-than-200-americans-tried-to-fight-for-isis-fbi-says>.

[32] Adam Goldman, Jia Lynn Yang, and John Muyskens, "The Islamic State's suspected inroads into America," *Washington Post*, September 8, 2015, available at <http://www.washingtonpost.com/graphics/national/isis-suspects/>.

[33] FBI News Blog, "Director Briefs Senate Committee on Current Threats to the Homeland," *FBI*, October 8, 2015, available at <https://www.fbi.gov/news/news_blog>.

[34] Pete Williams, "FBI: Dozens in U.S. in Secret Conversations With ISIS," *NBC News*, October 8, 2015, available at <http://www.nbcnews.com/storyline/isis-terror/fbi-dozens-u-s-secret-conversations-isis-n440946>.

[35] House Committee on Homeland Security, "Final Report of the Task Force on Combating Terrorist and Foreign Fighter Travel," September 29, 2015, available at <https://homeland.house.gov/wp-content/uploads/2015/09/FINAL_2pager1.pdf>.

[36] Homeland Security Committee Press Release, "Committee Unveils Foreign Fighter Task Force's Final Report," press release, *Homeland Security Committee*, September 29, 2015, available at <https://homeland.house.gov/press/committee-unveils-foreign-fighter-task-forces-final-report/>.

[37] Kevin Johnson, "Comey: Feds Have Roughly 900 Domestic Probes About Islamic State Operatives, Other Extremists," *USA Today*, October 23, 2015, available at <http://www.usatoday.com/story/news/politics/2015/10/23/fbi-comey-isil-domestic-probes/74455460/>.

[38] "Everything we know about the San Bernardino terror attack investigation so far," *LA Times*, December 14, 2015, available at <http://www.latimes.com/local/california/la-me-san-bernardino-shooting-terror-investigation-htmlstory.html>.

[39] Office of the Press Secretary, "FACT SHEET: The White House Summit on Countering Violent Extremism," press release, February 18, 2015, available at <https://www.whitehouse.gov/the-press-office/2015/02/18/fact-sheet-white-house-summit-countering-violent-extremism>.

[40] The White House, *Empowering Local Partners to Prevent Violent Extremism in the United States* (Washington, DC: White House, August 2011), available at <https://www.whitehouse.gov/sites/default/files/empowering_local_partners.pdf>.

[41] "Countering Violent Extremism," *Department of Homeland Security*, available at <http://www.dhs.gov/topic/countering-violent-extremism>.

[42] Ibid.

[43] "DHS' Approach to Countering Violent Extremism," *Department of Homeland Security*, available at <http://www.dhs.gov/dhss-approach-countering-violent-extremism>.

[44] Ibid.

[45] "Global Engagement Center," *U.S. Department of State*, available at <http://www.state.gov/r/cscc/>.

[46] Office of the Press Secretary, "FACT SHEET: Leaders' Summit to Counter ISIL and Violent Extremism," press release, September 29, 2015, available at <https://www.whitehouse.gov/the-press-office/2015/09/29/fact-sheet-leaders-summit-counter-isil-and-violent-extremism>.

[47] Greg Miller and Scott Hingham, "In a propaganda war against ISIS, the U.S. tried to play by the enemy's rules," *Washington Post*, May 8, 2015, available at <https://www.washingtonpost.com/world/national-security/in-a-propaganda-war-us-tried-to-play-by-the-enemys-rules/2015/05/08/6eb6b732-e52f-11e4-81ea-0649268f729e_story.html>.

[48] "Countering Violent Extremism," *U.S. Department of State*, available at <http://www.state.gov/j/cve/>.

[49] Office of the Press Secretary, "Remarks by the President at the Summit on Countering Violent Extremism," *White House*, February 19, 2015, available at <https://www.whitehouse.gov/the-press-office/2015/02/19/remarks-president-summit-countering-violent-extremism-february-19-2015>.

[50] Office of the Press Secretary, "FACT SHEET: Leaders' Summit to Counter ISIL and Violent Extremism."

[51] EdVenture Partners, "Peer 2 Peer: Challenging Extremism," *evp*, available at <http://edventurepartners.com/peer-to-peer/>.

[52] Alyssa Bereznek, "Marketing the anti-extremism message to millennials," *Yahoo News*, June 24, 2015, available at <https://www.yahoo.com/politics/marketing-the-anti-extremism-message-to-122267261971.html>.

[53] Greg Miller and Karen de Young, "Obama administration plans shake-up in propaganda war against ISIS," *Washington Post*, January 8, 2016, available at <https://www.washingtonpost.com/world/national-security/obama-administration-plans-shake-up-in-propaganda-war-against-the-islamic-state/2016/01/08/d482255c-b585-11e5-a842-0feb51d1d124_story.html>.

[54] Office of the Spokesperson, "A New Center for Global Engagement," press release, *U.S. Department of State*, January 8, 2016, available at <http://www.state.gov/r/pa/prs/ps/2016/01/251066.htm>.

16

Adapting to Today's Battlefield: The Islamic State and Irregular War as the "New Normal"

Sebastian Gorka

C lausewitz was, of course, correct: the first responsibility of any commander is to understand the nature of the war he is about to engage in. What is the reality of America's wars today, and how must we prepare for the future? How does the war with the Islamic State (IS) change our understanding of today's threats and those of tomorrow?

The Correlates of War Project at the Pennsylvania State University contains data from every war since the Napoleonic wars of 1815, to include Operation *Enduring Freedom* (OEF) and Operation *Iraqi Freedom* (OIF), and operations against IS today. David Kilcullen and I used this database for an article on America's experience of war in the post-9/11 world, to categorize the nature of war, and see how it has changed over the last two centuries.[1] The database declares any conflict in which there have been more than 1,000 casualties to be a "war;" as such, the dataset contains 460 wars in the last 200 years. We divided these wars into two categories: conventional war, characterized by state-on-state violence, versus irregular warfare or unconventional warfare, where at least one of the actors was a nonstate entity.

Figure 16.1. Typology of Conflict: The Reality of War

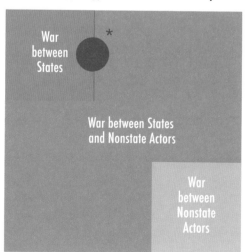

War between States

*

War between States and Nonstate Actors

War between Nonstate Actors

LEGEND

Less than 20% of war in last 2 centuries has been "conventional," state v. state, soldier versus soldier

More than half of all conflicts in the same period have been state v. nonstate actor (counterinsurgency, unconventional warfare, resistance warfare, counterterroism, counterpiracy, peace enforcement)

The remainder are conflicts between nonstate actors, "warrior versus warrior" (such as inter-tribal)

*Boundary conflicts include special cases such as civil war

The findings from our research (see Figure 16.1) shatter the common perspective of conflict in the 21st century. According to the empirical data, less than 20 percent of all wars, according to the Correlates of War Project, have been state-on-state, or what is deemed "conventional," such as World Wars I and II or Iran/Iraq, in which regular military units fought the regular military units of another nation or nations.

More than 80 percent of all wars since Napoleon—380 out of 460—fall into the "irregular" domain. The largest category of conflict in the last 200 years consists of instances where regular forces of one nation were fighting nonstate threat groups. This pattern is not driven purely by recent military endeavors, as both the U.S. Marines fighting the Barbary pirates off the shores of Tripoli over two centuries ago and Joint Special Operations Command (JSOC) elements today fighting against the Islamic State in Iraq or Syria are examples of such conflicts.

What that means is that the labels used inside the Pentagon, for doctrine and for analysis of the threat, are all diametrically opposed to reality on the battlefield. This prevents the United States from accurately assessing the threats we face and determining how best to defeat them. What we call "conventional" warfare should be considered anomalous given the relative scarcity of these conflicts throughout history. What we call "irregular warfare" (IW), which has been relegated to the purview of Fort Bragg and other special units, is really what war is all about. War is most often messy and nasty, without easily identified front lines. As a nation, we must admit sooner rather than later, that in the last 15 years, OEF and OIF are not the exceptions, but the rule and that "irregular war is war."

As a nation, we must move beyond outdated and Clausewitzian understandings of war as solely a functional operation of the nation-state. This is not to denigrate the Prussian's genius. However, his description of war as a continuation of politics by other means was an idealized description of state-on-state war and as such is fit fine for describing and understanding World War II or the Gulf War, but definitively lacking when we face groups that are not motivated by politics as we understand them, like al-Qaeda or IS. Whether fighting Shaka Zulu in Africa in the 19th century, the Taliban in Afghanistan, or IS today, our adversaries do not play by the Clausewitzian rule book. His concepts of friction and fog still apply, but the idea that our enemies make rational cost-benefit analyses about the reasons for going to war in ways that serve the *raison d'etat* really does not apply in the irregular domain, especially one in which our main enemy is transcendentally and apocalyptically motivated.

The primary conclusion that follows from this realization is that we must go beyond Westphalian and Clausewitzian understandings of war as they apply to the most prevalent form of armed conflict, which is undeniably what has been called "irregular." This demands adaptation not only for our armed services and our warfighters, but also for the way we think as a nation about war itself—what its purpose is and how our enemies will attempt to achieve their objectives in "unconventional" ways.

A secondary, and related conclusion, is that the defense community must help redefine IW as a core strategic function, not just an operational, tactical mission. To quote the former deputy director of Special Operations from the Pentagon, Major General Mike Nagata, all too often, the decisionmakers in D.C. look at our indirect options and at our special forces capabilities as the "easy button."[2] When a crisis strikes, our policymakers often default to an attitude of "Call Bragg, call Virginia Beach, and they'll sort it out." This is an utterly astrategic understanding of IW. Special Operations and Special Forces assets are, by their very nature, small-footprint investments that are meant to return strategic effects. Using them as a tactical hammer in an endless cycle of "whack-a-mole" denies the actual nature and purpose of such assets. It is our duty as national security professionals to convince the policymakers that "irregular war is war," and that our Special Forces and Special Operations units need to be understood as strategic assets rather than as a "quick fix" application for brush fires that break out in distant parts of the world.

The third significant conclusion from my research is that while America has, in the last 15 years, demonstrated that we are peerless in our capacity to apply kinetic force on target, the truth is that the victory in this current conflict will not come exclusively, or even primarily, through the use of force or kinetic solutions. It will come from identifying threats before they appear kinetically and deterring their full expression, neutralizing them in other indirect and unconventional ways, most especially in the counter-ideological domain.

That is not to say that applying violence to groups like IS is pointless. If we are relentlessly targeting their leaders, it makes it hard for them to think and act strategically. But counting Reaper hits against jihadi high-value targets is just as bad a metric of victory today as counting Viet Cong body bags was during the Vietnam War. Nonkinetic metrics will be the ultimate expression of victory in the sorts of conflicts that we face now and in the future. This does not mean that "Jobs for Jihadis," other programs focused on economic solutions, or nation-building are the solution.[3] Only a counter-ideological campaign will lead to ultimate victory. Along with our Sunni allies, we have to take down the ideology of global jihadism that has become far too appealing today, from the streets of Paris to Mosul, from Brussels to our very own San Bernardino. We have to undermine the ideology of jihadism just as we did with Communism during the Cold War.[4] Unfortunately, the bad news is that, in the last 15 years, the United States has not even begun to scratch the surface of the counter-ideological campaign for various reasons that can be the subject of another study.[5]

How Our New Enemy is Different and What This Means for the Prosecution of Future War

IS is much, much more dangerous than al-Qaeda for four major reasons, and these have severe ramifications for how America should wage war today and what it has to do to defeat similar enemies in the future.

Firstly, IS is the world's first transnational and multiregional insurgency, as opposed to al-Qaeda, which was just an international terrorist group. After 9/11, al-Qaeda only really functioned as a terrorist organization. In those theaters where there was an insurgency, al-Qaeda was actually a parasitic organization that attached itself to a preexisting insurgent force. Whether in Somalia, where it attached itself to al-Shabaab, or in Afghanistan, where al-Qaeda very deftly wormed its way into the Taliban and piggybacked its structure on top of the preexisting entity. Al-Qaeda never generated its own mass base of support, which is the mark of a true insurgency. Even where it supported insurgencies, al-Qaeda never generated its own cadre to populate a local insurgency. And, of course, the one key metric separating an insurgency from a terrorist organization is that an insurgency holds territory in daylight, thanks to the mass base of support it has successfully recruited.

IS, in contrast to al-Qaeda, is a full-fledged insurgency which has not had to exploit anybody else or piggyback on top of another indigenous fighting force to establish itself as such. Under Abu Bakr al-Baghdadi, IS has generated its own cadre, its own mass base of support across a number of countries. In fact, it has recruited tens of thousands of fighters.

Another crucial difference is that IS is a transnational insurgency. Every insurgency in modern history that was successful had some kind of international context, whether it was armed from the outside or had a place to rest and reorganize external to the battlefield. The international lines of effort for insurgencies are well-founded, but IS is not just international—it is transnational and multiregional.

Every insurgency of any merit in the last century, whether it was Mao in China or the Revolutionary Armed Forces of Colombia (FARC), had, as its strategic objective, the removal of the local government. Their strategic end state was always the same: capturing one nation-state's polity. IS is different; it is not interested in just capturing one nation's government and replacing it. It is interested in a global caliphate. Right now, IS holds territory in multiple countries in the Middle East and North Africa (MENA), including Iraq, Syria, and Libya, and has affiliates in another dozen countries. Not only that, it now has a fully integrated affiliate in West Africa since Boko Haram's pledge of *bayat* was accepted by Baghdadi, after which Boko Haram actually changed its name to "The West Africa Province of the Islamic State."

We have never seen in modern insurgent history a threat group that not only spans several nations in one region, but also has a significant presence in multiple regions. That, by itself, makes IS more important and much more dangerous than al-Qaeda ever was, and more importantly, than previous insurgencies or sundry irregular threats.

Moving beyond the unique qualities of the IS insurgency, the second metric is, very simply, wealth. As Jessica Stern documents in her contribution to this book, IS is the richest threat group of its kind in human history. The illicit oil sales through Turkey, the racketeering and kidnapping monies that are being raised by the new Caliphate, and even the sale of antiquities on the black market generated $500 million for IS in 2014. Even without these sources of funding, the group would still be incredibly wealthy. In the past three years, IS raided the Iraqi National Bank twice, netting $823 million in cash from the raids. To put that sort of financial power into perspective, consider that the 9/11 Commission Report estimated that al-Qaeda paid $500,000 to execute their 2001 attacks—the most destructive terrorist attack in our country's history.[6] That means that just from those two raids, IS netted enough to finance 1,600 attacks of the scale we experienced in September 2001. Of course, it will first have to consolidate its hold over the territory it has declared as the "New Caliphate" before it could consider sending sizeable teams to the United States to execute more attacks akin to 9/11, the Boston Bombing, or the San Bernardino massacre, but it has all the prerequisites: manpower, money, and the necessary skill sets.

Thirdly, and most obvious of all, comes their recruiting capability. The idea that IS—without piggybacking onto any other preexisting insurgent organization, has recruited at least 35,000 Syrians and Iraqis, and in the first 9 months of its renewed operations in Iraq, recruited more than 19,000 foreign fighters—is staggering. The United Nations' own estimate is that in the first year of renewed operations IS recruited 22,000 foreign fighters, including British, Americans, and Germans. To put that into perspective, before it became al-Qaeda in the 1980s, its predecessor was run by Abdullah Azzam under the name "The Arab Services Bureau for Mujahidin" (or the MAK in Arabic). For the whole decade of the Afghan War, from 1979 to 1989, the MAK recruited just 55,000 foreign fighters—IS has recruited 22,000 foreign fighters in just 12 months. IS's ability to attract supporters is what really makes al-Qaeda the "JV team," to use an analogy that President Obama most erroneously applied to IS.

The fourth and final metric is the one we really do not discuss adequately or fully understand within government circles: how IS has used religion to recruit and mobilize insurgents and terrorists in the Middle East, America, and elsewhere.

American policymakers still struggle to understand the strategic and tactical implications of Baghdadi climbing up to the pulpit of the Grand Mosque in Mosul in June 2014, dressed in the cloak of the religious authorities of Islam, declaring the reestablishment of the Caliphate, the theocratic empire of Islam, and declaring himself the new Caliph after nine decades of absence of the empire. The putative legitimacy of these claims is what makes the now "Islamic State" much more dangerous than any jihadi group we have seen in recent years.

Here we have to remember that the Caliphate is not just some idea of crazed extremists hiding out in remote parts of Central or South Asia; it was a real entity. The theocratic empire of Islam existed for more than a thousand years; first headquartered in Mecca, then in Damascus, then in Baghdad. And most importantly, less than a hundred years ago, the Caliphate still existed, though it was headquartered in Istanbul and run by Turks and not Arabs. The Ottoman Empire was the last formal version of the Caliphate, but was dissolved by then President Mustafa Kemal Atatürk after World War I, as he established and secularized the new Turkish Republic. When he formally dissolved the theocratic empire of Islam and "retired" the last Caliph in 1924, Atatürk unwittingly catalyzed a religiopolitical movement across the globe which has been demanding the reestablishment of the Caliphate ever since. Whether it was the original Muslim Brotherhood established in Egypt, or later organizations like al-Qaeda, the Islamists of this decentralized movement all believed that Atatürk had no authority to dissolve the Islamic empire, that within a caliphate is the only way for Muslims to live, and that Muslims must fight to see it rise again.

For 90 years, fundamentalists have been preaching for this to happen, political Islamists have been campaigning and conspiring to make it happen, and the most radical and dedicated members of this movement have been using terrorism to accomplish their shared goal. But Baghdadi did not talk about achieving the Caliphate, run for political office like the Brotherhood, or use disparate terrorist attacks to establish a caliphate—he simply went ahead and did it. He captured Mosul, the second largest city in Iraq, a city that Coalition soldiers and U.S. Marines used to patrol, and he declared the empire reborn with himself as its new head. This event should be understood as a watershed in modern IW for all the reasons mentioned above, and because IS is a fundamentally un-Clausewitzian enemy which typifies how the worldwide movement of violent actors who follow the ideology of "global jihad" understand war.

But the challenge for America is not simply one of understanding and dealing with different appreciations of war or strategy. We are also stymied by our legacy attitudes and our propensity to focus on that which is irrelevant and ignore or fail to appreciate that which is actually very important. And this is especially true in the U.S. intelligence community (IC).

For example, how many IC analysts or strategists can answer the question as to why did Osama bin Laden—or today, Ayman al-Zawahiri, the current head of al-Qaeda—not ever declare a caliphate, given that al-Qaeda was the brand leader of the international jihadi movement?

This, too, has to do with how war is "returning to type" and the way our enemies are motivated to fight in ways informed by pre-Clausewitzian mores. Here, one has to understand the history of Islam. Over the last 1,400 years, there have been two fundamental requirements before anyone could declare themselves the emperor of the Islamic State. The first is that the individual actually capture and hold territory, which, of course, is now the case for IS in Syria, Libya, and Iraq. The second requirement is that traditionally one cannot be the Caliph unless they belong to the same tribe as Mohammad, the founder of Islam. To bring about a revival of the Caliphate, you need to be of the tribe of Quresh. Neither bin Laden nor Zawahiri could claim to be Qureshis, but Baghdadi has managed to convince the populations of the territories under his control that he is a Qureshi. Not only has Baghdadi demonstrated his capacity to capture and hold the territory for a caliphate, he has marketed himself as meeting the Islamic criteria to do so. This has very little to do with Clausewitzian appreciations of war.

The Secret of the Success of the Islamic State: Realizing a Religious War

The question remains, however, how has IS so rapidly dethroned al-Qaeda and what does that mean for future IW campaigns involving America? The answer to this crucial question has to do with how Baghdadi has taken the concept of holy war, jihad, and given it a burning eschatological immanence.

IS is masterful in its understanding of psychological warfare and information operations (IO) in ways that al-Qaeda never was. IS controls more than 20,000 social media platforms and on average posts more than 50,000 pieces of propaganda for global consumption each and every day. These social media push the narrative of "global jihad" around the world 24/7, demanding a response from the United States that is simply not there. However, before the requisite response to IS IO can be designed and deployed, we must first understand what it is about their message that is so resonant and effective in recruiting both fighters to the theater and terrorists in the United States and elsewhere.

In the last two years in Washington, there has been a childish argument over what to call the threat. Some call it "ISIS," the Islamic State of Iraq in Syria. Others, like the president of the United States, call it "ISIL," the acronym for the Islamic State of Iraq and the Levant. Many commentators have criticized the Commander in Chief for this choice, saying that he uses ISIL because he wants to avoid referring to Syria and so remind audiences of the "red lines" he drew there and which were egregiously flouted by the Assad regime.

This is a churlish political debate because both of those names for our enemy are wrong. Both acronyms, from the point of view of rigorous intelligence preparation of the battlefield (IPB) are incorrect. Whenever you prepare for war what is the first thing you must do? Understand the enemy on his terms. We do not invent labels for them. We start by reading their statements, reading their doctrine, using their labels for themselves as the starting point for what we should call them.

We called the Soviet Union "the USSR" for a reason. We used the term "Third Reich" for a reason—not because we believed in Hitler's vision of a "thousand-year Aryan Empire," but because that is what they called themselves. Let us apply that to the threat we face today. Originally, of course, this group was al-Qaeda in Iraq—AQI, or the Iraqi franchise of al-Qaeda—which had been built up by the Jordanian thug, Abu Musab al-Zarqawi. After the pronouncement in June 2014 by Baghdadi of the reestablishment of the Caliphate, the correct designator for this threat is not ISIS, ISIL, or AQI. It is IS, or the Islamic State. They call themselves "the Caliphate," or "the Islamic State," and that is what we should call them.

But most importantly, we must appreciate what the group called themselves in the intervening years between being AQI and becoming IS, and what the evolution of their moniker tells us. As Baghdadi took his forces into Syria as the nation was collapsing and Nouri al-Maliki was ratcheting up his persecution of the Sunnis in Iraq, the group called themselves the "Islamic State of Iraq and al-Sham." (This is also the source of the derogatory name "DAESH," the Arabic acronym for the Islamic State of Iraq and al-Sham, which sounds much like the Arabic word, "to tread or step on," an act of disrespect in local culture and, thus, a slur to IS). This choice of name is a crucially important moment in the evolution of our enemy, because al-Sham is not simply a geographic designator; in Arabic, "al-Sham" can mean the "land that rises from the sea," which roughly equates in our terminology to Greater Syria or the Levant but, more importantly, it is a word associated directly with the End of Days, with the theology and eschatology of the religion of Islam.

Al-Sham is an extremely important concept in Islam. Like other religions, be it Buddhism, Hinduism, Zoroastrianism, Christianity, or Judaism, Islam has its own eschatology—a vision of how the world will end. For Christians, on the basis of tradition and the Book of Revelation, there is the expectation that there will be mighty battles before Judgment Day between the last true believers and the forces of the Antichrist, culminating on the plains of Megiddo in Northern Israel. In fact, the term "Armageddon" is derived from the town of Megiddo, the site of the last war between the believers and the nonbelievers before the end of the world.

Al-Sham is exactly that for Muslims; it is the Islamic Megiddo. All educated Muslims are taught that before the end of the world, there will come an earthly devil-like figure who will lead the forces of the *kuffar*, the infidels, in the last holy war, the final jihad, against the last true Muslim believers. According to Islam, this holy war will occur on al-Sham, according to the *sunna* of Islam.

This is a very powerful concept which goes to the heart of the staggering mobilizational capabilities of the IS. Baghdadi has not only named his organization after the territory that Muslims expect will be the site of the last holy war against the infidel, but he has also proceeded to capture said territory. Every time IS tweets, every time they put something on Snapchat, Telegram, or Facebook, they are sending a very clear and motivational message around the world. Each tweet or post can have a most powerful effect on teenage Muslims looking for a sense of purpose in life, for meaning in an increasingly vacuous and meaningless modern world. The message is: "If you want to save your soul, if you want to fight for the sovereignty of Allah, if you want to become a Mujahideen, this is your last chance. If you wish to cleanse your soul, if you wish to go to paradise, there will be no more jihads after this one. Just look at where we are fighting and that the Caliph is finally back. This is the very last pre-Judgment Day jihad, so buy that plane ticket to Turkey and walk across the border. We have initiated End Times. Just look at where we are: Al Sham!"

This leveraging of eschatological concepts is the only way to explain how IS has recruited over 35,000 foreign fighters in just months, with 6,000 of those, at least, being westerners from the United States and its partner or allied nations. Yet in our policy debates and national security analyses, we downplay the relevance of faith to our enemy, with top-level members of the administration going so far as to deny any relevance of religion to the actions of the IS.

Adapting to Understand Our New/Old Enemy

At this point we can conclude that:

- IW is back, and with a vengeance.
- IS is more dangerous and effective than all its predecessors.

As a result, we must reassess whether our legacy concepts about war, and our legacy capabilities, are applicable to the current and future threat environment. To do this we must follow the oft-quoted axiom of Sun Tzu, and understand properly the enemy we currently face.

When it comes to the *Art of War*, we in the West have our own strategic heavyweights, we study our Clausewitzes, our Mackinders, our Boyds, and so on. The enemy also has their own strategic powerhouses. Today's jihadists do not simply improvise their strategies; they are not random or capricious in their appreciation and application of war. They follow a plan based upon the strategic works of their most important authors. Everyone in the American national security establishment needs to be familiar with these jihadi strategists in order to understand the game plan of the enemies we face today, as well as tomorrow's threat groups, because IS will eventually be replaced by another, perhaps even more dangerous, Islamist threat group in the foreseeable future.

The first of the three most important grand strategic minds informing groups like IS today is Sayyed Qutb. Qutb is the most important strategic thinker of the Muslim Brotherhood. Qutb's book, *Milestones*, should be understood as the Field Manual (FM) of "global jihad." It is a book that has been found on high-value targets in every theater to which the United States has deployed in recent years. In this slim text, which has informed jihadi operations across the globe, including in America, Qutb is explicit on more than 20 occasions, when he writes that Muslims should understand that Islam is much more than a religion. For Qutb, and all salafists such as al-Qaeda and IS, Islam is, in fact, a supremacist political movement with a global mandate to reestablish the Caliphate for the greater glory of Allah. And all Muslims have a role to play in its glorious recreation.[7]

Second in importance to Qutb is bin Laden's former spiritual guide, Azzam. As he was creating the original al-Qaeda, Azzam wrote a 25-page *fatwa* called "Defense of Muslim Lands." That document was the most important call to jihad, the most important mobilizational document of international jihad in the 20th century, until IS declared its own caliphate after the fall of Mosul. Written after the Soviet invasion of Afghanistan, this *fatwa* declared jihad to be *fard 'ayn*, an individual and universal obligation for all Muslims. Azzam argued that since an infidel superpower had invaded historically Islamic territory, there must be a holy war—a jihad—in response. Yet, the Muslims of the world were waiting for a formal declaration of jihad; they were waiting to see if "deployment orders" would land in their letterboxes. But that would never happen, thanks to Atatürk's actions after World War I. After he retired the Caliph and dissolved the Caliphate, there was no one left to declare war or order Muslims into battle. As a result, it is now incumbent upon all believers, without waiting for orders, to deploy themselves, to grab a weapon and become mujahideen. This *fatwa* would become the most mobilizational document in international jihad until Baghdadi declared the Caliphate reborn in the summer of 2014 and himself the new defender of the faith.[8]

Most recently, another Abu Bakr, this time, Abu Bakr Naji (Mohammad Hasan Khalil al-Hakim), contributed to the operational strategy of IS through the e-book, *The Management of Savagery*. Naji was an Egyptian jihadi killed in Pakistan in 2008. Shortly before he died, he wrote the most practical guide on how to build the Caliphate today. This book represents the anti-FM 3-24 of the jihadi movement, if you will, their response to our own counterinsurgency (COIN) manual. The baseline of Naji's text is that the

Caliphate cannot be created through political reform. It must be achieved through violence, but this must be strategically exploited violence and savagery, and eventually the use of violence must lead to the capturing of territory which then must be governed effectively. This pragmatic approach to building the Caliphate should follow three distinct phases, according to Naji's book.[9]

The primary phase, which he calls the "vexation phase," is aimed at distracting and exhausting the infidel enemy and his allies by using dramatic operations, especially terrorist attacks. These attacks need not be on the scale of 9/11, but they need to sufficiently dramatic to prepare the battlefield for Phase Two, the "spreading of savagery." In this phase, the jihadis move to coordinated and synchronized attacks on a large scale, with the intent to dislodge governments from their capacity to actually govern their own territory. IS entered this stage two years ago. If the reports are true, that over 200 synchronized vehicle-borne improvised explosive devices (VBIED) were detonated on the day Ramadi fell, this means that IS has already gone through Phase Two. If they can actually use 200 VBIEDs in one day, they are following the Naji playbook and have already graduated to the last phase, which is called the "administer, consolidate, and expansion of savagery phase."

Here, just as in FM 3-24, the insurgent wants to stabilize held areas and to then unite the local populations into a fighting force. Simultaneously, this stage involves the implementation of governance in the form of shariah law and the use of this new stable area as a giant Forward Operating Base, a launching platform from which more Phase One and Two operations can be executed in new territories. This results in a "hybrid caliphate" contrasting strongly with the previous jihadi strategists who were purists when it came to how the Caliphate should be established and the role of force. Their earlier efforts relied on an almost automatic jump from violence to the emergence of a caliphate following the dislodging of infidel governments. Naji understood that regime decapitation does not lead by itself to a caliphate and that violence is only a means to an end. He outlined the importance of creating a quasi-nation-state base area before transmuting it into the final reality of the new Caliphate. It is his pragmatic approach to the quasi-nation-state, which would have been rejected by other strategists as a heretical, albeit temporary, aping of the West, that truly marks Naji apart. And unfortunately, this novel approach to creating a 21st-century caliphate is paying great dividends today for IS.

Lastly, the most important author and book of all is Brigadier S.K. Malik and his *Quranic Concept of War*. In this work, the Pakistani general officer effects an utter repudiation of Clausewitz and the Western way of war. In it, Malik states that jihad is something that all Muslims must do and that the only purpose for war is the reestablishment of the Caliphate—period. War has nothing at all to do with politics or *raison d'etat*, with the Clausewitzian need of a national government. War's sole purpose is the reestablishment of the theocratic empire of Islam, so Allah's writ can once again be sovereign here on earth.

Additionally, Malik writes, that the only center of gravity in war is not a physical target, but the soul of the *kuffar*, the infidel. The infidel must be made to convert to "the one true faith" or be killed. And he concludes, since the soul of the infidel is the only target that matters in war, if you wish to break his will to fight, the most effective mode of war is IW, most specifically, terror. If Islam is to be victorious, then the modes of attack preferred by groups such as al-Qaeda and IS, from the 9/11 hijackings to the pressure-cooker bombs of the Boston attack and the crucifixions and decapitations of the Caliphate today—that is the way to fight. This work is the "bible" of the global jihadist movement and must be read by anyone who works in the national security field, or who simply is concerned about the rise of threats like IS.[10]

To conclude, IS has been far more successful than al-Qaeda for identifiable reasons. Firstly, it has a correct understanding of IW. It realizes, as Mao did in China after World War II, that violence is but an instrument, not an end in itself. IS has set out to capture territory, declare the new empire and govern its population. Secondly, it has been highly effective in exploiting an extremely resonant and mobilizing "end times ideology" that the West has not even begun to counter at the strategic level. Because of that mobilizational ideology's resonance, in the long term, IO and psychological warfare will be the primary strategic weapons in this war, until IS and the broader global jihadist movement can be delegitimized with the help of our allies and Sunni Muslim partners.

As a result of these realities, the enemy threat doctrine of global jihadism must be built into our intelligence cycle and our IPB. We must study it and we must not allow politics into our intelligence preparations. Unfortunately, this does not seem to be the case at present.

Former U.S. military intelligence officer, Rick Francona, has written that there are now pressures upon intelligence analysts to not write reports wherein the attacks against IS are portrayed as being less than effective.[11] There appears to be censorship inside the system, with the Department of Defense actually launching an Inspector General investigation into the reports that show IS as resilient and growing, assessments that did not fit into a preapproved White House narrative of how IS is weak and "losing;" such reports have been doctored, censored, or lost.[12] Apparently we have allowed ideology or politics into the intelligence cycle and which has undermined our ability to understand our adversary. This will make winning the war against our new irregular enemy nigh impossible. It is impossible to cure a disease if you are not allowed to diagnose it objectively because facts become "inconvenient."

If we move beyond IS and look to the broader question of how America should respond to such IW now and in the future, we must address the issue of COIN's future. In recent years, this debate has boiled down to an unsophisticated standoff between the "pro-COIN cabal," often individuals personally connected to General David Petraeus and those involved in the rewriting of FM 3-24, versus the anti-COIN clique who wishes our armed services to return to the principles of fire and maneuver and so-called "conventional war" skills. However, both sides are wrong. COIN is a fundamentally un-American practice built on British and French colonial practices not applicable or even desirable today, and conventional warfare is the least likely scenario we face as a nation today or tomorrow, even when we look at nations like China or Russia, which clearly prefer unconventional and irregular warfare in the face of an overwhelming and conventionally defined military force such as America's.

America's Future Wars: What is to Be Done?

When it comes to defeating irregular threats, we should return to the beginning, to the types of missions and skills Army Special Forces were originally designed to hone and execute. The United States has historically excelled, and relied upon, Foreign Internal Defense (FID) as opposed to COIN when engaged in theaters where an insurgent force threatened our interests, or endangered the survival of our allies or partners. But FID has two non-negotiable requirements that we ignore at our peril. Number one, the indigenous government assisted by our trainers and advisors has to have a modicum of legitimacy with the local civilian population. They may not all love the government in power, but it must be understood by the local civilian population to be the government. You cannot do FID in a context where America's support of the regime makes the host government appear more illegitimate (as was the case with the Hamid Karzai government of Afghanistan). The second requirement is that the host nation's forces have to possess the capacity to absorb our support. It they do not, as was the case in both Afghanistan and Iraq, America is either wasting its money and expertise in a fruitless exercise, or worse, training cadres to fight more effectively, but most likely to fight against us or our allies, once they have learned what they wanted to learn from our trainers—about how to be a more effective fighting force, before abandoning the government to reinforce their own militias or tribes.

In addition to favoring FID-type missions over large-footprint COIN operations, we must have a strategic-level counter to the ideology of global jihadism, one that incorporates legitimate, moderate, local Muslim voices and authorities. America is not going to defeat IS acting unilaterally. U.S. taxpayers will not permit Washington to deploy troops at the levels needed to destroy IS unless there is a mass casualty attack on U.S. soil that is operationally linked to Baghdadi. As a result, it has to be the Iraqis, the Kurds, the Jordanians, and the Egyptians that close with and destroy IS forces on the ground. For multiple reasons, it has to be people like Egyptian President Abdel Fattah el-Sisi, who have the manpower, and leaders such as King Abdullah II of Jordan, who have the historic and lineage-based credentials, who must be the faces of this fight to take back the territory that needs to be recaptured from the insurgents.

The last part of the puzzle is the counter-ideological battlefront. Remember, Sun Tzu said the art of war is not to destroy the enemy's units—but to destroy the enemy's strategy. In order to defeat IS, we have to target their strategy, and that means countering the ideology of global jihad.[13]

The United States, in conjunction with our Sunni Muslim allies, must focus on those who have paid the greatest price in this war. The White House and politicians throughout the region should emphasize the number of Sunni Muslims killed in IS offensives—by demonstrating the toll that the ideology of jihad is having on Sunni Muslims in the region, we can begin to chip away the claim of authenticity made by Baghdadi and his murderous cohorts.

Ultimately, it will, however, have to be local actors that take the military fight to IS units; but they must have our support in this endeavor. No significant initiative is likely to be undertaken by the Sunni nations of the region unless they genuinely feel they have American support. And we must be a part of a global counter-ideological campaign, just as we were during the struggle against Communism in the Cold War.

Our Role in the War of Ideas

Any extremist ideology is most vulnerable in the delta between what it says is true and what the ground truth is. When the Soviet Union asserted that it was building a "workers' paradise" in which everyone will be equal—while Politburo members rode in limousines and the average Soviet citizen had to wait eight hours in line for a loaf of bread—they introduced a vulnerability into their ideological construct. Illuminating this delta is a critical part of countering ideologically motivated enemies.

In the speech he gave in June 2014, Baghdadi said he is the sword of Allah and the shield of Islam; our response should be to highlight that he is crucifying and immolating Sunni Muslims. Doing so will require incorporating discussions about religion at the highest policy levels in American national security circles. Additionally, we must commit ourselves to amplifying the voices of moderate Muslims without reducing their legitimacy in the eyes of local populations. The Muslims already using their own blogs, websites, and private organizations to counter the jihadi ideology of IS and the like, should be the tip of the nonkinetic spear, but they lack American support. We shy away from religious topics to our own detriment. If we do not lend support in the same way that we helped *Solidarnosc* (Solidarity) during the 1980s in Poland, then the United States will still be playing jihadi "whack-a-mole" a hundred years from now and we will be doomed to this kinetic approach as the threat continues to expand.

Both the Bush and Obama Administrations have argued that the United States cannot support the reformers, because American support will undermine them. I argue that this is lazy thinking; these reformers are already apostates in the eyes of the fundamentalists. Not lending them support merely reduces their capacity to spread a more tolerant Islamic message. And, of course, not all of our support need be overt. During the Cold War, the United States provided support to our ideological allies. Presidents made speeches about democracy and human rights in Poland and Hungary, and at the same time, there was enormous covert support provided to the dissidents best able to ideologically damage our totalitarian foe.[14]

If we do not send a message to the world that we are behind Egypt's President Sisi, that we are fully committed to the success of King Abdullah II of Jordan, they and the millions of peaceful Muslims they represent will not win this war against the "new totalitarians." Regardless of whether or not the United States lends it full-throated support, these groups and actors already have a target on their backs. We have to strengthen their hand so the modern, West-friendly versions of Islam can exorcise the fundamentalist and ultra-violent 7th-century version, which is currently in ascendance across the globe.

Public diplomacy must be a critical part of our IW strategy; we must engage in full-on strategic communications and IO at the national strategic-level, coupled with an aggressive covert counterpropaganda campaign amplifying local moderates. This will require us to have the capacity to map local communities in a nuanced and comprehensive manner, but right now we do not have these capabilities. They must be aggressively developed, along with a de-emphasis on technological modes of intelligence gathering, favoring instead more human intelligence.

Finally, we have to jettison the idea that that there are good Islamists and bad Islamists. If you believe in a caliphate—whether you want to create the Caliphate by decapitating people or through the ballot box, as former President Mohammed Morsi tried in Egypt—you are ultimately desirous of a theocratic state that is defined as the antithesis of everything that Western civilization holds dear, including freedom of religion, sexual equality, and minority rights.

And when it comes to the most dangerous jihadist threat today, we must understand that Baghdadi skillfully exploited the contingencies of the theater with the rising corruption of the Maliki regime, the collapse of Syria, and the execution of more than 200 unarmed Sunnis to create his insurgency. But removing Baghdadi will not solve this problem. Drone strikes only make sense if there is a limited pool of recruits to replace those that are killed and the enemy cannot replace them. We are not certain who is in the wings to replace Baghdadi—it may be someone with even greater organizational capacity than the current "Caliph." The limits of the kinetic application of force have to be internalized and we must work to actually make it less attractive for people to sign onto the global jihadist movement at all.

Conclusion: Back to the Future

Today's wars and tomorrow's campaigns can be understood as war returning to its normal state. The data is clear: IW is the most common form of war. John Keegan was right when he wrote that war is most often a social or cultural activity, as opposed to a cold and clinical activity done for "objective" reasons of the state.[15] If we realize that IS is but the expression of the most prevalent form of war, we will have taken the first step in being able to better win our current campaigns and future wars. But to do this, we will need to also:

- rebuild our national-level capabilities for IO and psychological warfare;
- invest in understanding the enemy threat doctrines much more than we do now;
- favor FID over COIN in our doctrine and policy; and
- remove politics and ideology from our intelligence analysis.

The West has demonstrated a capacity to do all these things individually at different times. In the 20th-century struggle against both fascism and communism, we built very effective counterpropaganda capabilities within both the IC and at the strategic level of national policy. During the Cold War, we invested heavily across both the government and private sectors to understand the political, economic, and military threat posed by the Soviet Union and so put our nation in a better place to win that confrontation. In both El Salvador and Colombia, we found a way to use very small numbers of U.S. forces to successfully strengthen the local host nation against an insurgent foe. And when we were at our best and the nation most secure, politics was not allowed to infect the intelligence function so as to distort reporting so that it comport with the preconceived expectations of the political elite. The question is, do we as a nation have the vision to see that we are once again faced with a new existential and totalitarian threat in the 21st century, and will we find the leadership, the courage, and the commitment to put all these pieces together? When the above short list becomes possible—which is really only a question of leadership—then we will be able to effectively assist our allies, defeat our shared enemies, and secure the Republic once more.

Notes

[1] Sebastian Gorka and David Kilcullen, "An Actor-centric Theory of War: Understanding the Difference Between *COIN* and *Counterinsurgency,*"*Joint Force Quarterly* 60, no. 1 (2011).

[2] Comment made to the author, September 2014.

[3] A concept made famous by Maria Harf, Deputy Spokesperson of the Department of State in an interview on national television.

[4] For a detailed analysis of how to do this, see Katharine Gorka and Patrick Sookhdeo, ed. *Fighting the Ideological War: Winning Strategies from Communism to Islamism* (Three Rivers, MI: Isaac Publishing, 2012).

[5] For a preliminary analysis of the strategic reasons for our failure as nation to engage in the counter-ideological war after 9/11, see Katharine Gorka, *The Flawed Science Behind U.S. Counterterrorism Strategy*, Council on Global Security White Paper (McLean, VA: The Council on Global Security, October 2014), available at <http://councilonglobalsecurity.org/wp-content/uploads/2014/10/Flawed-Science-Behind-US-CT-Strategy.pdf>.

[6] This estimate covers everything from flight schools to safe houses.

[7] Sayyed Qutb, *Milestones* (electronic version from *Kalamullah*), available at <http://www.kalamullah.com/Books/MILESTONES.pdf>

[8] Abdullah Azzam, "Defense of Muslim Lands," available at <https://archive.org/details/Defense_of_the_Muslim_Lands>

[9] Abu Bakr Naji, *The Management of Savagery*, available at <https://azelin.files.wordpress.com/2010/08/abu-bakr-naji-the-management-of-savagery-the-most-critical-stage-through-which-the-umma-will-pass.pdf>

[10] Unclassified English-language translations of all of the key jihadi strategic texts I have just summarized can be found at my personal website, TheGorkaBriefing.com. Detailed summaries of the works of the key jihadi strategists can also be found in my article: "Understanding Today's Enemy - The Grand Strategists of Modern Jihad,"o*Military Review* (May/June 2016): 32-39, available at <http://usacac.army.mil/CAC2/MilitaryReview/Archives/English/MilitaryReview_20160630_art001.pdf >.

[11] Lieutenant Colonel Rick Francona (Ret.), "Is your government lying to you about the war against ISIS?" *Middle East Perspectives by Rick Francona*, August 27, 2015, available at <http://francona.blogspot.co.uk/2015/08/is-your-government-lying-to-you-about.html>.

[12] Forty percent of USCENTCOM analysts recently answered "yes" to the question: "During the past year, do you believe that anyone attempted to distort or suppress analysis on which you were working in the face of persuasive evidence?" See Deb Riechmann, "House chairman: Military files, emails deleted amid ISIS intel assessment probe," *Military Times*, February 25, 2016, available at <http://www.militarytimes.com/story/military/capitol-hill/2016/02/25/house-chairman-military-files-emails-deleted-amid-isis-intel-assessment-probe/80933942/>.

[13] For our full report on the role of Information Operations in this and future wars, see the USASOC-sponsored report, Threat Knowledge Group, *The Islamic State and Information Warfare: Defeating ISIS and the Broader Global Jihadist Movement* (Threat Knowledge Group, January 2015), available at <http://threat-knowledge.org/wp-content/uploads/2015/11/TKG-Report-ISIS-Info-Warfare.pdf>.

[14] For a full exploration of how to win this war and the lessons we can use from the last conflict we won against a totalitarian foe, see my new book, Sebastian Gorka, *Defeating Jihad: The Winnable War* (Washington, DC: Regnery Publishing, 2016).

[15] John Keegan, *A History of Warfare* (New York, NY: Penguin Random House, 1993).

17

Networks at War: Organizational Innovation and Adaptation in the 21ˢᵗ Century

Christopher Fussell and D.W. Lee

The Evolving Landscape of Warfare

Recent observations of warfare clearly suggest that conflicts have become more transnational, longer, irregular, and network-centric.[1] Put differently, recent conflicts can be best described as protracted internal conflicts with multiple intervening state actors, networked with nonstate actors in a manner much like the multidimensional hybrid operational environment discussed in Army Special Operations (ARSOF) 2022.[2] The current conflicts in Iraq and Syria certainly meet this characterization; as do emerging crises in Ukraine, Yemen, and Libya, and longer-standing conflicts in Afghanistan, Somalia, and the Democratic Republic of the Congo. More state and nonstate actors support or sponsor movements in an intrastate conflict, making the termination of fighting very hard. For instance, the rapid resurgence of the Islamic State of Iraq and the Levant (ISIL) is largely attributed to the protracted Syrian civil war in which regional powers (including Iran, Saudi Arabia, and Turkey) as well as external nonstate actors such as ISIL, al-Nusra, and Hezbollah, to name just a few, sponsored local movements.

In essence, the complexity of warfare has increased due to the growing prevalence of networks utilized by states and nonstate actors who have found ways to countervail the kinetic superiority and hierarchical efficiency of big nation-states. This evolution demands a response from the United States and our allies, and requires a restructuring of our security apparatuses and a reframing of our definitions of preparedness and success.

Despite the changes that these threats demand, the changing nature of warfare is not a novel observation; the concept of "netwar" was coined by John Arquilla and David Ronfeldt in 1996.[3] Arquilla and Ronfeldt define "netwar" as "an emerging mode of conflict (and crimes) at societal levels, short of traditional military warfare, in which the protagonists use network forms of organization and related doctrines, strategies, and technologies attuned to the information age."[4] In their seminal paper, "The Advent of Netwar," the authors outline the defining characteristics of the netwar actor: it is necessarily "networked," containing nodes, clusters of nodes (i.e., cells), in a flat decentralized organizational structure. On offense, it is "adaptable, flexible, and versatile vis-à-vis opponents and challenges that arise."[5] Their now frequently coined phrase, "it takes networks to fight networks," was a clear foreshadowing of the transformation that state actors will need to go through if they hope to succeed in today's environment.[6]

This chapter explores the ways and means the United States can use to fight in this complex environment by harnessing the strategic utility of networks. Empirically, this objective is predicated upon the observation that an increasing number of external state actors overtly or covertly intervene in intrastate conflicts by exploiting various nonstate groups and networks in order to increase their respective strategic influence.[7] Similarly, nonstate actors also take advantage of interstate conflicts or political instability in their own countries and in neighboring regions.

Clearly, conflicts such as the Syrian civil war represent a sample of a larger shift in warfare toward more complex and hybridized dynamics. As of this writing, Uppsala University's world Conflict Data Program compiles data on 40 conflicts in the world for 2014. All but one of them are intrastate conflicts and 13 of them are internationalized.[8] The complexity of the current security environment is driven by an increasing number of state actors and nonstate actors who are networked to leverage and exploit the insurgent potential of multiple groups engaged in civil conflicts. Arguably ISIL, the most formidable terrorist movement of the 21st century, rose from this type of state and nonstate dynamics.[9] Defeating ISIL will require untangling the web of complex ties and competing interests between states and nonstate actors. This section explores ways and means of harnessing this complexity and suggests how such methods can be applied to help the United States fight more effectively against hybridized threats.

In order to fight in this complex hybrid environment, we argue that a deep understanding of network dynamics is critical.[10] Without understanding such dynamics, it becomes next to impossible to identify partnerships among disparate groups forming an alliance and coalition against American strategic interests. We also argue that fighting in the complex networked environment entails two interrelated innovative processes: 1) transforming our own organizations and communications to become more networked; and 2) mapping and illuminating the connective dynamics of adversarial networks.

Following the logic of netwar, we use two tales in this chapter to illustrate the sort of organizational revamping necessary to respond to modern conflicts. First we offer a narrative about the transformation of a highly specialized task force (TF) and how that organization had to force itself to adapt in the complex networked environment to safeguard the security of our nation. This is a story about fighting *ourselves* in order to become more agile and adaptive. The second story is about how we can fight more effectively against networked adversaries and strategic competitors. We argue that, within the context of organizational innovation, one effort is incomplete without the other.

Fighting Ourselves: Special Operations Task Force Transformation

The concept of organization-level change is easy in the abstract, but in most groups, real change remains strictly theoretical. Without a genuine imperative, human nature will drive each of us to resist fundamental, systems-level changes. There is a threshold of change beyond which the system can no longer revert to the known, and organizations (and the individual decisionmakers that comprise the organization) are generally incentivized to avoid crossing this line. When they are forced to do so, they often return as quickly as possible to the previous state of perceived stability. The military is as guilty of this pattern as any large bureaucracy. As with any industry, this can be both a strength and a vulnerability.

The U.S. Army's Airborne School has gone without systematic changes for generations of paratroopers. While parachute, aircraft, and individual equipment technology continues to improve, the step-by-step approach to preparing an individual soldier for his or her first parachute jump looks largely the same in 2015 as it would have in 1970. It is, ultimately, a linear process that has reached a very high level of efficiency and effectiveness, managing appropriate throughput while controlling for expected levels of capability and safety. The so-called "Jump School" is an excellent, and highly optimized, system.

The challenge in a bureaucratic system as big as the U.S. military (or any other hierarchical, global enterprise), however, is to encourage a differentiation between problems that can be solved through linear optimization of the current organizational model, and those problems that necessitate a rewiring of the fundamental way in which the organization functions. If the Department of Defense suddenly identified the need for twice as many paratroopers, the solution would be complicated. Along with high resource costs, the solutions would map to a series of second- and third-order impacts. It would demand the focus and intellectual power of experienced leaders and planners who understand the multitude of implications of the transition. The execution of the plan would require excellent cross-functional leadership and project management skills, ideally from leaders with deep knowledge of the training environment and existing relationships that would allow them to move as a team. But ultimately, the answer to "how can we create twice as many paratroopers?" can be known, within a few degrees of accuracy, from the outset of the execution of the plan, assuming it was properly produced. Indeed, this is one of the core strengths of the U.S. military dating, arguably, to lessons of military industrialization that were learned and ingrained over generations of industrial-age warfare. These lessons have driven advances in training, programs, and growth for decades inside the ordered industrial-age military environment.

But, as the Special Operations community learned in the early years following the attacks of September 11, the same organizational muscle that drives effective organizations toward optimization and efficiency gains can lead to organizational uncertainty and inertia around the development and implementation of viable solutions when subjected to a new military environment. The reality that the Special Operations TF encountered was that the arrival of the information age led to a level of interconnectedness between individuals around the globe that traditional systems were simply not designed to cope with. In practice, this meant that networks of al-Qaeda leaders, fighters, influencers, and financiers could connect globally and in real time, maintaining strategic context and broad alignment with the goals of al-Qaeda while constantly adapting to the demands of local conditions. The TF was facing an organic network, able to create leaderless action, and thereby quickly negate the effectiveness of predictive analysis that the American military had relied upon for decades. Al-Qaeda's actions did not come from a strategic plan, but emerged bottom-up from the real-time thinking and planning of its numerous, highly autonomous, individual nodes. While some of the TF's tactical-level leaders could sense that conditions were changing on the ground, traditional military approaches incentivized members to attempt to frame the disorder of wholly unpredictable problems into a linear solution set; this breeds, inevitably, ill-fitting solutions to misunderstood problems. If the only thing you have is traditional enterprise thinking, you become the infamous hammer in search of a nail and your actions run the risk of creating as many problems as they solve.

None were guiltier of this than the Special Operations community in the early days of the fight with al-Qaeda in Iraq (AQI). In late 2003, it was clear that the momentum of AQI was outpacing our efforts in Iraq; by the end of the year, there had been more terrorist attacks in Iraq alone than there had been in the entire world in 2003.[11] And it only got worse. By the end of 2005, terrorism claimed 8,300 Iraqi lives; by early 2006, more than a thousand Iraqis died each month.[12] What was unclear was why this was happening. On paper, and through the accepted view of the world in 2004, this made no sense. The forces that comprised the Special Operations TF had clear and undeniable points of superiority, which included:

- capability level and training of the individual operator;
- cohesion and tactical effectiveness of small team units;
- advanced weapon systems at every level (individual operator to overhead strike assets);
- full spectrum intelligence collection and dissemination capabilities;
- ability to dominate night operations; and
- highly refined, global-reaching logistics and supply chain operations.

This list could go on. In short, the TF could move exponentially more capable forces around the globe, with superior equipment, at unparalleled levels of speed and efficiency, and place them in tactical scenarios where they held nearly every advantage. These points of superiority led to the measurable fact that, when elements of the TF were able to lock members of AQI in time and place, and then close in on them with a tactical ground force, these special operations teams demonstrated a near-perfect record of winning the engagement. However, victory in the moment in each tactical engagement was not the issue.

Despite all of these advantages, it was clear by 2004 that AQI was somehow outpacing some of the world's most highly trained and well-funded units. We were winning tactical engagements, but losing the overall war. The entire U.S. military system had, in retrospect, failed to properly weight the new variable of global interconnectivity. There was a time when state-run organizations controlled information flow, as there was a significant barrier of entry to pass information on a global scale. The TF knew conceptually, early in the fight, that modern technology was creating globally connected networks, but it did not realize that this new reality had nearly instantaneously changed the face of the battlefield.

Traditional bureaucratic systems designed around the control of information flows were not designed to handle this new reality. As in most hierarchies, the TF was grounded in the fundamental belief that information is power and that the ability to gather information from multiple silos and synthesize it into knowledge and insight that others could not produce is the source of ultimate power and respect. But playing an industrial-age game while the external system is operating by the rules of the information age adds incredible risk and is destined for failure.

By 2004, it had become clear to the senior leadership of the Special Operations community that the existing bureaucratic model was not going to allow the TF to move fast enough to keep pace with al-Qaeda. That same senior leadership also made it clear that the solution would involve a fundamental shift in how the TF operated as an organization.

In the early days of the fight with AQI, the problem was not that the TF needed exponentially more operators, helicopters, or weapon systems. The TF did not need new agencies to be invented to solve the problem, and it certainly did not need more individual data points (raw intelligence) to be collected from the field; if anything, the information-age battlefield was already overwhelming the system with data that could not be sorted and acted upon fast enough. Simply put, there was no linear solution, organization chart redesign, or any single silo within the enterprise that would somehow solve the problem. The U.S. military, quite understandably, was locked in a collective cognitive bias (more specifically, a classic status quo bias) that forced many of the TF to take complex information from the battlefield and create ways to explain it through preestablished ways of thinking that made sense according to the old norms of the organization. For example, if a detained member of al-Qaeda offered information about the person he reported to, it was difficult for the TF to understand that the person being referenced was part of a fluid network, not part of a crisp hierarchical organization chart. By the time the intelligence had become actionable, the person being described may have already moved multiple times, both in position and influence, within the self-rearranging network. But the organization's status quo bias drove the TF, at a systems level, to fixate on targeting the next person up the perceived hierarchy. Hours of planning and energy would go toward locating the reported "boss," only to find, after targeting that individual, that the new target was no more or less important that the person initially captured. And the cycle would repeat. This led, of course, to days, weeks, and months of head-scratching and to post-it note hierarchies covering the walls of outstations throughout Iraq, fruitlessly attempting to tell a story about al-Qaeda's structure that simply was not there. The TF could no more create an organization chart that defined the totality of al-Qaeda than one could create a similar document for all Facebook users.

The bias toward seeking a linear solution is obvious and reasonable, and it would have been ideal for the design of the enterprise. Much like the problem of creating twice the number of Jump School graduates, if we could have solved the math problem and declared with certainty—there are x number of al-Qaeda fighters, and we need to produce y amount of actions to defeat them—then a complicated solution would have started with: assets + personnel = y (actions). And like the Jump School plan, the TF would have increased the first two variables until its output was moving faster than the growth of AQI. It would have been costly and difficult, but knowable and measurable. And this, predictably, is what the TF tried to achieve in the early part of the conflict. But by late 2004, there were no other assets or personnel to add to the mix, and stretched to the limit, the TF's actions could move no faster.

In retrospect, the ultimate problem was a not uncommon organizational bias against recognizing a massive shift in the external environment. The TF had been created in a complicated, but, ultimately, ordered environment where nation-states and enterprise-level systems controlled the flow of information and action. Individual actors in any space (on the battlefield, in business, etc.) could certainly exercise their free will and step outside the norms, but the risks were incredibly high and the likelihood of strategic impact very low. A soldier in World War II could sneak through enemy lines with relative ease and engage a less defended position, but the odds of returning to safe territory were low, as were the chances of engaging an enemy soldier of any strategic importance. The rational player who wanted to have significant impact, therefore, was incentivized to understand and master the rules of the system. The general officers on the battlefield, CEOs of the corporate world, or the dignitaries in nation-state interactions were the positions that consistently provided the opportunity for strategic impact. But the TF had, as a direct result of the reduced barrier of entry for global communications and the subsequent interconnectivity of billions, entered an environment dominated by disorder and complexity. No longer could the scale advantage of large systems control the entire environment. Suddenly, the individual who had not mastered the system or reached any traditional position of power or influence was able to become a strategic player based simply on their ability to connect, influence, and create action within a networked organization.

But these types of players exist outside of the ordered systems that *other* ordered systems are biased toward looking for. In the absence of seeing an ordered system in the fight against al-Qaeda, the TF worked diligently to create structure where it did not exist. This, of course, proved impossible and ultimately drove the TF to shift to an entirely new operating framework. The TF, and the large U.S. military, had moved into a world of disorder, where environmental conditions can no longer be dominated or controlled by large systems. On the battlefield, the coupling of the interconnectedness of individual nodes in the al-Qaeda network with the speed at which the global information system allows for the flow of information and ideas led to the creation of a complex system; that is, a system whose output can no longer be effectively predicted based on input variables. The output of a complex system emerges as more than a simple sum of input variables, a reality that fundamentally undermines the predictive nature of traditional military systems.

In this complicated battlefield scenario, those closest to the problem (e.g., the "front line" of the fight) were looking for a relatively small number of large data points, such as the number of enemy fighters, types of weapon systems, and estimates on supply chain capability. These large chunks of data could be effectively synthesized at the top of the hierarchy, predictive analysis done on possible and likely enemy courses of action, and orders disseminated back to the frontline elements. These orders, for operational security reasons, could be kept relatively compartmentalized, as synchronization and de-confliction between the various frontline elements could be centrally controlled. This worked, in large part, because both sides were following the same basic set of rules even if their end states were radically different. Therefore, the side that was able to optimize their system most effectively was likely to come out victorious. This was true in the kinetic environment of World War II, and in the nonkinetic, proxy environment of the U.S.-Soviet Cold War. But when the opposition has no allegiance to a traditional system, leveraging old systems in hopes of creating a predictive analysis capability is fruitless. When the frontline elements are looking not at a few large pieces of intelligence, but instead at thousands of highly nuanced pieces of data ranging from individual relationships to shifts in tribal allegiances to community members that may float in a single day between family interactions, legitimate business work, U.S. military partnerships, and al-Qaeda relationships, synthesizing data at the top of the hierarchy and distributing useful guidance is an impossible task. When the TF tried this approach, its individual actions were accurate and successful, but their sum total was exponentially slower than was necessary.

Looking at this environment through an adapted version of the Cynefin model, first designed by David Snowden and Mary Boone, the battlefield environment had progressed from the complicated and ordered environment to the complex and disordered space (with occasional upshots into chaos):[13]

Figure 17.1. Adapted Version of the Cynefin Model

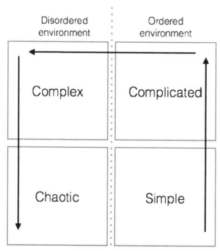

The initial reaction of the TF was to push and strain systems designed for a complicated world as far as it possibly could, but to no avail. The organization pushed assets harder than ever. Operators reviewed plans and intelligence and briefed the chain of command; elite

Figure 17.2. Adaptation and Extension of the Cynefin Model

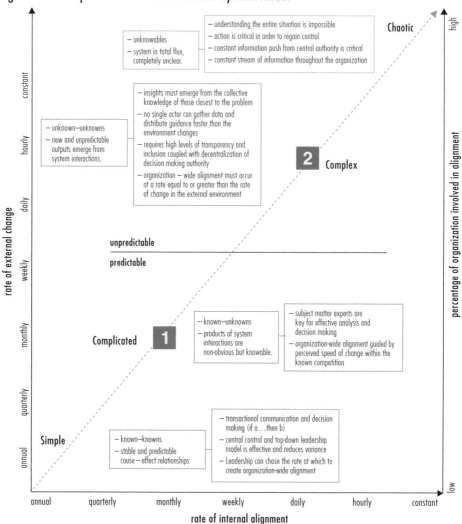

small teams went to work in the dark hours of night, hitting multiple targets, meticulously planning and executing each operation, sleeping for only a few hours after returning to base each morning. This process that drove the TF became known as "F3EA:" Find—Fix—Finish—Exploit—Analyze.[14]

But it was not enough. Component parts were being optimized, but the overall system was misaligned to the speed of information flow on the battlefield. In the disordered and complex environment, the TF's approach to communication, intelligence sharing, decentralization of decisionmaking rights, and ability to take autonomous action was mismatched relative to the speed and complexity of the enemy network. For its system to be effective, the TF needed to understand the pace of the problem it was facing, then index that against the speed with which it was able to move information and create action.

Through that optic, it was clear that the system needed to change at a very fundamental level in order for a large, global enterprise to outpace the agile al-Qaeda network.

This level of change was an easy thing to conceptualize, but difficult to execute, as the TF realized that it was not designing a temporary measure to account for a sudden change in the enemy's approach; rather, that it had stumbled upon a fundamental shift in the operating environment, so its change needed to be permanent, not a patch. An early realization was that its pace of communication would need to increase exponentially if it hoped to match the speed of a complex environment. The graph below, an adaptation and extension of the Cynefin model, shows an x-axis representing increasing time separation between moments of organization-wide strategic alignment, a $y1$-axis showing an increasing rate of change in the external environment, and a $y2$-axis showing the percentage of the organization that must be pulled into broad communication forums in order to match the speed of the external environment.

This graph maps the entire construct, but the majority of the problems in the counterterrorism fight against AQI sat at the transition point between "complicated" and "complex." The new reality that the arrival of the information age brought with it, for the foreseeable future, is an external environment in a regular state of disorder. Speed and interconnectivity of millions of individual nodes have created a constant state of complexity. The crossing of that horizontal divide from order to disorder is what the TF had to accept as the new reality of the world we all now live in. The implication was that the organizational norms and leadership behaviors of the optimization- and efficiency-based models that dominate an ordered environment were no longer relevant.

The requirement was to move the organization from point 1 to point 2, as shown in the graph provided, shifting from a structure capable of dealing with complicated and predictable problems to one comfortable handling a complex and unpredictable environment at the enterprise level. Long a trait of small Special Operations teams, the ability to adapt in near real time to an ever-shifting landscape had never been seen as a requirement at the global-system level, where scale and efficiency have always seemed to matter most. Consider any place along the red line in the graph shown, the point at which a sufficient amount of the enterprise is aligned on strategy to allow those closest to the problem set to take independent action without creating strategic-level errors. The Special Operations TF's optimization/efficiency model did this well, but far too slowly. The TF was stuck in the ordered/complicated space, while AQI was operating in the disordered/complex world that was the actual battlefield.

To move from point 1 to point 2 above, the TF needed to execute two major change functions. First, it needed to acknowledge that it was now in a fundamentally unpredictable environment, which meant building a network of all stakeholders involved in the fight. Without a physical center of gravity in Iraq, it was necessary to interconnect parts of the organization that would have preferred to remain siloed from one another, even within the Special Operations community itself (i.e., cultural and operational divisions between Navy SEALs, Army Rangers, and other specialized units). In a complicated environment, each of these components would own a piece of the battlefield and the section of the problem that was contained therein; but in the complex environment, the al-Qaeda problem flowed quickly and rapidly between these self-constructed boundaries. Therefore, the TF needed its operational elements to be interconnected at an unprecedented level. But al-Qaeda was a global network that was in a constant state of change, so it was necessary to expand the network model well beyond the military units involved in the ground war. An al-Qaeda network involved in recruiting, radicalizing, and moving foreign fighters into Iraq demanded of the Department of Defense a real-time relationship with the Department of State, as the problem would extend through multiple countries before reaching the declared theater of war. Understanding the interconnection between intelligence source networks and the al-Qaeda fighters being targeted by tactical units would require seamless connectivity with the various intelligence agencies involved in the fight. The list of key relationships would grow over time, and ultimately, it became a true interagency effort on a global scale (arguably the only time the United States has successfully established and maintained such a model for an extended period).

Second, the TF needed to significantly increase its speed of communication, minimizing the amount of time needed to cultivate organization-wide alignment. Al-Qaeda was not adjusting its plans based on any master plan; rather, it was a complex system whose actions emerged bottom-up from the input variables in the environment. Those input variables were immeasurable, from local conditions on the ground to tribal politics, number of new recruits to disputes within the network, and from the raids the TF was conducting against them on a regular basis to the success or failure of their own operations. The AQI system changed daily, not based on any centralized strategy, but by the very nature of it being an organic network. Therefore, the TF needed to match the agility of the enemy if it hoped to eventually overtake the speed of AQI. As the TF's own network grew, it simultaneously increased the speed at which it ensured organization-wide alignment. This was accomplished by steadily increasing the number of participants included in, and the precision with which it was executed, a daily global video teleconference meeting titled, "Operations and Intelligence Update" (O&I).

During the peak periods of violence in Iraq between 2006 and 2007, daily attendance on this video teleconference numbered in the thousands; attendees were scattered around the globe, and every interagency organization critical to the fight was represented. Because of the tempo of this forum (roughly an hour and a half everyday), it was critical to have large numbers in attendance. If the forum had been attended only by the top level of leadership, the conversation would have been limited to mostly strategic discussion, and the organization would have spent many of the remaining hours of the day simply cascading information down through a series of small, siloed meetings with inconsistent context being distributed at varying levels of effectiveness (normal bureaucratic behavior that resembles a giant game of telephone). But by the time important strategic information hit the tactical level of the organization, many layers of filters would have interpreted it and passed down their view on what was and was not relevant, and the time remaining in a 24-hour operational cycle would be severely constrained. The system simply could not have moved as fast as the al-Qaeda network under that constraint. The ultimate decision was simple in theory, challenging in practice. Everyone needed to hear the same truth, simultaneously, at a rate that moved as fast as, or faster than, changes within the al-Qaeda network. In retrospect, we refer to this change as the creation of "shared consciousness" within the organization.

Creating shared consciousness was a critical step in establishing enterprise-level alignment across the TF with consistency and speed. By pulling together large numbers of individuals representing all of the critical geographic locations, intelligence organizations, leadership teams, and tactical arms of the organization, the TF was able to create broad awareness around critical changes ranging from the strategic to the tactical.

Two aspects in particular were essential to this transformation. First, instead of the traditional well-rehearsed brief by a junior member followed by black-and-white questions (e.g., "How many x?"), our dialogue became broader and more participatory in order to glean the context (e.g., "Why are you thinking x?"). Second, until the revamping of the O&I meeting, the work done by operators and intelligence analysts was inextricably linked, yet siloed—the two groups had on organizational blinders in the name of efficiency and compartmentalization of information, which hampered attempts to build relationships and effectively cooperate in real time. The O&I meeting tore down these barriers, fused operations and intelligence—O and I—for the first time, and connected the purpose of both groups' efforts.[15]

With shared consciousness (driven by the O&I) in place, the organization developed a consistent level of what we will now refer to as "empowered execution." That is, the expectation was established that operational- and tactical-level elements would move with high levels of speed and autonomy during the periods between organization-wide synchronization (O&I sessions). In real terms, this meant that the elite small teams within the global enterprise were now interconnected as part of a network, given access to a daily forum to gain near complete levels of strategic alignment, then empowered, expected, and held accountable for their ability to move quickly and autonomously for the remaining 22 hours of the operating cycle. It was the informed autonomy that high performance organizations and aggressive leaders naturally desire, but rarely put in place.

As the Greek philosopher Heraclitus tells us, you cannot step in the same river twice. That became an apt way to describe the operations cycle once the enterprise was locked in shared consciousness and driven by empowered execution. What happened inside the remaining 22 hours of the operating cycle was never the same thing twice, as the organization had become more than the sum of its parts. As an interconnected network with decentralized authorities to create and adjust relationships, shift assets between units, share new intelligence in real time, and drive autonomous action, a global force of thousands was suddenly moving faster in its analysis-decisionmaking-action cycle than the small pockets of fighters in the al-Qaeda network. Goliath had maintained his strength, but was now more agile than David, and the scales began to shift.

Fighting the Enemy: The Human Domain Approach

Adapting in the complex environment of multiple interacting components is a daunting challenge for nation-states. The previous tale demonstrates how challenging it can be to defeat highly networked and adaptive adversaries. This section demonstrates that fighting in this highly dynamic and potentially chaotic environment requires a deep understanding of both the context and the structure of node, ties, and boundaries.[16] Following this logic, this section explores how to understand complex networks without losing sight of their inherent dynamics.

Illuminating the complexity of networked adversaries begins with a broad collection of information about the conditions that underpin the networks, key influencers that keep the networks connected, and affiliated social and political organizations. Insurgent or terrorist movements do not emerge just because of economic grievances or governance vacuums. Rather, they emerge by manipulating existing conditions of grievance and ties between and across cohesive networks.[17]

Following this observation, there is a growing consensus on the necessity of mapping complex social dynamics to understand the social foundations of insurgent movements as early as possible. For instance, the new Army Operating Concept emphasizes the need to understand the human dynamics prior to the outbreak of violence.[18] It acknowledges that such an understanding is a must, in order to be able to shape and manipulate the operational environment during the conflict.[19] In other words, harnessing the complexity of the modern security environment can begin with understanding the relational dynamics of comprising entities.

United States Special Operations Command (USSOCOM) recently published a concept paper that emphasizes cultivating soldiers "with the knowledge, skills, and abilities to understand and influence human actions and activities."[20] In addition, the same concept stresses the need to link these activities to creating desired effects in the human domain.[21] In other words, understanding how to operate in the human domain will be a critical part of future fighting. Much like one would not think about going to war without detailed physical maps of the terrain and topology of the operational environment, human domain mapping can provide the initial social, cultural, and political dynamic understanding that can help the United States fight networked and complex adversaries. What remains largely unexplored is the type of methodology needed to fight effectively in this complex human environment.

A good analogy to describe the utility of these processes would be countering vehicle-borne improvised explosive devices (VBIED). In order to counter and prevent attacks from VBIEDs, intelligence work typically focuses on the overall terrain, roads, and key junctions. Instead of solely tracking only potential vehicles that may carry explosives, it makes sense to factor in the local road system with possible entry and exit points from known areas under insurgent control. In addition, one can also overlay known IED-manufacturing sites and insurgent activities onto the overall topography of the environment. By integrating the road system, known insurgent sites, and potential support elements, one can achieve a better understanding of how to deploy the Intelligence, Surveillance, and Reconnaissance (ISR) element more surgically, thus, improving the probability of early detection and interdiction. In fact, most counter-IED strategies do factor in environmental and topological variables.[22]

Operating in the complex human environment can be conceptualized in a similar vein. In other words, it entails collecting and analyzing information pertaining to three broad categories: mapping key conditions of insurgency development, pathways of interaction, and nodes of influence.[23] These are well-supported in the literature of social movement theory and social network analysis (SNA). For instance, Doug McAdam shows that robust insurgencies emerge from existing ties. ISIL's organizational growth shows similar patterns where preexisting ties of old Baathists and Republican Guards comprised the infrastructure of the movement to expand in Syria. In essence, AQI's weakened leadership developed ties with other inmates at Camp Bucca, which later became the overall leadership structure of ISIL.[24] Expanding in Syria, ISIL exploited the local conditions of lawlessness and divided opposition groups by first identifying key influencers of major fighting groups and then either co-opting existing social structures or replacing them. By establishing *Dawahs*, or charity groups, first, ISIL embedded itself within existing social networks before it began to exert its social and political control. From this brief description, it becomes clear how conditions, pathways of interaction, and influential nodes are critical categories of information in fighting in the human domain.

The Conditions: Context of Network Development

The first step of mapping the human domain begins with the identification of conditions that are highly associated with the emergence and development of radical collective action.[25] As David Kilcullen points out, powerful insurgent movements emerge from a coalition of disparate groups when certain conditions compel them to work together.[26] In order to anticipate and counter emerging insurgent movements, it is critical that intelligence preparation of the battlefield begin by analyzing what conditions exist in the operational environment.

Social movement theory suggests four major types of antecedent conditions: political, economic, social, and ideological. Political conditions can be factions within the regime or the existence of political opposition groups. Going back to the expansion of the ISIL, Nouri al-Malaki's systematic persecution of the Iraqi Sunni population created a very permissive political environment that allowed the remnants of AQI to mobilize a great many Sunni groups in Iraq.

Certain economic conditions are highly associated with the onset of radical movements. Typically, these conditions include income inequalities, underemployment, unemployment, inflation, or income stagnation. Note that it is often external shocks that trigger the exacerbation of these conditions.[27]

Ungoverned or underregulated economics can also provide a fertile ground for insurgent groups to generate resources to sustain themselves. These unsanctioned economic areas typically have built-in informal or autonomous channels of resource extraction and redistribution. The autonomy of the bazaar in Iran was a major factor of success during the Iranian Revolution of 1979.[28] The bazaar provided much needed resources to key organizers of the movement when the regime had cut subsidies and stipends to students and academics.[29]

It must be stressed that economic conditions themselves are rarely sufficient for resistance to emerge or take hold. For instance, while the overall economic conditions of the Middle Eastern states were generally comparable in the 1980s and 1990s, insurgent movements emerged in only a select few countries.[30] While all major macroeconomic indicators were comparable in Algeria, Egypt, Jordan, Morocco, and Tunisia between 1980 and 1992, only the first two countries experienced major insurgent movements. This observation is not uncommon in the literature.[31] Assessing individual grievances is an important part of understanding how radical elements become networked to form a broader movement. What is critical to understand is the process through which individual grievances are transformed into a group narrative. In other words, grievances become instrumental when they are exploited and framed by groups or networks actively seeking to create opportunities for collective mobilization.

Social divides and existing dissident networks provide great potential for resistance. In particular, external actors can leverage these social conditions to establish a robust organizational platform. It is no coincidence that most robust resistance movements emerge from preexisting ties and networks. These preexisting ties typically have built-in mechanisms to coordinate information and action across civil society. Ethnic divides can be a powerful location from which collective action emerges. The cohesiveness of existing socio-ethnic divides can also generate resources to create a broad coalition of insurgent movements as opposed to just focusing on one cohesive group.

Ideological conditions are based on existing grievances stemming from economic disparities or structural strains such as income inequalities, unemployment, underemployment, or discrimination. In essence, these conditions often stem from social, economic, or political strains. They also include existing norms of collective action and violence that can be utilized to justify mobilizing large groups for resistance. For instance, a sense of victimization is often used by Islamists to justify jihad.[32] Typically, insurgents will try to align their ideology with socially accepted themes of expressing dissent.[33] Instead of treating resistance ideology as a monolithic worldview, it is more useful to approach it as a set of grievances specifically framed to motivate and justify collective action.[34]

The Pathways

Once the context of a network is analyzed, then we can start mapping the internal components. The second process of illuminating the human domain begins by understanding how various groups are connected with other groups as well as external actors. Social movement theory suggests that existing ties and channels between social and political organizations play a critical role for the growth and expansion of insurgent movements.[35] In essence, existing relations function as the path of least resistance between groups and actors. As individuals try to disseminate new information and find others to join them in collective action, informal ties are instrumental to collecting and distributing resources and ideas. Figure 17.3 illustrates the importance of understanding existing ties to harness the complexity of networked entities. Figure 17.3 provides a graphic representation of observed individuals and relations among them. In SNA, this type of visualization is called a "sociogram."[37] In essence, the sociogram depicts a map of existing ties between several political factions in Iraqi Kurdistan. It captures four types of ties between the members of the Iraqi parliament and major Kurdish political movements: ties with nonpolitical organizations, political movement affiliations, shared government organizations, and shared military ties. Each political affiliation is color-coded.[38] The Kurdistan Democratic Party (KDP) and the Patriotic Union of Kurdistan (PUK) are well-known to the outside world, whereas Gorran is a new political movement within Iraqi Kurdistan.

Figure 17.3. Network Map of the Power Structures of Iraqi Kurdistan[36]

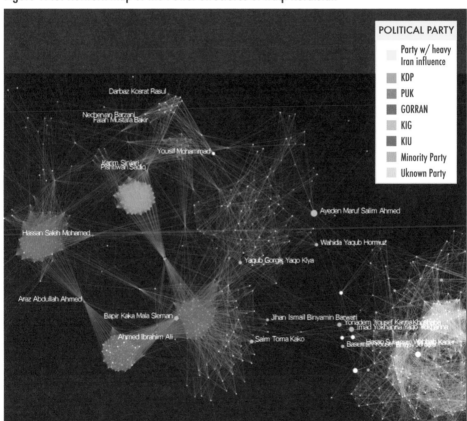

The sociogram includes both current and past interactions, such as working in the same committee or for the same project.[39] Understanding and mapping past relations is a critical part of mapping a complex network with multiple subgroups, as movements typically emerge by activating old ties.[40] Especially for political influence, it frequently follows existing human relations as building new rapport can be too conspicuous or costly. From the sociogram, we can see that the overall political landscape in Iraqi Kurdistan is characterized by identifiable and cohesive political parties connected by less conspicuous connective tissue.

By following existing pro-Iranian individuals within the Iraqi government and their prior associations with Kurdish politicians, we can illuminate existing relational pathways between and across major political groups in Iraqi Kurdistan. It is worth noting that Iranians are leveraging existing interpersonal and interorganizational ties within the Iraqi government in an attempt to extend their access and placement in Iraqi Kurdistan. While the PUK has had a stronger relationship with the Iranians and the Shia militias they back, such as the Badr Brigades, the actual pathways utilized by the Iranian Revolutionary Guard Corps (IRGC) to exert influence in Kurdistan indicate that Iran is also diversifying its ties within Iraqi Kurdistan. Pathways of influence can often be fluid and dynamic. In this regard, mapping the key pathways that connect external actors and organic networks should also focus on temporal changes in the network to anticipate how their interconnectivity changes over time.

The Influential Nodes

Once the overall structure of key pathways is analyzed, we start identifying critical brokers and influencers between and across subgroups in order to understand key individuals who sustain the network. The purpose of this process is twofold. First, it illuminates and identifies those who are important in sustaining the network but not well-known to the outside world; they are often called "emergent nodes."[41] Understanding who the emergent nodes are can broaden options to leverage and shape the network. Second, it establishes quantitative social network metrics to assess each key individual's type of influence as well as relative measure of influence. Once the emergent nodes are estimated with specific metrics, then outside actors can prioritize and select a few to optimize the process of building access and placement within the network.

In Iraqi Kurdistan, it is true that traditional power brokers and leaders of the PUK and KDP still wield an enormous amount of power. However, the perceived concentration of power among these few leaders makes influencing the regions quite challenging. Upon a careful network analysis of the sociogram, we can see that there are other individuals who occupy key locations within the network. Many of those individuals, who have long worked various groups while representing Iranian interests, are not widely known influencers. However, their relative influence can be measured by computing their activities and relations with well-established social network metrics. This is where the quantitative aspect of SNA can be increasingly insightful. For instance, the IRGC seems exceedingly pragmatic in that they have established ties with minority leaders who have organic relations with other major parties, such as the Kurdistan Regional Government (KRG) and PUK in the region.

Table 17.1 summarizes the highest valued agents according to several important metrics, as well as overall key influencers in the overall network. It is worth noting that Yousif Mohammad and Bapir Kaka Mala Sleman are both members of minority parties, who function as brokers within Kurdistan's government.[42] In other words, SNA can reveal highly influential nodes who are not necessarily the most obvious.

Table 17.1. Key Actor Values [43]

Rank	Node	Key Actor Value
1	Ayeden Maruf Salim Ahmed	0.048
2	Sabah Jalloub Faleh Hami Al-Sa'idi	0.043
3	Imad Yokhanna Yaqo Yokhanna	0.037
4	Mathhar Khader Naser	0.032
5	Bapir Kaka Mala Sleman	0.031
6	Safiyah Taleb Ali Alsouhail	0.028
7	Yonadem Yousef Kanna Khoshaba	0.027
8	Mahmous Ali Othman Omar	0.022
9	Salim Toma Kako	0.021
10	Kathem Atiyah Alshammari	0.021
11	Yousif Mohammad	0.02
12	Yaqub Gorgis Yaqo Klya	0.02

It is worth noting that the minority groups have ideally positioned themselves to be able to access major political factions, indicated by their high boundary spanner potential. Boundary spanner potential indicates structural positions that allow access or influence across multiple groups.[44] Put differently, boundary spanners have great potential to bridge and connect multiple subnetworks.[45] It is also noteworthy that Iran seems willing to work with those with the most potential for connecting multiple groups as opposed to those who are just politically aligned with Iranian interests.

Given that these minority groups' leaders have shown a great deal of pragmatism by working with various major factions, these potential brokers provide great opportunities for the United States to enhance its influence in the region as well. Being able to identify and locate these key brokers can be a critical step toward minimizing and undermining adversarial influence in Iraqi Kurdistan. For example, Yousif Mohammad can potentially provide multiple avenues of approach to the KDP or Gorran to counter Iranian influence within the KRG.

Table 17.2. Key Influential Brokers[46]

Rank	Node	Boundary Spanning Value
1	Kardo [last name redacted]	0.034
2	Mawlood Murad Mohyeldin	0.033
3	Yousif Mohammad	0.027
4	Mustafa Sayid Qadir	0.025
5	Kamal [last name redacted]	0.024
6	Pishtiwan Sadiq	0.022
7	Farsat Ahmad Abdullah	0.022
8	Salar Mahmud Murad Ali	0.021
9	Suzan Shahab Nuri	0.019
10	Abdullah Mahmud Muhammad	0.019
11	Arez Abdullah Ahmed	0.018
12	Bapir Kaka Mala Sleman	0.017

Mapping the Kurdish political landscape in Iraq yields powerful insights relevant to how to operate effectively in the complex networked environment. It demonstrates that mapping relational dynamics is critical to lifting the fog of complex operational environment. SNA, when properly executed, can shed much light on not only the pathways used for coordinating operations and narratives, but also inconspicuous yet key individuals who perform such functions. Key is understanding both the underlying concepts and techniques that reveal the hidden structures and interactions of networks.

Understanding the complexity of networked adversaries is a daunting task. Part of the methodological and conceptual challenge is to reduce the complexity without degrading our ability to cope with its contextual nuance. As shown in the Kurdistan example, it is possible to understand how state actors actively try to leverage existing or emerging networks to increase their influence in that environment. However, it does require a conceptual understanding of what comprises such networks and what methodological tools are available to illuminate key pathways and key brokers that keep that connective tissue functioning.

SNA, combined with a deep understanding of the sociopolitical conditions that enable network emergence, can be particularly powerful. It should be noted that mapping social relations has its limitations, such as a high dependency on data fidelity and availability and a potential temporal lag between observation and changes in network dynamics. However, such weaknesses typically apply to all analytic procedures, and not exclusively to SNA. What should be stressed instead is the need to develop an intelligence system that analyzes interconnectivity of networked adversaries by continuously mapping and updating relational data. Just as the TF transformation is characterized by extensive interconnectivity and distributed decisionmaking, the same logic must be applied to how we understand and analyze the dynamics of network connectivity within a broad human domain, characterized by both transnational actors and localized influencers. As rapid and routinized communication helped the TF form a collective consciousness of the mission and the battlespace, it is also critically important to map and analyze how individual and parochial narratives are aligned to support the ideological appeal of networked adversaries. Again, such mapping and analysis is not infeasible. The CORE Lab at the Naval Postgraduate School has repeatedly utilized various mapping techniques from social media outlets and shown in detail how ISIL has adapted its narrative to align itself with various warring factions in Syria.[47] Much like physical relational ties, insurgent and terrorist narratives can be understood as a system of themes, idioms, and resonating cultural norms.[48] This is not a trivial observation. Part of fighting in the complex networked environment is waging "battles of the story" and understanding how stories of battles are composed and aligned with political and strategic interests.[49]

Conclusion

As the TF transformation and the Kurdish example of mapping the human domain suggest, the fog of complex networks can be lifted; change is not necessarily followed by chaos. Our ability to harness this complexity can be enhanced by mapping and analyzing the conditions that underpin network dynamics, interaction pathways, and key influencers. This is in essence a two-prong race. On one hand, it hinges on how we internally network our organizations and collective awareness in order to increase our adaptive agility against networked threats. On the other hand, it also requires a concerted effort to map and illuminate the environment, comprising nodes, ties, and boundaries of networked adversaries.

The TF transformation strongly indicates that it is entirely possible, albeit often painfully slow and opposed, for the United States to function and fight as a networked force pursuing a unified objective through shared consciousness. Our adversaries have harnessed the strategic utility of this organizational innovation. The sheer complexity of various alliances between nonstate actors and external sponsors is just a symptom of a much larger pattern of warfare characterized by increasingly transnationalized, protracted, movement-centric, and networked conflicts.[50] There is no doubt that we are late to the game. Powerful states typically do not have the luxury of agile adaptation frequently associated with less powerful states leveraging nonstate actors. "The (b)end of history," as John Arquilla described it, where states and networks coexist for strategic competition, cannot be won by linear thinking and hierarchical execution.[51] We hope that the TF transformation and the human domain approach we have discussed in this chapter pave the way forward for our nation to harness and utilize the strategic utility of networks. The era of organizational innovation will favor those who outpace others in learning how to interface and interact with networks toward their strategic objectives.[52]

Notes

[1] Glenn Johnson and Doowan Lee, "Revisiting the Social Movement Approach to Unconventional Warfare," *Small Wars Journal* 7, no. 11 (2014).

[2] U.S. Army Special Operations Command, *ARSOF 2022* (Fort Bragg, NC: United States Army John F. Kennedy Special Warfare Center and School, 2013), 3.

[3] John Arquilla and David Ronfeldt, *The Advent of Netwar* (Montclair, NJ: RAND Corporation, 1996), available at <http://www.rand.org/pubs/monograph_reports/MR789.html>.

[4] Ibid., 6.

[5] Ibid., 11.

[6] Ibid., 81-82.

[7] Note that internal conflicts can be a civil war or an internal political confrontation, or both.

[8] Department of Peace and Conflict Research, "Uppsala Conflict Data Program," *Uppsala Universitet,* available at <http://www.pcr.uu.se/research/ucdp/>.

[9] Christopher Reuter, "The Terror Strategist: Secret Files Reveal the Structure of Islamic State," *Spiegel*, April 18, 2015, available at <http://www.spiegel.de/international/world/islamic-state-files-show-structure-of-islamist-terror-group-a-1029274.html>.

[10] In the SOF community, this is conceptualized as "operating in the human domain."

[11] Found in: General Stanley McChrystal, Tantum Collins, David Silverman, and Chris Fussell, *Team of Teams: New Rules of Engagement for a Complex World* (New York, NY: Penguin, 2015), 22, referencing the U.S. State Department annual terrorism reports, which reported 198 "significant" terrorist incidents in Iraq in 2004, compared with the worldwide total of 175 in 2003, 22 of which occurred in Iraq.

[12] Of the 20,000 deaths worldwide from terrorist attacks, 65 percent were in Iraq (13,000 deaths). Found in: McChrystal et al., *Team of Teams,* 22; referencing: National Counterterrorism Center, *Report on Terrorist Incidents - 2006,* April 30, 2007, available at <http://www.fbi.gov/stats-services/publications/terror_06.pdf>.

[13] David Snowden and Mary Boone, "A Leader's Framework for Decision Making," *Harvard Business Review*, November 2007, available at <https://hbr.org/2007/11/a-leaders-framework-for-decision-making>.

[14] F3EA was derived from similar targeting and decisionmaking processes, such as the well-known OODA loop (Observe—Orient—Decide—Assess) that was associated with fighter pilots. Found in: McChrystal et al., *Team of Teams,* 50.

[15] Ibid., 169-170.

[16] Johnson and Lee, "Revisiting the Social Movement Approach to Unconventional Warfare."

[17] David Kilcullen, *The Accidental Guerilla: Fighting Small Wars in the Midst of a Big One* (New York, NY: Oxford University Press, 2011).

[18] Emphasis added by author. TRADOC Pamphlet 525-3-1, *The U.S. Army Operating Concept* (Washington, DC: Department of the Army, 2014).

[19] Derek Raymond, "Human Domain Mapping in the 21st Century," *Small Wars Journal* (2015); TRADOC Pamphlet 525-3-1, *The U.S. Army Operating Concept.*

[20] United States Special Operations Command, *Operating in the Human Domain, Version 1.0* (Tampa, FL: USSOCOM, August 2015), 38.

[21] Ibid.

[22] For a detailed comparison of counter-IED strategies, see: Lora Weiss, Elizabeth Whitaker, Erica Briscoe, and Ethan Trewhitt, "Evaluating Counter-IED Strategies," *Defense and Security Analysis* 27, no. 2 (2011).

[23] These processes are mostly informed by social network analysis and social movement theory.

[24] Derek Raymond, "Combating Daesh: A Socially Unconventional Strategy" (Master's thesis, Naval Postgraduate School, June 2015).

[25] An antecedent condition can be defined as "a phenomenon whose presence activates or magnifies the action of a causal law or hypothesis." Stephen Van Evera, *Guide to Methods for Students of Political Science* (Ithaca, NY: Cornell University Press, 1997), 9-10.

[26] Kilcullen, *The Accidental Guerilla.*

[27] Doowan Lee, "A Social Movement Approach to Unconventional Warfare," *Special Warfare Magazine* 26, no. 3 (2013): 30.

[28] Benjamin Smith, "Collective Action with and without Islam: Mobilizing the Bazaar in Iran," in *Islamic Activism: A Social Movement Theory Approach,* ed. Quintan Wiktorowicz (Bloomington, IN: Indiana University Press, 2004), 187-9.

[29] Ibid.

[30] Mohammed M. Hafez, *Why Muslims Rebel* (Boulder, CO: Lynne Rienner Publishers, 2004), 10-16.

[31] Doug McAdam, *Political Process and the Development of Black Insurgency, 1930-1970* (Chicago, IL: University of Chicago Press, 1982), 11-16.

[32] David A. Snow and Scott C. Byrd, "Ideology, Framing Processes, and Islamic Terrorist Movements," *An International Quarterly Review* 12, no. 1 (2006).

[33] David A. Snow, E. Burke Rochford, Jr., Steven K. Worden, and Robert D. Benford, "Frame Alignment Processes. Micromobilization, and Movement Participation," *American Sociological Review* 51, no. 4 (1986): 467-76.

[34] Ibid.

[35] Mario Diani and Doug McAdam, ed. *Social Movements and Networks, Relational Approaches to Collective Action* (New York, NY: Oxford University Press, 2003), 7.

[36] The nodes are sized by each person's potential to bridge different groups. In this case, we are using a specific social network measure called "boundary spanning," that captures intergroup and intragroup ties. Research was conducted by Christopher Couch and Doowan Lee at the Naval Postgraduate School in Monterey, California. This project was initiated to support U.S. Special Operations Command Central in Tampa, Florida. The sociogram was generated with ORA Software, Kathleen M. Carley, Center for Computational Analysis of Social and Organizational Systems, and Carnegie Mellon University. Copyright 2001-2011. A full analysis can be found in Christopher Couch, "Aghas, Sheiks, and Daesh in Iraq, Kudish Robust Action in Turmoil" (Master's thesis, Naval Postgraduate School, June 2015).

[37] Stanley Wasserman and Katherine Faust, *Social Network Analysis: Methods and Applications,* (Cambridge: Cambridge University Press, 1994), 12.

[38] The yellow nodes represent the political entities closely aligned with Iran, the red nodes represent members of the KDP, the blue nodes show members of the PUK, the purple nodes denote members of the newly formed Gorran Party, the green nodes represent the two Islamist Kurdish parties, and the light blue nodes show minority political parties. The light purple nodes denote an unknown party affiliation. The resulting network is composed of 919 individuals and 32,030 ties.

[39] The dataset is coded from every article on Kurdistan in the Al-Monitor newspaper written in 2014 and various other online sources. A full list of sources is available upon request.

[40] Donatella della Porta, "Recruitment Processes in Clandestine Political Organizations: Italian Left-Wing Terrorism," *International Social Movement Research* 1 (1988).

[41] Kathleen Carley, "Destabilization of Covert Networks," *Computational and Mathematical Organization Theory* 12, no. 1 (2003).

[42] ORA calculates the top-ranked agents in terms of the following metrics: Emergent Leader, In_the_ Know, Clique Count, Eigenvector Centrality, Eigenvector Centrality Per Component, Hub Centrality, Authority Centrality, Betweenness Centrality, How they Connect Groups, and Group Awareness.

[43] Data Generated by ORA. Reproduced with author's permission from Couch, "Aghas, Sheiks, and Daesh in Iraq, Kudish Robust Action in Turmoil." Full names can be obtained by contacting the authors.

[44] For a detailed discussion of how brokers shape and sustain networks, see: Mario Diani, "'Leaders' or Brokers? Positions and Influence in Social Movement Networks," in *Social Movements and Networks, Social Movements and Networks, Relational Approaches to Collective Action*, ed. Mario Diani and Doug McAdam (New York, NY: Oxford University Press, 2003); Shin-Kap Han, "The Other Ride of Paul Revere: The Brokerage Role in the Making of the American Revolution," *Mobilization: An International Quarterly* 14, no. 2 (2009).

[45] BSP technically is a calculation of the betweenness centrality of a node divided by the degree centrality of that node.

[46] Data generated by ORA. Reproduced with author's permission from Couch, "Aghas, Sheiks, and Daesh in Iraq, Kudish Robust Action in Turmoil." Full names can be obtained by contacting the authors.

[47] Gregeory Freeman and Robert Schroeder, *Social Media Exploitation: An Assessment* (Monterey, CA: Naval Postgraduate School CORE Lab, September 2014).

[48] Snow and Byrd, "Ideology, Framing Processes, and Islamic Terrorist Movements."

[49] John Arquilla, "The (B)end of History," *Foreign Policy*, December 15, 2011, available at <http://foreignpolicy.com/2011/12/15/the-bend-of-history/>; Lee, "A Social Movement Approach to Unconventional Warfare."

[50] Johnson and Lee, "Revisiting the Social Movement Approach to Unconventional Warfare."

[51] Arquilla, "The (B)end of History."

[52] Ibid.

This chapter was written with assistance from Jessica Craige.

Contributors

Hilary Matfess is a research associate at the Institute for Defense Analyses (IDA), a freelance journalist, and a contributor to the Nigeria Social Violence Project at the Johns Hopkins University School of Advanced International Studies. She has conducted fieldwork in Tanzania, Rwanda, Nigeria, and Ethiopia. Her current research focuses on social violence and the Boko Haram insurgency in Nigeria. Prior to joining IDA, she was a research analyst at the Center for Complex Operations at the National Defense University.

Michael Miklaucic, Director of Research, Information, and Publications at the Center for Complex Operations (CCO) at National Defense University, is also the editor of *PRISM*, the journal of CCO. Prior to this assignment, he served in various positions at the U.S. Agency for International Development (USAID) and the Department of State, including chief operating officer for the USAID Office of Democracy and Governance, and rule of law specialist in the Center for Democracy and Governance. From 2002 to 2003, he served as the Department of State deputy for War Crimes Issues. He later returned to State as a USAID representative on the Civilian Response Corps Inter-Agency Task Force.

Scott Atran, a French American anthropologist, is the director of research at the Centre National de la Recherche Scientifique in Paris, an adjunct professor at the University of Michigan, a presidential scholar in sociology at John Jay College of Criminal Justice, and a co-founder of ARTIS International and of the Centre for the Resolution of Intractable Conflict at Oxford University. He has studied and written extensively about terrorism, violence, and religion, and has also done fieldwork revolving around terrorists, Islamic fundamentalists, and political leaders. His book, *Talking to the Enemy: Religion, Brotherhood, and the (Un)Making of Terrorists* (New York, NY: Ecco, 2011), delves into such topics, and his work has been featured around the world—in *The New York Times*, *Reuters*, *Agence France-Presse*, *Wall Street Journal*, *Newsweek*, *Time*, *Discover*, *The Guardian*, *Financial Times*, *La Recherche* (France), *BBC World Service*, and *CNN*, to name a few.

Jay Chittooran is a policy advisor in the Economic Department at Third Way. Chittooran's portfolio includes trade policy, with particular attention to the Trans-Pacific Partnership and Trade Promotion Authority, and works on cross-commodity strategy as a part of trade analysis. Chittooran previously served in positions focused on energy economics at the Council on Foreign Relations, Google Ideas, Goldman Sachs, and West Point. He also recently served as a policy advisor and speechwriter to Ray Kelly, the former commissioner of the New York City Police Department. His research has appeared or been cited in the *Washington Post, The Hill, Politico, Roll Call, Wall Street Journal*, the *Financial Times*, the *Houston Chronicle*, and *Time*, among other publications, and he has briefed government officials, media outlets, and other interested parties. He holds an M.A. from Seton Hall University and a B.A. from Loyola University Chicago.

Christopher Dishman is the South Central region director for the Department of Homeland Security's (DHS) Field Operations Division. Prior to joining Field Operations Division, Dishman served as the chief of the Border Security Branch in the Office of Intelligence and Analysis (I&A) of the DHS. Prior to I&A, Dishman served as a policy analyst at the Office of National Drug Control Policy, where he led a team charged with understanding the economics behind drug trafficking and proposed innovative ways to disrupt the global drug trafficking industry. Dishman has written extensively in academic journals, newspapers, and magazines about the relationship between terrorism and organized crime and other terrorist-related issues. He has lectured on these topics at the National Defense University and the National War College, among other venues. He is currently pursuing a Ph.D. degree in public affairs at the University of Texas, Dallas.

Douglas Farah is the president of IBI Consultants, LLC and senior visiting fellow at the Center for Complex Operations at the National Defense University. He is also senior non-resident associate at the Americas Program at the Center for Strategic and International Studies. Farah, who specializes in field research, works as a consultant and a subject matter expert on security challenges, terrorism, and transnational organized crime in Latin America, both for the U.S. government (Department of Defense, Department of Homeland Security, and others) and the private sector. Farah is the author of two books, *Blood from Stones: The Secret Financial Network of Terror* (New York, NY: Doubleday, 2004) and *Merchant of Death: Money, Guns, Planes, and the Man Who Makes War Possible* with Stephen Braun (Hoboken, NJ: J. Wiley, 2007). He has also written dozens of articles and monographs for peer-reviewed journals and the media.

Francis Fukuyama received his B.A. from Cornell University in classics and his Ph.D. from Harvard in political science. He was a member of the Political Science Department of the RAND Corporation, and of the Policy Planning Staff of the U.S. Department of State. He previously taught at the Paul H. Nitze School of Advanced International Studies (SAIS) of Johns Hopkins University and at George Mason University's School of Public Policy. He served as a member of the President's Council on Bioethics from 2001 to 2004. He is the chairman of the editorial board of *The American Interest*, which he helped found in 2005. He is a senior fellow at the Johns Hopkins SAIS Foreign Policy Institute, and a non-resident fellow at the Carnegie Endowment for International Peace and the Center for Global Development. He holds honorary doctorates from Connecticut College, Doane College, Doshisha University (Japan), Kansai University (Japan), Aarhus University (Denmark), and the Pardee RAND Graduate School. He is a member of the Board of Governors of the Pardee RAND Graduate School, the Board of Directors of the National Endowment for Democracy, and a member of the advisory board for the *Journal of Democracy*. He is also a member of the American Political Science Association, the Council on Foreign Relations, and the Pacific Council for International Affairs. He is married to Laura Holmgren and has three children.

Christopher Fussell, a former U.S. Navy SEAL officer, has spent the past four years adapting strategies he learned in the military to the corporate world. He is a managing partner at the McChrystal Group, where he oversees the McChrystal Group Leadership Institute. He is a co-author of *Team of Teams*, and currently authoring the Group's next book about the key processes involved in creating team-based models for organizations.

Nils Gilman, Associate Chancellor at the University of California, Berkeley, is a strategic advisor to the Chancellor and responsible for the effective administrative organization and daily functioning of the Office of the Chancellor. Gilman works closely with the Chancellor and his senior leadership team, across a range of day-to-day and long-range responsibilities. He also represents the Office of the Chancellor to a variety of internal and external constituencies, providing leadership and overall project management for key strategic projects, as well as coordinating activities and communications. Gilman also provides the Chancellor with briefings, administrative and policy support, and strategic advice.

Sebastian Gorka is a national security professional specializing in irregular warfare, including counterinsurgency and counterterrorism. He is the vice president of the Institute of World Politics in Washington, D.C.; chairman of Threat Knowledge Group; and author of the *New York Times* bestseller, *Defeating Jihad*. Previously, he served as the Major General Matthew C. Horner Distinguished Chair of Military Theory at the Marine Corps University. He is a founding member of the Council for Emerging National Security Affairs and has served as the associate dean for Congressional Affairs and Relations to the Special Operations Community at the National Defense University. He is also currently affiliated with U.S. Special Operations Command's Joint Special Operations University, and is a regular instructor at the John F. Kennedy Special Warfare Center and School in Fort Bragg, as well as the Federal Bureau of Investigation's Counterterrorism Division. He has testified before the U.S. Congress on the threat of IS and global jihadism and briefed the Office of the Director of National Intelligence, the Central Intelligence Agency, the Defense Intelligence Agency, the National Intelligence Council, the National Counterterrorism Center, and the Commandant of the Marine Corps. Born in the United Kingdom to Hungarian parents, Gorka became an American citizen in 2012.

Scott Helfstein is a global market strategist at BNY Mellon Investment Management, focused on multi-asset class investing, global equities, and geopolitical investment opportunities. He is also a non-resident fellow at West Point's Combating Terrorism Center (CTC) and a senior fellow at George Washington University's Center for Cyber and Homeland Security. Prior to joining BNY Mellon, Helfstein helped run a small entrepreneurial organization as the director of Research and Strategic Initiatives at West Point's CTC and was an assistant professor in the Department of Social Science at the United States Military Academy. He also worked at the Federal Reserve Board of Governors and as an investment banker focusing on mergers and acquisitions at Credit Suisse First Boston. He holds a B.A. in finance from George Washington University, an M.A. in war studies from King's College, London, and a Doctorate in public policy from the University of Michigan. Helfstein is a

Term Member in the Council on Foreign Relations (CFR), a member of the CFR Term Member Advisory Committee, and a member of the Economic Club of New York.

Karl Lallerstedt is a co-founder of Black Market Watch and leads the program on illicit trade at the Global Initiative Against Transnational Organized Crime. Formerly he worked as anti-illicit trade strategy director at a leading multinational corporation, where he also served as a steering committee member of the International Chamber of Commerce's Business Action to Stop Counterfeiting and Piracy (BASCAP). Lallerstedt has a background as a political and economic analyst for the Department of State, Oxford Analytica, and the Economist Intelligence Unit. He is a member of the Organisation for Economic Co-operation and Development Task Force on Countering Illicit Trade.

D. W. Lee is an educator who teaches social revolution and unconventional warfare at the Naval Postgraduate School, and unconventional warfare operational design and special warfare network development at the U.S. Army Advanced Special Operations Training Center. His current research focuses on social movement theory, civil resistance, and comparative politics. He is the author of numerous publications, and completed graduate studies at the University of Chicago.

Matthew Levitt is the Fromer-Wexler fellow and director of The Washington Institute's Stein Program on Counterterrorism and Intelligence. From 2005 to early 2007, he served as deputy assistant secretary for intelligence and analysis at the U.S. Department of the Treasury. In that capacity, he served both as a senior official within the department's terrorism and financial intelligence branch and as deputy chief of the Office of Intelligence and Analysis, one of 16 U.S. intelligence agencies coordinated under the Office of the Director of National Intelligence. From 2008 to 2009, he served as a State Department counterterrorism advisor to the Special Envoy for Middle East Regional Security, General James L. Jones. Previously, he served as a counterterrorism intelligence analyst at the Federal Bureau of Investigation, where he provided tactical and strategic analytical support for counterterrorism operations, focusing on fundraising and logistical support networks for Middle Eastern terrorist groups.

Clare Lockhart is the co-founder and CEO of the Institute for State Effectiveness (ISE), founded in 2005 to find and promote approaches to building good governance. She, and ISE, now work in countries around the world to support leaders and managers find paths for their countries to stability and prosperity, and work with networks globally to rethink the balance between state, market and civil society for the 21st century. She served in Afghanistan as an adviser to the UN during the Bonn Process and to the Afghan Government from 2001 to 2005, designing a number of national initiatives including a program that provides a block grant to every village in Afghanistan, now present in 23,000 villages. She is co-author with Ashraf Ghani of *Fixing Failed States* (New York, NY: Oxford University Press, 2008) and contributes to the media on issues of peace and state-building.

Celina Realuyo is a professor of practice at the William J. Perry Center for Hemispheric Defense Studies at the National Defense University, where she focuses on U.S. national security, illicit networks, transnational organized crime, counterterrorism, and threat finance issues in the Americas. As a former U.S. diplomat, international banker with Goldman Sachs, U.S. foreign policy advisor under the Clinton and Bush Administrations, and professor of international security affairs at the National Defense, Georgetown, George Washington, and Joint Special Operations Universities, Realuyo has over two decades of international experience in the public, private, and academic sectors. She speaks regularly in English and Spanish on "Managing U.S. National Security in the New Global Security Environment," "Following the Money Trail to Combat Terrorism, Crime, and Corruption," "Combating Illicit Networks in an Age of Globalization," and "Designing Strategies to Counter Terrorism." Realuyo appears and comments regularly in the international media, including *CNN en Español*, *Foreign Policy*, *Reuters*, *Voice of America*, and *Univision Radio*.

Tuesday Reitano is the head of the Secretariat at the Global Initiative Against Transnational Organized Crime, the director for an independent policy and monitoring unit for the European Union's programmes in counterterrorism, and a senior research advisor at the Institute for Security Studies in Pretoria, South Africa. Reitano has 12 years of experience as a policy specialist in the UN System, including with the UN Development Programme, the UN Development Group and the UN Office on Drugs and Crime, as well as a number of years in a boutique consulting firm as an advisor on justice, security, and governance issues. In this time, she has amassed a wealth of experience in fragile states and development working both with states, civil society, and at the community level to strengthen resilience to transnational threats, promote sustainable development and the rule of law. Reitano has Master's degrees in business administration (MBA) and public administration (MPA), and a Master of Science in security, conflict, and international development (MSc). She has also published extensively in leading academic journals and policy institutions. Reitano is based in Beirut, Lebanon, with her family.

Raj Samani is an active member of the information security industry, through involvement with numerous initiatives to improve the awareness and application of security in business and society. He is currently working as the EMEA chief technical officer for Intel Security, having previously worked as the chief information security officer for a large public sector organization in the United Kingdom. He was inducted into the Infosecurity Europe Hall of Fame (2012), won the Virus Bulletin Péter Ször Award for the paper/investigation he co-authored on the takedown of the Beebone Botnet, and was named in the UK's top 50 data leaders and influencers by Information Age. He is also the special advisor for the European CyberCrime Centre, also on the advisory council for the Infosecurity Europe show, *Infosecurity Magazine*, and expert on both searchsecurity.co.uk and Infosec portal, and a regular columnist on *Help Net Security*. He has had numerous security papers published, and regularly appears on television commenting on computer security issues.

Mark Shaw is the director of the Global Initiative Against Transnational Organized Crime, an international network of experts and think tank headquartered in Geneva, which is focused on catalyzing new responses to organized crime. He is also the National Research Foundation professor of Justice and Security and director of the Centre of Criminology at the University of Cape Town. Shaw previously worked for 10 years at the United Nations Office on Drugs and Crime, including as inter-regional adviser and chief of the Criminal Justice Reform Unit, with extensive fieldwork in fragile states. He holds a Ph.D. from the University of the Witwatersrand, Johannesburg, and has wide experience with both governmental, nongovernmental, and private sector organizations working on issues of transnational threats, governance, and conflict.

Jessica Stern is a research professor at Boston University's Pardee School of Global Studies. She is also a visiting fellow at Hoover Institution and an advanced academic candidate at the Massachusetts Institute of Psychoanalysis. She is a consultant on a Department of Defense-Minerva and National Institute of Justice-funded project on Somali immigrant children at the Boston Children's Hospital. She is a co-author of *ISIS: The State of Terror* (New York, NY: Ecco, 2015) and the author of *Denial: A Memoir of Terror* (New York, NY: Ecco, 2011), *Terror in the Name of God: Why Religious Militants Kill* (New York, NY: Harper Perennial, 2004), and *The Ultimate Terrorists* (Cambridge, MA: Harvard University Press, 2001). She was selected as a 2014-2015 Fulbright Scholar. In 2009, she was awarded a Guggenheim Fellowship for her work on trauma and terror. She served on President Clinton's National Security Council Staff from 1994 to 1995. Stern advises a number of government agencies on issues related to terrorism and has taught courses for government officials. She is a member of the Council on Foreign Relations (CFR) and was named a CFR International Affairs Fellow, a National Fellow at Stanford University's Hoover Institution, a Fellow of the World Economic Forum, and a Harvard MacArthur Fellow. Stern has a B.A. from Barnard College in chemistry, an M.A. from Massachusetts Institute of Technology in technology policy, and a Doctorate from Harvard University in public policy.

Andrew Trabulsi is a consultant, author, and entrepreneur, focusing on technology, geopolitics, and economic development policy. Based in San Francisco, his work and research has included technology capacity building with indigenous communities in the Amazon rainforest, economic development with the Federal Reserve Bank, innovation consulting with Deloitte LLP, and geopolitical analysis of transnational criminal organizations. Andrew advises public, private, and social sector clients on issues of strategy, geopolitics, forecasting, and policy development. His first book, *Warlords, Inc.: Black Markets, Broken States, and the Rise of the Warlord Entrepreneur* (Berkeley, CA: North Atlantic Books), was published in 2015.

Phil Williams is holder of the Wesley W. Posvar Chair and director of the Matthew B. Ridgway Center for International Security Studies at the University of Pittsburgh. His previous assignments included Visiting Professor at the Strategic Studies Institute at the

U.S. Army War College and Visiting Scientist at the Computer Emergency Response Team of the Carnegie Mellon University, where he worked on cybercrime and infrastructure protection. He has worked extensively on transnational criminal networks, terrorist networks, terrorist finances, the rise of drug trafficking violence in Mexico, and has focused most recently on issues of violence and governance in Central America. He has published extensively in the field of international security.